Integral, Probability, and
Fractal Measures

Springer
New York
Berlin
Heidelberg
Barcelona
Budapest
Hong Kong
London
Milan
Paris
Santa Clara
Singapore
Tokyo

Gerald A. Edgar

Integral, Probability, and Fractal Measures

With 36 Figures

 Springer

Gerald A. Edgar
Department of Mathematics
The Ohio State University
Columbus, OH 43210-1174
USA

Mathematics Subject Classification (1991): 28A80, 60D05

Library of Congress Cataloging-in-Publication Data
Edgar, Gerald A., 1949–
    Integral, probability, and fractal measures/Gerald A. Edgar.
        p.   cm.
    Includes index.

    1. Fractals.   2. Measure theory.   3. Probability measures.
    I. Title.
    QA614.86.E335   1997
    514′.742 – dc21   97-6236

Printed on acid-free paper.

ISBN 978-1-4419-3112-2

Production managed by Francine McNeill; manufacturing supervised by Jeffrey Taub.
Photocomposed copy prepared using the author's AMS-TeX files.

Printed in the United States of America.

9 8 7 6 5 4 3 2 1

Preface

This book may be considered a continuation of my Springer-Verlag text *Measure, Topology, and Fractal Geometry*. It presupposes some elementary knowledge of fractal geometry and the mathematics behind fractal geometry. Such knowledge might be obtained by study of *Measure, Topology, and Fractal Geometry* or by study of one of the other mathematically oriented texts (such as [13] or [87]). I hope this book will be appropriate to mathematics students at the beginning graduate level in the U.S. Most references are numbered and may be found at the end of the book; but *Measure, Topology, and Fractal Geometry* is referred to as [*MTFG*].

One of the reviews of [*MTFG*] says that it "sacrific[es] breadth of coverage for systematic development"[1]—although I did not have it so clearly formulated as that in my mind at the time I was writing the book, I think that remark is exactly on target. That sacrifice has been made in this volume as well. In many cases, I do not include the most general or most complete form of a result. Sometimes I have only an example of an important development. The goal was to omit most material that is too tedious or that requires too much background.

In this volume, the reader will again learn some of the mathematical background to be used in our study of fractal topics. Chapter 2 deals with integration in the modern sense. Since [*MTFG*] dealt with measures and had no theory of integrals, we have here used the knowledge of measures to aid in the discussion of integrals. Chapter 4 deals with mathematical probability. A mathematician may sometimes be inclined to think of probability as a branch of measure theory, but some of its motivations and techniques are quite different from the ones commonly seen in measure theory. In both of these chapters, only parts of the complete theory (of integrals, or of probability) are included—emphasis is on those particular results that are used elsewhere in the book for our discussion of fractals.

What is a **fractal**? There is, as yet, no widely accepted definition.

In [176, p. 15], Benoit Mandelbrot writes: *A fractal is by definition a set for which the Hausdorff-Besicovitch dimension strictly exceeds the topological dimension.* In the notation of [*MTFG*], a "fractal" is a set E with ind $E <$ dim E. I will sometimes call this a **fractal in the sense of Mandelbrot**. But note that in the second printing of 1982 (and more explicitly in [178, p. 8])

---

[1] Alec Norton's review in the *American Mathematical Monthly*, April 1992.

Mandelbrot states that he now prefers *to leave the term "fractal" without a pedantic definition* [176, p. 548].

S. James Taylor [255] has proposed that a fractal is a set for which the Hausdorff dimension coincides with the packing dimension. That is, a "fractal" is a set E with dim $E = $ Dim E. I will sometimes call this a **fractal in the sense of Taylor**. But in fact, Taylor also is no longer promoting his definition.

Thus we are left with no precise definition of the term "fractal"; so of course there can be no theorems about "fractals" as such. Our theorems will be stated about sets E with certain properties, appropriate for the purpose we have in mind.

The term **fractal measure** in the title of the book has three possible meanings. All three will be seen in the course of the book:

(1) A "fractal measure" could be one of the measures (like $\mathcal{H}^s$ or $\mathcal{P}^s$) associated with the measurement of the s-dimensional "size" of a set and thus associated with the definition of various fractal dimensions (like the Hausdorff dimension or the packing dimension). We begin with such fractal measures in Chapter 1.

(2) M. Barnsley [13, §IX.5] suggests that a "fractal" really is an element of the space $\mathfrak{P}(S)$ of probability measures on a metric space S. Just as elements of $\mathfrak{K}(S)$ are "fractal sets," so elements of $\mathfrak{P}(S)$ are "fractal measures." Dimensions may be associated with measures in many of the same ways that they are associated with sets. Such fractal measures, in particular "self-similar" fractal measures, are found in Chapter 3.

(3) B. Mandelbrot [81] has proposed the term "fractal measure" for a kind of decomposition arising when a natural measure on a fractal set is not completely uniform: rather than a single number (the fractal dimension), the set may be decomposed into parts, each exhibiting its own dimension. Another term used for this decomposition is "multifractal." An example of this is in §5.7.

When I was first planning to write this book, I intended to include more material on random fractals; in particular, more material on the dimensions of sets associated with Brownian motion. But during the time that this book was under preparation there appeared a fine book by P. Mattila [186]. It contains much more along those lines than I could hope to include. So in this book I have contented myself with a brief introduction to the possibilities (Chapter 5). Mattila's book is more demanding on the reader in terms of the required background, but in return the results are much more complete.

In a similar way, there appeared a book by P. Massopust [185]; it contains material on fractal functions that will not be duplicated here, except for a few special cases in Chapter 3.

Here are some remarks on notation. Usually, we follow the notation of [*MTFG*]. But because of my use of uppercase script letters for measures, I have replaced some of my former uses of script letters by other notation. We will use Fraktur (German) letters to represent certain spaces: $\mathfrak{K}(S)$ the space of nonempty compact subsets of a metric space S with Hausdorff metric [*MTFG*,

p. 66]; $\mathfrak{C}(S,T)$ the space of continuous functions from S to T with uniform metric (when S is compact) [*MTFG*, p. 61]; $\mathfrak{P}(S)$ the space of probability measures on S with Hutchinson metric. We write $\mathbb{R}$ for the set of real numbers, $\mathbb{Q}$ for the set of rational numbers, $\mathbb{Z}$ for the set of integers, and $\mathbb{N}$ for the set $\{1,2,3,\cdots\}$ of natural numbers. In a metric space, diam A is the diameter of a set A, $B_r(x)$ is an open ball, $\overline{B}_r(x)$ is a closed ball, ∂A is the boundary of a set A, and $\overline{A}$ is the closure of a set A. The maximum of two real numbers a,b is written $a \vee b$, and the minimum is written $a \wedge b$. If a sequence x_n increases to a limit x, we write $x_n \nearrow x$. In $\mathbb{R}^d$ we write $|x|$ for the Euclidean norm of x, and $\mathcal{L}^d$ for d-dimensional Lebesgue measure. The Dirac measure at a point a is the measure $\mathcal{E}_a$ defined such that $\mathcal{E}_a(A) = 1$ if $a \in A$ but $\mathcal{E}_a(A) = 0$ if $a \notin A$. The indicator function of a set A is the function $\mathbb{1}_A$ defined by $\mathbb{1}_A(x) = 1$ if $x \in A$ and $\mathbb{1}_A(x) = 0$ if $x \notin A$. (Thus $\mathbb{1}_A(a) = \mathcal{E}_a(A)$.)

A metric space S is **totally bounded** iff for every $n \in \mathbb{N}$ and every $\varepsilon > 0$ there is a finite set A such that every point of S is within distance ε of a point of A. (A metric space is compact if and only if it is complete and totally bounded. This was essentially proved in passing in [*MTFG*].)

There are two different uses of the Greek letter ω in this book—I hope they will not be confused. One use is to represent a sample point in probability theory (Chapter 4). The other is to represent the least infinite ordinal. That is the meaning in the superscript $E^{(\omega)}$. It indicates that the letters in the strings $\sigma \in E^{(\omega)}$ are to be understood as ordered in that way (a first letter, a second letter, and so on).

Thanks are due to Edmund Mullins and Jeffrey Golds, who read parts of the text and found many errors.

Exercises will be found throughout the text. The reader is invited to provide proofs or investigate a topic further. Some of them are easy, some of them are hard, and a few I do not know how to solve. They should always be understood in the sense of "prove or disprove"; when an assertion turns out to be wrong, try to see what you can do to salvage it. Some of the exercises have hints (or even solutions) elsewhere in the text.

The ninety-nine figures in the text were drawn by the author on a Power Macintosh using these programs: Maple V, Photoshop, MacDraw Pro, Logo-Mation, and Fractal Attraction. The text was typeset using $\mathcal{A}_{\mathcal{M}}\mathcal{S}$-TEX, with XY-pic and some macros provided by Springer-Verlag.

Columbus, Ohio   Gerald A. Edgar

Textbooks and Heaven only are Ideal.
—John Updike, *Scientific American* (January 1969)

Contents

---

* Asterisks indicate optional sections.

1. Fractal Measures

We will begin with a review of some measure theory, with an eye to the most common measures used in the study of fractals: the Hausdorff measures and the packing measures. In [MTFG] we reached the Hausdorff measure only near the end of the book, because of the prerequisites that had to be taken care of first. Here we start off with it immediately. In [MTFG] the packing measures were barely mentioned. I think if I were starting to write that book today, I would give them a more prominent place, almost as prominent as the Hausdorff measure.

This chapter will deal briefly with fractal measures other than these two. It will also discuss a few (more or less) "geometric" properties of fractal measures.

1.1 Measure Theory

For this chapter, some basic knowledge of measure theory is required, such as [MTFG, Chapter 5]. Let us begin with some review of these basics, and at the same time fix some notation and terminology.

We will do computations in the extended real number system, including ∞ and $-\infty$. In addition to the obvious conventions, it will be useful to adopt a convention that is not obvious: We define the product $0 \cdot \infty$ as 0. For example, a line

$$\{ (x, y) \in \mathbb{R}^2 : x = 0 \}$$

in the plane has area (two-dimensional Lebesgue measure or two-dimensional Hausdorff measure) zero. It can be thought of as the product of an interval $[0, 0]$ of length zero by an interval $(-\infty, \infty)$ of infinite length. The area of such a "rectangle" is the product of its two dimensions. But be careful with the convention $0 \cdot \infty = 0$. Making such a definition does not eliminate the fact that $0 \cdot \infty$ is an indeterminate form. Multiplication is *not* continuous at that point.

We will write $\mathcal{L}$ for Lebesgue measure on $\mathbb{R}$, and $\mathcal{L}^d$ for d-dimensional Lebesgue measure on $\mathbb{R}^d$.

Carathéodory's Outer Measures. Let X be a set. An **outer measure** on X is a set-function $\overline{M}$ that assigns to every subset $A \subseteq X$ an element $\overline{M}(A) \in [0, \infty]$ and satisfies

(a) $\overline{\mathcal{M}}(\varnothing) = 0$.

(b) Monotone: If $A \subseteq B$, then $\overline{\mathcal{M}}(A) \leq \overline{\mathcal{M}}(B)$.

(c) Countably subadditive: If $A_1, A_2, \cdots$ are disjoint, then

$$\overline{\mathcal{M}}\left(\bigcup_{n=1}^{\infty} A_n\right) \leq \sum_{n=1}^{\infty} \overline{\mathcal{M}}(A_n).$$

(1.1.1) Exercise. Show that (b) and (c) may be replaced by:

(c*) If A_n are any subsets of X, and $A \subseteq \bigcup_{n=1}^{\infty} A_n$, then

$$\overline{\mathcal{M}}(A) \leq \sum_{n=1}^{\infty} \overline{\mathcal{M}}(A_n).$$

Note that this combines (b) and (c), as well as dropping the requirement of disjointness.

A set $E \subseteq X$ is called $\overline{\mathcal{M}}$-**measurable** iff every set $A \subseteq X$ satisfies

$$\overline{\mathcal{M}}(A) = \overline{\mathcal{M}}(A \cap E) + \overline{\mathcal{M}}(A \setminus E).$$

The collection $\mathcal{F}$ of all measurable sets is a σ-algebra of subsets of X, and the restriction of $\overline{\mathcal{M}}$ to $\mathcal{F}$ is a measure. We will often write $\mathcal{M}$ for this restriction.

(1.1.2) Exercise. An outer measure $\overline{\mathcal{M}}$ is **continuous from below** provided that if $A_1 \subseteq A_2 \subseteq \cdots$, then

$$(*) \qquad \overline{\mathcal{M}}\left(\bigcup_{n=1}^{\infty} A_n\right) = \lim_{n \to \infty} \overline{\mathcal{M}}(A_n).$$

   (a) Show by example that an outer measure need not be continuous from below.

   (b) Show that $(*)$ is true when the sets A_n are measurable.

An outer measure $\overline{\mathcal{M}}$ on a set X is called **regular** iff for every set $A \subseteq X$, there is a measurable set $E \supseteq A$ with $\overline{\mathcal{M}}(A) = \overline{\mathcal{M}}(E)$.

(1.1.3) Theorem. *A regular outer measure is continuous from below.*

Proof. Let $\overline{\mathcal{M}}$ be a regular outer measure, and let $A_1 \subseteq A_2 \subseteq \cdots$ be an increasing sequence of sets. Write $A = \bigcup_{n=1}^{\infty} A_n$. I must show that $\overline{\mathcal{M}}(A) = \lim_{n \to \infty} \overline{\mathcal{M}}(A_n)$. Since $A_n \subseteq A$ for each n, we have $\overline{\mathcal{M}}(A) \geq \lim_{n \to \infty} \overline{\mathcal{M}}(A_n)$. So I must prove the opposite inequality.

   For each n, choose a measurable set $E_n \supseteq A_n$ with $\overline{\mathcal{M}}(E_n) = \overline{\mathcal{M}}(A_n)$. Then let

$$C_n = \bigcap_{k=n}^{\infty} E_k.$$

Then C_n is measurable, and $A_n \subseteq C_n \subseteq E_n$, so $\overline{\mathcal{M}}(C_n) = \overline{\mathcal{M}}(A_n)$. But also $C_1 \subseteq C_2 \subseteq \cdots$. Now let $C = \bigcup_{n=1}^{\infty} C_n$. We have $\overline{\mathcal{M}}(C) = \lim_n \overline{\mathcal{M}}(C_n)$ since the sets C_n are measurable. But also $A \subseteq C$, so

$$\overline{\mathcal{M}}(A) \leq \overline{\mathcal{M}}(C) = \lim_{n \to \infty} \overline{\mathcal{M}}(C_n) = \lim_{n \to \infty} \overline{\mathcal{M}}(A_n),$$

as required.   ☺

Borel Space. A set X together with a σ-algebra of subsets will be called a **Borel space** or a **measurable space**. The fact that there are two different terms stems from the fact that the concept generalizes two different settings. Some texts use the term "Borel structure" to refer to a σ-algebra. One good example of a Borel space is a metric space S and the σ-algebra $\mathcal{B}(S)$ of Borel sets in S [MTFG, §5.2]. Another example is a set X together with the σ-algebra of sets measurable with respect to a given outer measure $\overline{\mathcal{M}}$ on X.

If X is a set and $\mathcal{A}$ is some family of subsets of X, then there is a smallest σ-algebra $\mathcal{F}$ on X that contains $\mathcal{A}$. This σ-algebra is called the σ-algebra generated by $\mathcal{A}$ [MTFG, Theorem (5.2.1)]. This definition is not constructive; but because of its form, it is adapted to an "inductive" kind of proof. To show that every set $A \in \mathcal{F}$ satisfies a proposition $P(A)$, it suffices to show

(1) every set $A \in \mathcal{A}$ satisfies $P(A)$;
(2) the collection of all sets A satisfying $P(A)$ is a σ-algebra.

(1.1.4) Exercise. Let $\mathcal{A}$ be a finite family of subsets of X. Then there is a good description of the σ-algebra generated: First add the complements, $\mathcal{A}_1 = \mathcal{A} \cup \{X \setminus A : A \in \mathcal{A}\}$. Then take intersections, $\mathcal{A}_2 = \{\bigcap \mathcal{C} : \mathcal{C} \subseteq \mathcal{A}_1\}$. Finally take unions, $\mathcal{A}_3 = \{\bigcup \mathcal{C} : \mathcal{C} \subseteq \mathcal{A}_2\}$. Show that $\mathcal{A}_3$ is the σ-algebra generated by $\mathcal{A}$.

Let $\mathcal{F}$ be a σ-algebra. A minimal nonempty element of $\mathcal{F}$ is called an **atom** of $\mathcal{F}$; that is, $A \in \mathcal{F}$ is an atom iff $A \neq \varnothing$, but for all $B \in \mathcal{F}$, if $B \subseteq A$, then either $B = \varnothing$ or $B = A$.

(1.1.5) Exercise. Let $\mathcal{F}$ be a finite σ-algebra. Write $\mathcal{A} = \{A_1, A_2, \cdots, A_n\}$ for the set of all atoms of $\mathcal{F}$. Show that $\mathcal{F}$ consists of exactly 2^n distinct elements, the unions of subcollections of $\mathcal{A}$.

Measures. Let $(X, \mathcal{F})$ be a Borel space. A **measure** on $(X, \mathcal{F})$ is a function $\mathcal{M} \colon \mathcal{F} \to [0, \infty]$ such that

(1) $\mathcal{M}(\varnothing) = 0$;
(2) If $A_n \in \mathcal{F}$ is a disjoint sequence of sets, then $\mathcal{M}(\bigcup A_n) = \sum \mathcal{M}(A_n)$.

Suppose $(X, \mathcal{F})$ is a Borel space. If $E \subseteq X$, then there is a natural Borel structure on E induced by the Borel structure $\mathcal{F}$. We will call it the **restriction** of $\mathcal{F}$ to E and write

$$\mathcal{F}|E = \{A \cap E : A \in \mathcal{F}\}.$$

If $E \in \mathcal{F}$, then it may be written also as

$$\mathcal{F}|E = \{A \in \mathcal{F} : A \subseteq E\}.$$

Let $(X, \mathcal{F}, \mathcal{M})$ be a measure space, and let $E \in \mathcal{F}$. The **restriction** of the measure $\mathcal{M}$ to the set E is the set-function $\mathcal{M}|E$ defined by

$$\mathcal{M}|E(A) = \mathcal{M}(A \cap E) \quad \text{for all } A \in \mathcal{F}.$$

The restriction is a measure on X, but it assigns measure zero to the complement of E. Thus we may, in some cases, consider it to be a measure on the subset Borel space $(E, \mathcal{F}|E)$.

The measure space $(X, \mathcal{F}, \mathcal{M})$ (or the measure $\mathcal{M}$) is called **finite** iff $\mathcal{M}(X) < \infty$. A **probability measure** is a measure $\mathcal{M}$ on a set X with $\mathcal{M}(X) = 1$. (See §4.1.)

Suppose $(X, \mathcal{F}, \mathcal{M})$ is a measure space. If we say that something is true **almost everywhere** on X, we mean that the set A of points x where it fails is a set of measure zero, $\mathcal{M}(A) = 0$. We may also say that something is true for **almost all** points x. In integration theory, we can often ignore a set of measure zero, so what happens almost everywhere will often be all that matters. If more than one measure is being considered, we may say "$\mathcal{M}$-almost everywhere" or "$\mathcal{M}$-almost all" to specify which measure is meant. In Euclidean space $\mathbb{R}^d$, when we say "almost everywhere" without mentioning a measure, it is to be understood that Lebesgue measure $\mathcal{L}^d$ is intended.

If A and B are sets, their **symmetric difference** consists of those points that belong to exactly one of the sets:

$$A \triangle B = (A \setminus B) \cup (B \setminus A).$$

(1.1.6) Exercise. Let $\mathcal{F}$ be a σ-algebra of subsets of X. Let $\mathcal{M}$ be a finite measure on $\mathcal{F}$. Define $\rho(A, B) = \mathcal{M}(A \triangle B)$. Then for all $A, B, C \in \mathcal{F}$:

(a) $\rho(A, A) = 0$;
(b) $\rho(A, B) = \rho(B, A)$;
(c) $\rho(A, C) \leq \rho(A, B) + \rho(B, C)$.

So this ρ is almost a metric (it lacks only the assertion that $\rho(A, B) = 0$ only if $A = B$).

Measurable Functions. If $(X_1, \mathcal{F}_1)$ and $(X_2, \mathcal{F}_2)$ are Borel spaces, we will often want to discuss functions (mappings, transformations) from one to the other. It may be useful for such a function to be compatible with the Borel structures. A function

$$f: X_1 \rightarrow X_2$$

is called a **Borel function** (from $\mathcal{F}_1$ to $\mathcal{F}_2$) iff $f^{-1}[E] \in \mathcal{F}_1$ for all $E \in \mathcal{F}_2$. Such a function will also be called "measurable" with respect to $\mathcal{F}_1$ and $\mathcal{F}_2$.

There are some elementary ways to combine Borel functions. Suppose $(X_i, \mathcal{F}_i)$ for $i = 1, 2, 3$, are three Borel spaces. If $f \colon X_1 \to X_2$ is a Borel function from $\mathcal{F}_1$ to $\mathcal{F}_2$ and $g \colon X_2 \to X_3$ is a Borel function from $\mathcal{F}_2$ to $\mathcal{F}_3$, then clearly the composite function $g \circ f \colon X_1 \to X_3$ is a Borel function from $\mathcal{F}_1$ to $\mathcal{F}_3$.

Suppose $X_1 = \bigcup_{i=1}^{\infty} E_i$ is a disjoint union, where $E_i \in \mathcal{F}_1$ for all i. Suppose functions $f_i \colon E_i \to X_2$ are all Borel functions from $\mathcal{F}_1 | E_i$ to $\mathcal{F}_2$. Then a new function $f \colon X_1 \to X_2$ may be defined by

$$f(x) = f_i(x) \qquad \text{for } x \in E_i, i = 1, 2, \cdots.$$

The function f defined in this way will be measurable from $\mathcal{F}_1$ to $\mathcal{F}_2$: if $B \in \mathcal{F}_2$, then

$$f^{-1}[B] = \bigcup_{i=1}^{\infty} f_i^{-1}[B],$$

a countable union of $\mathcal{F}_1$-measurable sets.

Pi and Lambda Systems. The definition of a σ-algebra may be divided into two parts in a useful way. Let X be a set. A collection $\mathcal{A}$ of subsets of X is called a **pi system** iff it is closed under finite intersections: If $A, B \in \mathcal{A}$, then $A \cap B \in \mathcal{A}$. A collection $\mathcal{A}$ of subsets of X is called a **lambda system** iff it contains the whole space and is closed under proper differences and increasing countable unions:

(1) $X \in \mathcal{A}$;
(2) if $A, B \in \mathcal{A}$ and $A \subseteq B$, then $B \setminus A \in \mathcal{A}$;
(3) if $A_n \in \mathcal{A}$ $(n = 1, 2, \cdots)$ with $A_n \subseteq A_{n+1}$, then $\bigcup_n A_n \in \mathcal{A}$.

Note that a collection $\mathcal{A}$ is a σ-algebra if and only if it is both a pi system and a lambda system.

(1.1.7) Pi–Lambda Theorem. *Let X be a set, let $\mathcal{A}$ be a pi system on X, and let $\mathcal{B}$ be a lambda system on X. If $\mathcal{B} \supseteq \mathcal{A}$, then $\mathcal{B}$ contains the σ-algebra generated by $\mathcal{A}$.*

Proof. Let $\mathcal{D}$ be the smallest lambda system containing $\mathcal{A}$: There is at least one lambda system containing $\mathcal{A}$ (namely $\mathcal{B}$), and the intersection of any family of lambda systems containing $\mathcal{A}$ is again a lambda system containing $\mathcal{A}$; so there is a smallest lambda system $\mathcal{D}$ containing $\mathcal{A}$. Clearly, $\mathcal{A} \subseteq \mathcal{D} \subseteq \mathcal{B}$.

I claim that $\mathcal{D}$ is also a pi system. If this is proved, then $\mathcal{D}$ will be a σ-algebra, and therefore $\mathcal{D}$ will be the σ-algebra generated by $\mathcal{A}$, so this will complete the proof.

For a set $A \in \mathcal{D}$, let

$$\mathcal{G}_A = \{ B \in \mathcal{B} : A \cap B \in \mathcal{D} \}.$$

Since $\mathcal{D}$ is a lambda system, it follows easily that $\mathcal{G}_A$ is a lambda system. If $A \in \mathcal{A}$, then (since $\mathcal{A}$ is a pi system) $\mathcal{G}_A \supseteq \mathcal{A}$; so $\mathcal{G}_A$ is a lambda system that contains $\mathcal{A}$, and therefore (by minimality) $\mathcal{G}_A \supseteq \mathcal{D}$.

Now we know that if $A \in \mathcal{A}$ and $B \in \mathcal{D}$, then $A \cap B \in \mathcal{D}$. So if $B \in \mathcal{D}$, then $\mathcal{A} \subseteq \mathcal{G}_B$; but $\mathcal{G}_B$ is a lambda system, so (by minimality) $\mathcal{D} \subseteq \mathcal{G}_B$. Thus, if $A, B \in \mathcal{D}$, then $A \cap B \in \mathcal{D}$. This shows that $\mathcal{D}$ is a pi system, as required. ☺

Here is a simple application of the Pi–Lambda Theorem. The collection of sets where two measures agree is not, in general, a σ-algebra, so the result is not immediate from the definitions.

(1.1.8) Proposition. *Let $(X, \mathcal{F})$ be a Borel space. Let $\mathcal{A} \subseteq \mathcal{F}$ be a pi system. Let $\mathcal{M}_1$ and $\mathcal{M}_2$ be two probability measures on $\mathcal{F}$. If $\mathcal{M}_1(A) = \mathcal{M}_2(A)$ for all $A \in \mathcal{A}$, then also $\mathcal{M}_1(A) = \mathcal{M}_2(A)$ for all A in the σ-algebra generated by $\mathcal{A}$.*

Proof. To apply the Pi–Lambda Theorem, it suffices to verify that the set of A satisfying $\mathcal{M}_1(A) = \mathcal{M}_2(A)$ is a lambda system. Since $\mathcal{M}_1$ and $\mathcal{M}_2$ are both probability measures, $\mathcal{M}_1(X) = \mathcal{M}_2(X)$. If $A \subseteq B$, $\mathcal{M}_1(A) = \mathcal{M}_2(A)$, and $\mathcal{M}_1(B) = \mathcal{M}_2(B)$, then $\mathcal{M}_1(B \setminus A) = \mathcal{M}_1(B) - \mathcal{M}_1(A) = \mathcal{M}_2(B) - \mathcal{M}_2(A) = \mathcal{M}_2(B \setminus A)$. And if A_n increases to A, with $\mathcal{M}_1(A_n) = \mathcal{M}_2(A_n)$ for all n, then $\mathcal{M}_1(A) = \lim \mathcal{M}_1(A_n) = \lim \mathcal{M}_2(A_n) = \mathcal{M}_2(A)$. ☺

(1.1.9) Algebra Approximation Theorem. *Let $(X, \mathcal{F}, \mathcal{M})$ be a finite measure space. Let $\mathcal{R} \subseteq \mathcal{F}$ be an algebra of sets and let $\mathcal{G}$ be the σ-algebra generated by $\mathcal{R}$. Then $\mathcal{R}$ is dense in $\mathcal{G}$ according to the symmetric difference. That is, for every $G \in \mathcal{G}$ and every $\varepsilon > 0$, there is $R \in \mathcal{R}$ such that $\mathcal{M}(R \triangle G) < \varepsilon$.*

Proof. Let

$$\mathcal{G}' = \{ G \in \mathcal{G} : \text{for every } \varepsilon > 0, \text{ there is } R \in \mathcal{R} \text{ such that } \mathcal{M}(R \triangle G) < \varepsilon \}.$$

Then certainly $\mathcal{R}$ is a pi system and $\mathcal{R} \subseteq \mathcal{G}'$. It only remains to show that $\mathcal{G}'$ is a lambda system.

Certainly $X \in \mathcal{G}'$.

Suppose $A, B \in \mathcal{G}'$ and $A \subseteq B$. Given $\varepsilon > 0$, there are $R, S \in \mathcal{R}$ with $\mathcal{M}(R \triangle A) < \varepsilon/2$ and $\mathcal{M}(S \triangle B) < \varepsilon/2$. But $(B \setminus A) \triangle (S \setminus R) \subseteq (B \triangle S) \cup (A \triangle R)$, so we may conclude that $\mathcal{M}\big((B \setminus A) \triangle (S \setminus R)\big) < \varepsilon$. Therefore, $B \setminus A \in \mathcal{G}'$.

Suppose that $A_n \in \mathcal{G}'$ and $A_n \nearrow A$. Given $\varepsilon > 0$, there is n such that $\mathcal{M}(A \setminus A_n) < \varepsilon/2$. Then there is $R \in \mathcal{R}$ such that $\mathcal{M}(R \triangle A_n) < \varepsilon/2$. Then $\mathcal{M}(A \triangle R) \leq \mathcal{M}(A \triangle A_n) + \mathcal{M}(A_n \triangle R) < \varepsilon$. Therefore, $A \in \mathcal{G}'$. ☺

Measures in Metric Space. If S is a metric space, the **Borel sets** in S constitute the σ-algebra $\mathcal{B}(S)$ generated by the open sets [*MTFG*, §5.2]. If S is separable, then by the Lindelöf property, $\mathcal{B}(S)$ is generated by the open balls of S. When we deal with a metric space, this natural Borel structure will be understood.

Suppose S_1 and S_2 are metric spaces. A function $f\colon S_1 \to S_2$ is called a **Borel function** iff it is measurable from $\mathcal{B}(S_1)$ to $\mathcal{B}(S_2)$. That is, $f^{-1}[B] \in \mathcal{B}(S_1)$ for all $B \in \mathcal{B}(S_2)$. Now, $\mathcal{B}(S_2)$ is generated by the open sets in S_2, so f is a Borel function if and only if $f^{-1}[B] \in \mathcal{B}(S_1)$ for all open sets $B \subseteq S_2$. This will be true in particular if $f^{-1}[B]$ is open in S_1 for all open sets $B \subseteq S_2$; that is, if $f\colon S_1 \to S_2$ is continuous.

An outer measure $\overline{\mathcal{M}}$ on a metric space S is called a **metric outer measure** iff it satisfies

$$\overline{\mathcal{M}}(A \cup B) = \overline{\mathcal{M}}(A) + \overline{\mathcal{M}}(B)$$

for any sets A, B with positive separation $\mathrm{dist}(A, B) > 0$. The interest of metric outer measures is that all Borel sets are measurable [*MTFG*, (5.2.6)]. The converse is true as well:

(1.1.10) Exercise. Let S be a metric space, and let $\overline{\mathcal{M}}$ be an outer measure on S. If all Borel sets are $\overline{\mathcal{M}}$-measurable, then $\overline{\mathcal{M}}$ is a metric outer measure.

Let (S, ρ) be a metric space, and let $\mathcal{M}$ be a Borel measure on S. We say that $\mathcal{M}$ is **supported** by the closed set $F \subseteq S$ iff $\mathcal{M}(S \setminus F) = 0$. If S is separable, then the intersection of all closed sets that support $\mathcal{M}$ is again a set that supports $\mathcal{M}$. This smallest closed set supporting $\mathcal{M}$ is known as the **support** of $\mathcal{M}$.

Here is a consequence of the Pi–Lambda Theorem.

(1.1.11) Exercise. Let $\mathcal{M}_1, \mathcal{M}_2$ be two Borel measures on the metric space S. If $\mathcal{M}_1(F) = \mathcal{M}_2(F)$ for all closed sets $F \subseteq S$, then $\mathcal{M}_1(E) = \mathcal{M}_2(E)$ for all Borel sets.

A result more informative than the preceding exercise says that Borel sets can be approximated by closed sets or by open sets.

(1.1.12) Theorem. *Let (S, ρ) be a metric space, and let $\mathcal{M}$ be a finite Borel measure on S. Suppose $A \subseteq S$ is a Borel set. Then for any positive number ε, there exist a closed set F and an open set U with $F \subseteq A \subseteq U$ and $\mathcal{M}(U \setminus F) < \varepsilon$.*

Proof. Let $\mathcal{A}$ be the collection of all sets A with the approximation property specified. I will show that $\mathcal{A}$ is a σ-algebra containing the open sets.

Suppose first that A is an open set. The sets

$$F_n = \left\{ x \in S : B_{1/n}(x) \subseteq A \right\}$$

increase to A in the sense that $F_1 \subseteq F_2 \subseteq \cdots$, and $\bigcup_n F_n = A$. Therefore, $\lim_n \mathcal{M}(F_n) = \mathcal{M}(A)$, so there is n so large that $\mathcal{M}(A \setminus F_n) < \varepsilon$. The sets F_n are closed: if $y \notin F_n$, then $B_{1/n}(y) \not\subseteq A$, so there is $x \notin A$ with $\rho(x, y) < 1/n$. If r satisfies $r < 1/n - \rho(x, y)$, then $B_r(y) \cap F_n = \varnothing$. Thus the complement of F_n is open. The sets $F = F_n$ and $U = A$ show that $A \in \mathcal{A}$.

Now suppose $A \in \mathcal{A}$, and consider the complement $X \setminus A$. There exist closed F and open U with $F \subseteq A \subseteq U$ and $\mathcal{M}(U \setminus F) < \varepsilon$. But $F' = X \setminus U$ is closed, $U' = X \setminus F$ is open, $F' \subseteq X \setminus A \subseteq U'$, and $\mathcal{M}(U' \setminus F') = \mathcal{M}(U \setminus F) < \varepsilon$. So also $X \setminus A \in \mathcal{A}$.

Suppose $A_n \in \mathcal{A}$ for $n \in \mathbb{N}$. Let $\varepsilon > 0$. There exist open sets U_n and closed sets F_n with $F_n \subseteq A_n \subseteq U_n$ and $\mathcal{M}(U_n \setminus F_n) < \varepsilon/2^{n+1}$. Now

$$\bigcup_n F_n \subseteq \bigcup_n A_n \subseteq \bigcup_n U_n$$

and

$$\mathcal{M}\left(\left(\bigcup U_n\right) \setminus \left(\bigcup F_n\right)\right) \le \mathcal{M}\left(\bigcup(U_n \setminus F_n)\right) \le \sum \mathcal{M}(U_n \setminus F_n)$$
$$\le \sum \frac{\varepsilon}{2^{n+1}} = \frac{\varepsilon}{2}.$$

The set $U = \bigcup U_n$ is open. Choose m large enough that

$$\mathcal{M}\left(\bigcup_{n=1}^{\infty} F_n\right) - \mathcal{M}\left(\bigcup_{n=1}^{m} F_n\right) < \varepsilon/2.$$

The set $F = \bigcup_{n=1}^{m} F_n$ is closed. Finally, $F \subseteq \bigcup A_n \subseteq U$ and $\mathcal{M}(U \setminus F) < \varepsilon$. Therefore, $\bigcup A_n \in \mathcal{A}$.

Therefore, $\mathcal{A}$ is a σ-algebra containing the open sets. So $\mathcal{A}$ contains all Borel sets.   ☺

A measure space $(X, \mathcal{F}, \mathcal{M})$ is said to be σ-**finite** iff the space is a countable union of measurable sets of finite measure: there exist $X_n \in \mathcal{F}$ with $\mathcal{M}(X_n) < \infty$ and $X = \bigcup_{n=1}^{\infty} X_n$. Lebesgue measure $\mathcal{L}^d$ in $\mathbb{R}^d$ is σ-finite, since the whole space is the countable union of the cubes X_n, centered at the origin, with side n.

(1.1.13) Exercise. Does Theorem 1.1.12 remain correct for a σ-finite Borel measure on a metric space S?

A complete separable metric space admits many compact subsets. The following theorem is one way to exploit this fact.

(1.1.14) Theorem. *Let S be a complete separable metric space, and let $\mathcal{M}$ be a finite Borel measure on S. Then*

$$\mathcal{M}(S) = \sup\{\mathcal{M}(K) : K \subseteq S, K \text{ compact}\}.$$

Proof. If $\mathcal{M}(S) = 0$, there is nothing to prove, so suppose $\mathcal{M}(S) > 0$. Fix a number $a < \mathcal{M}(S)$. I will show that there is a compact set $K \subseteq S$ with $\mathcal{M}(K) \ge a$.

Begin by defining recursively a sequence T_n of closed sets. $T_0 = S$, so T_0 is closed and $\mathcal{M}(T_0) > a$. Suppose $n \ge 1$ and T_{n-1} has been defined, T_{n-1} is a closed set, and $\mathcal{M}(T_{n-1}) > a$. The open balls of radius 2^{-n} cover T_{n-1}. By the Lindelöf property, countably many of them also cover T_{n-1}, say

$$\bigcup_{k=1}^{\infty} B_{2^{-n}}(x_k) \supseteq T_{n-1}.$$

Thus

$$\mathcal{M}\left(\bigcup_{k=1}^{\infty} B_{2^{-n}}(x_k) \cap T_{n-1}\right) = \mathcal{M}(T_{n-1}) > a,$$

so by the countable additivity of $\mathcal{M}$, there is a finite number r_n with

$$\mathcal{M}\left(\bigcup_{k=1}^{r_n} B_{2^{-n}}(x_k) \cap T_{n-1}\right) > a.$$

Let

$$T_n = \bigcup_{k=1}^{r_n} \overline{B}_{2^{-n}}(x_k) \cap T_{n-1}.$$

Because the balls used are closed, T_n is a closed set, $T_n \subseteq T_{n-1}$, but still $\mathcal{M}(T_n) > a$. Because of the construction, T_n is covered by finitely many closed balls of radius 2^{-n}.

Continue in this way to define all sets T_n ($n = 1, 2, \cdots$). Let $K = \bigcap_{n=1}^{\infty} T_n$. Then K is a closed set. For each n, the set K is covered by finitely many balls of radius 2^{-n}. That is, K is totally bounded. Also, S is complete and K is closed, so K is also complete. Therefore, K is compact. Finally, the sets T_n decrease, so $\mathcal{M}(K) = \lim_n \mathcal{M}(T_n) \geq a$. ☺

Note the requirement that $\mathcal{M}$ be finite. In general, if $\mathcal{M}(S) = \infty$, there need not be any compact sets of positive measure.

(1.1.15) Corollary. *Let S be a complete separable metric space, and let $\mathcal{M}$ be a finite Borel measure on S. For any Borel set $A \subseteq S$ and any $\varepsilon > 0$, there exist compact sets K_1 and K_2 with $K_1 \subseteq A \subseteq S \backslash K_2$ and $\mathcal{M}(S \backslash (K_1 \cup K_2)) < \varepsilon$.*

Proof. Approximate A from inside by a closed set F using Theorem 1.1.12. F itself is a complete metric space. Then approximate F from inside by a compact set K_1 using Theorem 1.1.14. Similarly, approximate the complement from inside by a compact set K_2. ☺

Methods I and II. The construction of measures is an important part of measure theory, and various constructions will be used in this book.

Let X be a set and E a subset. A collection $\mathcal{A}$ of subsets of X is called a **cover** of E iff every point of E belongs to some set $A \in \mathcal{A}$; we will also say that $\mathcal{A}$ **covers** E. If a set-function $C \colon \mathcal{A} \to [0, \infty]$ is given, there is a greatest outer measure $\overline{\mathcal{M}}$ on X with $\overline{\mathcal{M}}(A) \leq C(A)$ for all $A \in \mathcal{A}$ [MTFG, Theorem (5.2.2)]. This measure may be constructed by a procedure that we will call **method I**: for any set $E \subseteq X$,

$$\overline{\mathcal{M}}(E) = \inf \sum_{A \in \mathcal{D}} C(A),$$

where the infimum is over all countable families $\mathcal{D} \subseteq \mathcal{A}$ that cover E. By convention, inf $\varnothing = \infty$, so if no such cover exists, $\overline{\mathcal{M}}(E) = \infty$. The empty set $E = \varnothing$ is covered by the empty family $\mathcal{D} = \varnothing \subseteq \mathcal{A}$, and the empty sum $\sum_{A \in \varnothing} C(A)$ is 0, so $\overline{\mathcal{M}}(\varnothing) = 0$.

Note that if all elements of $\mathcal{A}$ turn out to be $\overline{\mathcal{M}}$-measurable, then $\overline{\mathcal{M}}$ will be a regular measure.

Let S be a metric space. A **Vitali cover** of a set $E \subseteq S$ is a family $\mathcal{A}$ of subsets of S such that for every $x \in E$ and every $\varepsilon > 0$, there is $A \in \mathcal{A}$ with $x \in A$ and diam $A \leq \varepsilon$. Given a Vitali cover $\mathcal{A}$ of S and a set function $C \colon \mathcal{A} \to [0, \infty]$, there is a construction of an outer measure called **method II**: For each $\varepsilon > 0$, let $\overline{\mathcal{M}}_\varepsilon$ be the method I outer measure constructed from the cover $\{ A \in \mathcal{A} : \text{diam } A \leq \varepsilon \}$ and the set-function C. Then $\overline{\mathcal{M}}(E)$ is the limit of $\overline{\mathcal{M}}_\varepsilon(E)$ as $\varepsilon \to 0$. As ε decreases, $\overline{\mathcal{M}}_\varepsilon(E)$ increases, so this limit exists in $[0, \infty]$. An outer measure $\overline{\mathcal{M}}$ constructed by method II is a metric outer measure [*MTFG*, Theorem (5.4.2)]. So all Borel sets are $\overline{\mathcal{M}}$-measurable.

Measures Constructed as Variations. There are other methods that may be used to construct metric outer measures. Two "variation" constructions will be discussed here. I will basically follow B. Thomson.

Thomson's construction may be done with any "differentiation basis" in a metric space S, but we will be content to use the basis defined by the closed balls $\overline{B}_r(x)$, where $x \in S$ and $r > 0$. Now, in most familiar metric spaces (such as Euclidean space $\mathbb{R}^d$), if we know a closed ball $\overline{B}_r(x)$ as a point-set, then the center x and radius r are uniquely determined. But in "strange" metric spaces, such as ultrametric spaces, this may not be true: It may happen that

$$\overline{B}_r(x) = \overline{B}_{r'}(x')$$

even if $x \neq x'$ and/or $r \neq r'$. So the **constituents** in our constructions will be taken to be ordered pairs (x, r), with $x \in S$ and $r > 0$. Of course, the closed ball $\overline{B}_r(x)$ is associated with the pair (x, r).

There are two "variations" to be defined, called the **full variation** and the **fine variation**.

Let $E \subseteq S$ be a set in the metric space S. A **gauge** on E is a function $\Delta \colon E \to (0, \infty)$. (It is not required to be continuous.) It is intended that we associate to each point x a maximum radius $\Delta(x)$. That radius may be small at some points x and large at other points x. A constituent (x, r) is Δ-**fine** iff $r < \Delta(x)$ (that is, the ball $\overline{B}_r(x)$ stays within the maximum allowed radius $\Delta(x)$). A set β of constituents is Δ-**fine** iff $r < \Delta(x)$ for all $(x, r) \in \beta$. A finite set π of constituents is a **packing** iff its elements are disjoint—that is, $\overline{B}_r(x) \cap \overline{B}_{r'}(x')$ if $(x, r), (x', r') \in \pi$ and $(x, r) \neq (x', r')$. We say that π is a packing of E iff it is a packing such that $x \in E$ for all $(x, r) \in \pi$. (More technically, this may be called a packing of E by centered closed balls.)

We begin with a **constituent function** $C \colon S \times (0, \infty) \to [0, \infty)$. It associates to each constituent (x, r) a nonnegative real number $C(x, r)$. Let $E \subseteq S$ be a set. Let Δ be a gauge on E. Then we define

$$V_\Delta^C(E) = \sup \sum_{(x,r)\in\pi} C(x,r),$$

where the supremum is over all Δ-fine packings π of E. Note that when we decrease the gauge Δ, the value $V_\Delta^C(E)$ decreases. We define the **full variation** of C on E as the limit of $V_\Delta^C(E)$ as $\Delta \to 0$:

$$V^C(E) = \inf V_\Delta^C(E),$$

where the infimum is over all gauges Δ on E.

(1.1.16) Theorem. *Let S be a metric space, and let C be a constituent function. Then the full variation V^C is a metric outer measure.*

Proof. (a) The only packing of $\varnothing$ is the empty packing, and the value of the empty sum is 0, so $V_\Delta^C(\varnothing) = 0$ and thus $V^C(\varnothing) = 0$.

(b) Suppose $E \subseteq F$. If Δ is a gauge on F, then the restriction $\Delta{\restriction}E$ is a gauge on E. Any $(\Delta{\restriction}E)$-fine packing of E is also a Δ-fine packing of F, so $V_{\Delta{\restriction}E}^C(E) \leq V_\Delta^C(F)$. By the definition of V^C, this yields $V^C(E) \leq V_\Delta^C(F)$. Take the infimum over Δ to get $V^C(E) \leq V^C(F)$.

(c) Suppose $E = \bigcup_{n=1}^\infty E_n$. We claim that $V^C(E) \leq \sum V^C(E_n)$. Now, if the right-hand side is ∞, there is nothing to prove, so suppose it is finite. Fix $\varepsilon > 0$. For each n, choose a gauge Δ_n on E_n such that $V_{\Delta_n}^C(E_n) \leq V^C(E_n) + \varepsilon/2^n$. Define Δ on E by $\Delta(x) = \Delta_n(x)$ if $x \in E_n \setminus \bigcup_{k=1}^{n-1} E_k$. Then Δ is a gauge on E. Now let π be a Δ-fine packing of E. Then for each n,

$$\pi_n = \left\{ (x,r) \in \pi : x \in E_n \setminus \bigcup_{k=1}^{n-1} E_k \right\}$$

is a Δ_n-fine packing of E_n. So

$$\sum_{(x,r)\in\pi} C(x,r) = \sum_n \sum_{(x,r)\in\pi_n} C(x,r) \leq \sum_n V_{\Delta_n}^C(E_n) \leq \varepsilon + \sum_n V^C(E_n).$$

Take the supremum on π to get $V_\Delta^C(E) \leq \varepsilon + \sum V^C(E_n)$. So by the definition of V^C, we have $V^C(E) \leq \varepsilon + \sum V^C(E_n)$. Then take the infimum on ε to conclude that $V^C(E) \leq \sum V^C(E_n)$.

(d) Suppose E and F have positive separation, $\mathrm{dist}(E,F) = \varepsilon > 0$. Let Δ be any gauge on $E \cup F$ with $\Delta(x) < \varepsilon/2$. Then the constituents of any Δ-fine packing of E are disjoint from the constituents of any Δ-fine packing of F. So the union of two such packings is a Δ-fine packing of $E \cup F$. Therefore, $V_\Delta^C(E \cup F) = V_\Delta^C(E) + V_\Delta^C(F)$. This is true for all sufficiently small gauges, so taking the limit yields $V^C(E \cup F) = V^C(E) + V^C(F)$.   ☺

(1.1.17) Exercise. In the metric space $S = \mathbb{R}$, define $\underline{C}(x,r) = 2r$. Show that the full variation V^C is the Lebesgue outer measure $\overline{\mathcal{L}}$.

With the preceding exercise as a warm-up, try this one:

(1.1.18) Exercise. Let S be a metric space, and let $\mathcal{N}$ be a finite Borel measure on S. Let a constituent function C be defined by $C(x,r) = \mathcal{N}(\overline{B}_r(x))$. Show that $V^C(F) \le \mathcal{N}(F)$ for all Borel sets F.

We will see later (Proposition 1.3.18) that in many situations there is equality: $V^C(F) = \mathcal{N}(F)$. But it is not always true ([54], [186, p. 42]).

Next we turn to the definition of the fine variation. Let $E \subseteq S$. A set β of constituents is called a **fine cover** of E iff β contains arbitrarily small balls centered at each point of E; that is, $x \in E$ for all $(x,r) \in \beta$, and for every $x \in E$ and every $\delta > 0$ there is $(x,r) \in \beta$ with $r < \delta$. As before, let C be a constituent function. Now, a fine cover is likely to be an infinite set (or even an uncountable set), but we may define

$$v_\beta^C = \sup \sum_{(x,r) \in \pi} C(x,r),$$

where the supremum is over all packings $\pi \subseteq \beta$. The **fine variation** of C on E is

$$v^C(E) = \inf v_\beta^C,$$

where the infimum is over all fine covers β of E.

(1.1.19) Theorem. *Let S be a metric space, and let C be a constituent function. Then the fine variation v^C is a metric outer measure.*

Proof. (a) The only fine cover of $\varnothing$ is the empty cover $\varnothing$, which contains only the empty packing, so $v_\varnothing^C = 0$ and $v^C(\varnothing) = 0$.

(b) Suppose $E \subseteq F$. Let β be a fine cover of F. Then

$$\beta' = \{ (x,r) \in \beta : x \in E \}$$

is a fine cover of E. If $\pi \subseteq \beta'$ is a packing of E, then also $\pi \subseteq \beta$ is a packing of F, so $v_{\beta'}^C \le v_\beta^C$. Therefore, $v^C(E) \le v_\beta^C$. Taking the infimum over β, we get $v^C(E) \le v^C(F)$.

(c) Suppose $E = \bigcup_{n=1}^\infty E_n$. We must show that $v^C(E) \le \sum_{n=1}^\infty v^C(E_n)$. Because of the monotonicity proved in (b), we may assume that the E_n are disjoint. If $\sum v^C(E_n) = \infty$, we are done, so assume it is $< \infty$. Let $\varepsilon > 0$ be given. Now let β_n be a fine cover of E_n such that $v_{\beta_n}^C < v^C(E_n) + \varepsilon/2^n$. Then $\beta = \bigcup_n \beta_n$ is a fine cover of E. Consider a packing $\pi \subseteq \beta$. We may write $\pi = \bigcup \pi_n$, where each $\pi_n \subseteq \beta_n$. Then

$$\sum_{(x,r) \in \pi_n} C(x,r) \le v_{\beta_n}^C \le v^C(E_n) + \frac{\varepsilon}{2^n},$$

and thus

$$\sum_{(x,r) \in \pi} C(x,r) \le \varepsilon + \sum_n v^C(E_n).$$

Take the supremum over π to get $v_\beta^C \leq \varepsilon + \sum v^C(E_n)$. Take the infimum over ε to get $v^C(E) \leq \sum v^C(E_n)$.

(d) Suppose E and F are positively separated: $\mathrm{dist}(E, F) = \varepsilon > 0$. Let β be any fine cover of $E \cup F$. Then

$$\beta' = \left\{ (x, r) \in \beta : r < \frac{\varepsilon}{2} \right\}$$

is also a fine cover of $E \cup F$, and $v_{\beta'}^C \leq v_\beta^C$. Now

$$\beta_E = \{ (x, r) \in \beta' : x \in E \},$$
$$\beta_F = \{ (x, r) \in \beta' : x \in F \}$$

are fine covers of E and F, respectively, and $v_{\beta'}^C = v_{\beta_E}^C + v_{\beta_F}^C$. Now

$$v^C(E) + v^C(F) \leq v_{\beta_E}^C + v_{\beta_F}^C = v_{\beta'}^C \leq v_\beta^C.$$

Take the infimum over β to get $v^C(E) + v^C(F) \leq v^C(E \cup F)$.   ☺

The two variations (defined from the same constituent function C) are related.

(1.1.20) Theorem. *Let S be a metric space, and let C be a constituent function. Then $v^C(F) \leq V^C(F)$ for all $F \subseteq S$.*

Proof. If $V^C(F) = \infty$, there is nothing to prove, so assume that $V^C(F) < \infty$. Let $\varepsilon > 0$ be given. There is a gauge Δ on F such that $V_\Delta^C(F) \leq V^C(F) + \varepsilon$. Now,

$$\beta = \{ (x, r) : x \in F, r < \Delta(x) \}$$

is a fine cover of F. If $\pi \subseteq \beta$ is a packing, then it is Δ-fine, so

$$\sum_{(x,r) \in \pi} C(x, r) \leq V_\Delta^C(F) \leq V^C(F) + \varepsilon.$$

This is true for all π, so $v_\beta^C \leq V^C(F) + \varepsilon$. Therefore, $v^C(F) \leq V^C(F) + \varepsilon$. Take the infimum on ε to get $v^C(F) \leq V^C(F)$.   ☺

1.2 Hausdorff and Packing Measures

When we say that a surface in space is "two-dimensional," one of the characteristics that we have in mind is the possibility of measuring its "surface area." Similarly, for a "one-dimensional" curve, we may try to measure its "arc length"; for a "three-dimensional" solid figure, we may try to measure its volume. In the same way, when we discuss "fractal" sets, with "fractal dimension" s that is possibly not an integer, it may be possible to measure the

"s-dimensional content" of the set. Roughly speaking, this is what the **fractal measures** that are discussed here will do for us.

We will consider **local** fractal measures, in the sense that the computation of the measure of a set depends only on how the very small parts of the set behave, not on how the overall large-scale parts of the set behave. So we are interested in taking limits as the measuring scale decreases to zero.[1]

In this section we define the two fractal measures that will be most used in this book: the Hausdorff measure $\mathcal{H}^s$ and the packing measure $\mathcal{P}^s$. Some other local fractal measures will be considered in Section 1.4.

Hausdorff Measure. Let S be a metric space, with metric ρ. The **diameter** of a set $A \subseteq S$ is

$$\operatorname{diam} A = \sup \{ \rho(x,y) : x, y \in A \}.$$

A set with finite diameter is called **bounded**. Let s be a positive real number. The s-**dimensional Hausdorff outer measure** is the method II outer measure $\overline{\mathcal{H}}^s$ defined using the Vitali cover of all bounded subsets of S and the set-function

$$C(A) = (\operatorname{diam} A)^s.$$

So the outer measure may be computed as a limit,

$$\overline{\mathcal{H}}^s(A) = \lim_{\varepsilon \to 0} \overline{\mathcal{H}}^s_\varepsilon(A),$$

where $\overline{\mathcal{H}}^s_\varepsilon$ is the method I measure constructed using the cover consisting of of all sets A with diameter $< \varepsilon$, and the same set-function $C(A)$.

It will be convenient to define a 0-**dimensional Hausdorff outer measure** $\overline{\mathcal{H}}^0$ in the same way, where we interpret $(\operatorname{diam} A)^0$ as 1 for nonempty bounded sets A, but $(\operatorname{diam} \varnothing)^0$ as 0.

As usual, when we restrict the outer measure $\overline{\mathcal{H}}^s$ to its measurable sets, we get a measure $\mathcal{H}^s$, called the s-**dimensional Hausdorff measure**. As with all method II outer measures, $\overline{\mathcal{H}}^s$ is a metric outer measure, so all Borel sets are measurable.

(1.2.1) Proposition. *The 0-dimensional Hausdorff measure is a* **counting measure**. *Every subset $A \subseteq S$ is measurable; if A is an infinite set, then $\mathcal{H}^0(A) = \infty$; if n is a nonnegative integer and A is a set with n elements, then $\mathcal{H}^0(A) = n$.*

Proof. Suppose $A = \{a_1, a_2, \cdots a_n\}$ is a set with n elements. For any $\varepsilon > 0$, the set is covered by

---

[1] For real-world fractals, this may not be what is desired. There, it may be that we should consider a range of length scales, not too large and not too small. But for mathematical purposes, we consider length scales that approach zero. This will let us recover arc length, area, and volume. Even in the classical cases, in order to measure the exact value of the arc length, we need to consider a limit as the measuring scale goes to zero.

$$A \subseteq \bigcup_{i=1}^{n} B_{\varepsilon/2}(a_i),$$

so that

$$\overline{\mathcal{H}}^0_\varepsilon(A) \leq \sum_{i=1}^{n} \left(\text{diam } B_{\varepsilon/2}(a_i)\right)^0 = n.$$

On the other hand, if ε is smaller than the minimum of all distances $\rho(a_i, a_j)$, then any set with diameter $< \varepsilon$ contains at most one of the points a_i, so any cover of A by sets with diameter this small contains at least n sets. Thus $\overline{\mathcal{H}}^0_\varepsilon(A) \geq n$. When we let $\varepsilon \to 0$ we get $\overline{\mathcal{H}}^0(A) = n$.

If A is an infinite set, then (for each n) it contains a subset with n elements, so $\overline{\mathcal{H}}^0(A) \geq n$. This is true for each positive integer n, so $\overline{\mathcal{H}}^0(A) = \infty$.

I claim that any set $A \subseteq S$ is measurable. So suppose $E \subseteq S$ is another set. The set E is the union of two disjoint sets

$$E \cap A \quad \text{and} \quad E \setminus A.$$

If E is infinite, then at least one of these sets is infinite, so $\overline{\mathcal{H}}^0(E) = \infty = \overline{\mathcal{H}}^0(E \cap A) + \overline{\mathcal{H}}^0(E \setminus A)$. If E is finite, then the two sets have positive separation, so again $\overline{\mathcal{H}}^0(E) = \overline{\mathcal{H}}^0(E \cap A) + \overline{\mathcal{H}}^0(E \setminus A)$. This completes the proof that A is measurable. ☺

Now suppose $\mathcal{B}$ is some Vitali cover of S by bounded sets. We may use the set-function $C(A) = (\text{diam } A)^s$ to define a method II outer measure. This is closely related to the construction of Hausdorff measure. Let us write

$$\overline{\mathcal{H}}^s_{\mathcal{B}}(A) = \lim_{\varepsilon \to 0} \overline{\mathcal{H}}^s_{\mathcal{B}, \varepsilon}(A).$$

The resulting measures clearly satisfy

$$\overline{\mathcal{H}}^s_\varepsilon(A) \leq \overline{\mathcal{H}}^s_{\mathcal{B}, \varepsilon}(A),$$
$$\overline{\mathcal{H}}^s(A) \leq \overline{\mathcal{H}}^s_{\mathcal{B}}(A).$$

In many cases there is equality.

(1.2.2) Exercise. Consider the Vitali covers $\mathcal{B}$ specified below. In which cases is it true (for all positive real numbers s) that $\overline{\mathcal{H}}^s = \overline{\mathcal{H}}^s_{\mathcal{B}}$?

(a) $\mathcal{B}$ is the collection of all bounded closed sets.
(b) $\mathcal{B}$ is the collection of all bounded open sets.
(c) $\mathcal{B}$ is the collection of all closed balls.
(d) $S = \mathbb{R}^d$, and $\mathcal{B}$ is the collection of all bounded closed convex sets.

There are other cases where the outer measures $\overline{\mathcal{H}}^s$ and $\overline{\mathcal{H}}^s_{\mathcal{B}}$ are within a constant factor of each other. See Section 1.7.

Given a set A, there is a critical value $s_0 \in [0, \infty]$ such that

$$\overline{\mathcal{H}}^s(A) = \begin{cases} \infty, & \text{for all } s \text{ with } s < s_0, \\ 0, & \text{for all } s \text{ with } s > s_0. \end{cases}$$

This critical value is the **Hausdorff dimension** of the set A [MTFG, p. 149], which will be written dim $A = s_0$ in this book.

Suppose A is a Borel set. If $\mathcal{H}^s(A) > 0$, then dim $A \geq s$. If $\mathcal{H}^s(A) < \infty$, then dim $A \leq s$. If A is σ-finite for $\mathcal{H}^s$, then dim $A \leq s$.

(1.2.3) Exercise. Give examples of Borel sets $A \subseteq \mathbb{R}^2$ such that:

(a) dim $A = 1$ and $\mathcal{H}^1(A) = 0$.
(b) dim $A = 1$ and $0 < \mathcal{H}^1(A) < \infty$.
(c) dim $A = 1$, $\mathcal{H}^1(A) = \infty$, and A is σ-finite for $\mathcal{H}^1$.
(d) dim $A = 1$, $\mathcal{H}^1(A) = \infty$, and A is not σ-finite for $\mathcal{H}^1$.

(1.2.4) Exercise. Let S and T be metric spaces. Let p be a positive number. A function $f\colon S \to T$ satisfies a **Hölder condition of order p** iff there is a constant M such that

$$\rho\big(f(x), f(y)\big) \leq M\rho(x,y)^p$$

for all $x, y \in S$. In that case, what can be said about the relation between the Hausdorff dimensions dim A and dim $f[A]$ for $A \subseteq S$?

Increasing Sets Lemma. Recall that an outer measure $\overline{M}$ on a space X is said to be **regular** iff for every set $A \subseteq X$ there is a measurable set $C \supseteq A$ with $\overline{M}(A) = \overline{M}(C)$.

(1.2.5) Proposition. *Let S be a metric space, and let $s > 0$. Then the Hausdorff outer measure $\overline{\mathcal{H}}^s$ is regular.*

Proof. Let $A \subseteq S$ be a (possibly nonmeasurable) set. For each positive integer n, choose closed sets E_{in}, $i = 1, 2, \cdots$, such that $A \subseteq \bigcup_i E_{in}$, diam $E_{in} < 1/n$, and $\sum_i (\text{diam } E_{in})^s < \overline{\mathcal{H}}^s_{1/n}(A) + 1/n$. Let

$$E = \bigcap_n \bigcup_i E_{in}.$$

Then $A \subseteq E$. Since the sets E_{in} are closed, certainly E is a Borel set. The covers E_{in} show that

$$\overline{\mathcal{H}}^s_{1/n}(E) \leq \overline{\mathcal{H}}^s_{1/n}(A) + \frac{1}{n},$$

so that $\overline{\mathcal{H}}^s(E) \leq \overline{\mathcal{H}}^s(A)$, and therefore $\overline{\mathcal{H}}^s(E) = \overline{\mathcal{H}}^s(A)$.   ☺

Combining this with (1.1.3), we see that Hausdorff measure is continuous from below. It is important that the sets A_n need not be measurable.

(1.2.6) Corollary. *Let S be a metric space, and let $s, \varepsilon > 0$. Let $A_1 \subseteq A_2 \subseteq \cdots$ be an increasing sequence of sets. Then*

$$\overline{\mathcal{H}}^s\left(\bigcup_{n=1}^\infty A_n\right) = \lim_{n \to \infty} \overline{\mathcal{H}}^s(A_n).$$

It will be useful to have a similar result for the approximations $\overline{\mathcal{H}}_\varepsilon^s$. Because so few sets are $\overline{\mathcal{H}}_\varepsilon^s$-measurable, it may not be reasonable to expect $\overline{\mathcal{H}}_\varepsilon^s$ to be regular.

(1.2.7) Exercise. Is $\overline{\mathcal{H}}_1^1$ a regular outer measure on $\mathbb{R}^2$?

Even if $\overline{\mathcal{H}}_\varepsilon^s$ is not regular, it is still continuous from below in certain cases. Let S be a compact metric space, and let $s, \varepsilon > 0$. Let $A_1 \subseteq A_2 \subseteq \cdots$ be an increasing sequence of sets. Then

$$\overline{\mathcal{H}}_\varepsilon^s \left(\bigcup_{n=1}^\infty A_n \right) = \lim_{n \to \infty} \overline{\mathcal{H}}_\varepsilon^s (A_n).$$

For our purposes, it will be enough to prove a slightly weaker result:[2]

(1.2.8) Increasing Sets Lemma. *Let S be a compact metric space, $s > 0$, and $0 < \varepsilon < \alpha$. Let $A_1 \subseteq A_2 \subseteq \cdots$ be an increasing sequence of sets. Then*

$$\overline{\mathcal{H}}_\alpha^s \left(\bigcup_{n=1}^\infty A_n \right) \leq \sup_{n \in \mathbb{N}} \overline{\mathcal{H}}_\varepsilon^s (A_n).$$

Proof. Write $A = \bigcup_n A_n$ and $\lambda = \sup_n \overline{\mathcal{H}}_\varepsilon^s(A_n)$. We may assume $A \neq \varnothing$ and $\lambda < \infty$.

Let $\eta > 0$ be given. For any positive integer n, choose closed sets F_{in} with

$$A_n \subseteq \bigcup_i F_{in},$$

$$\text{diam } F_{in} < \varepsilon,$$

$$\sum_i (\text{diam } F_{in})^s < \lambda + \eta.$$

Since $\lambda + \eta < \infty$, these diameters approach zero; thus we may number them in decreasing order of size:

$$\varepsilon > \text{diam } F_{1n} \geq \text{diam } F_{2n} \geq \text{diam } F_{3n} \geq \cdots.$$

Now, the space $\mathcal{K}(S)$ is a compact metric space in the Hausdorff metric [MTFG, §2.4]. So by taking a subsequence, we may assume that the sequence F_{in} converges as $n \to \infty$ for each i, say $F_{in} \to F_i$. Write $d_i = \text{diam } F_i$. Now, $\text{diam } F_{in} \to d_i$, so $\varepsilon \geq d_1 \geq d_2 \geq \cdots$ and

$$l = \sum_i d_i^s \leq \lambda + \eta.$$

This is finite, so $d_i \to 0$ as $i \to \infty$.

---

[2] The following is an optional result. It is used only in Sections 1.6 and 1.7.

Now choose numbers $c_i > d_i$ such that $c_i < \alpha$ and $c_i^s < d_i^s + \eta/2^i$. Let

$$V_i = \left\{ x : \operatorname{dist}(x, F_i) \le \frac{1}{2}(c_i - d_i) \right\}.$$

Now, V_i is closed (since F_i is compact) and diam $V_i \le$ diam $F_i + (c_i - d_i) = c_i$, so

$$\sum_i (\operatorname{diam} V_i)^s < \sum \left(d_i^s + \frac{\eta}{2^i} \right) \le l + \eta.$$

For each i, there is a large enough n that $F_{in} \subseteq V_i$, but that n may depend on i. Write $V = \bigcup_i V_i$.

Consider sets $A_n \setminus V$. I claim that $\overline{\mathcal{H}}^s(A_n \setminus V)_\downarrow \le \lambda - l + 3\eta$. Fix n. Let $\varepsilon^* > 0$ be given. There is i^* such that $d_{i^*} < \varepsilon^*$ and $\sum_{i=i^*+1}^\infty d_i^s < \eta$. Then choose $n^* > n$ such that

$$F_{in^*} \subseteq V_i \qquad (i = 1, 2, \cdots, i^*),$$

$$(\operatorname{diam} F_{in^*})^s \ge d_i^s - \frac{\eta}{2^i} \qquad (i = 1, 2, \cdots, i^*),$$

$$\operatorname{diam} F_{i^*n^*} < \varepsilon^*.$$

Thus $A_n \subseteq A_{n^*} \subseteq \bigcup_{i=1}^\infty F_{in^*}$ and $\bigcup_{i=1}^{i^*} F_{in^*} \subseteq \bigcup_i V_i = V$, so $A_n \setminus V \subseteq \bigcup_{i=i^*+1}^\infty F_{in^*}$. Further, for $i \ge i^* + 1$,

$$\operatorname{diam} F_{in^*} \le \operatorname{diam} F_{i^*n^*} < \varepsilon^*.$$

Finally,

$$\sum_{i=i^*+1}^\infty (\operatorname{diam} F_{in^*})^s = \sum_{i=1}^\infty (\operatorname{diam} F_{in^*})^s - \sum_{i=1}^{i^*} (\operatorname{diam} F_{in^*})^s$$

$$\le \lambda + \eta - \sum_{i=1}^{i^*} \left(d_i^s - \frac{\eta}{2^i} \right)$$

$$< \lambda + 2\eta - \sum_{i=1}^{i^*} d_i^s$$

$$< \lambda - l + 3\eta.$$

Hence $\overline{\mathcal{H}}^s_{\varepsilon^*}(A_n \setminus V) \le \lambda - l + 3\eta$. Let $\varepsilon^* \to 0$ to conclude that $\overline{\mathcal{H}}^s(A_n \setminus V) \le \lambda - l + 3\eta$.

Next, note that $\overline{\mathcal{H}}^s(A \setminus V) = \overline{\mathcal{H}}^s(\bigcup_n (A_n \setminus V)) = \sup_n \overline{\mathcal{H}}^s(A_n \setminus V) \le \lambda - l + 3\eta$. Now, V is covered by $\{V_i\}$, so $\overline{\mathcal{H}}^s_\alpha(V) \le \sum_i (\operatorname{diam} V_i)^s < l + \eta$. So $\overline{\mathcal{H}}^s_\alpha(A) \le \overline{\mathcal{H}}^s_\alpha(A \cap V) + \overline{\mathcal{H}}^s_\alpha(A \setminus V) \le l + \eta + \lambda - l + 3\eta = \lambda + 4\eta$. Since η was arbitrary, we have $\overline{\mathcal{H}}^s_\alpha(A) \le \lambda$, as required.   ☺

Packing Measure. The second fractal measure to be considered is the "packing measure." In an approximate sense, it is a counterpart of the Hausdorff measure.

Let S be a metric space, and let $A \subseteq S$ be a set. A **centered-ball packing** of A is a countable disjoint collection of closed balls with centers in A:

$$\left\{ \overline{B}_{r_1}(x_1), \overline{B}_{r_2}(x_2), \cdots \right\},$$

where $x_i \in A$ and $\overline{B}_{r_i}(x_i) \cap \overline{B}_{r_j}(x_j) = \varnothing$ for $i \neq j$. Let s be a positive number. For $\varepsilon > 0$, define

$$\widetilde{\mathcal{P}}^s_\varepsilon(A) = \sup \sum_i (2r_i)^s,$$

where the supremum is over all packings $\{\overline{B}_{r_i}(x_i)\}$ of A by centered balls with $r_i \leq \varepsilon$. (Of course, in Euclidean space diam $B_r(x) = 2r$, but in a general metric space this need not be true.) The **s-dimensional packing premeasure** of A is

$$\widetilde{\mathcal{P}}^s(A) = \lim_{\varepsilon \to 0} \widetilde{\mathcal{P}}^s_\varepsilon(A).$$

The **s-dimensional packing outer measure** is the measure $\overline{\mathcal{P}}^s$ defined from the set-function $\widetilde{\mathcal{P}}^s$ by method I. That is,

$$\overline{\mathcal{P}}^s(A) = \inf \left\{ \sum_{D \in \mathcal{D}} \widetilde{\mathcal{P}}^s(D) : \mathcal{D} \text{ is a countable cover of } A \right\}.$$

Then $\overline{\mathcal{P}}^s$ is a metric outer measure. The **s-dimensional packing measure** $\mathcal{P}^s$ is the restriction of $\overline{\mathcal{P}}^s$ to its measurable sets.

(1.2.9) Exercise. Show that each $\widetilde{\mathcal{P}}^s_\varepsilon$ is an outer measure. Give an example showing that $\widetilde{\mathcal{P}}^s$ need not be countably subadditive.

(1.2.10) Exercise. Finite packings suffice. That is,

$$\widetilde{\mathcal{P}}^s_\varepsilon(A) = \sup \sum_i (2r_i)^s,$$

where the supremum is over all finite packings $\{\overline{B}_{r_i}(x_i)\}$ of A by centered balls with $r_i \leq \varepsilon$.

An open ball $B_r(x)$ is the increasing union of a sequence $\overline{B}_{r-1/n}(x)$ of closed balls, so the same value of $\widetilde{\mathcal{P}}^s_\varepsilon$ is obtained if we use packings by open balls instead of closed balls.

If $A \subseteq S$ is a set and $\overline{A}$ is the closure of that set, then $\widetilde{\mathcal{P}}^s(A) = \widetilde{\mathcal{P}}^s(\overline{A})$. Indeed, on the one hand, any centered-ball packing of A is also a centered-ball packing of $\overline{A}$, so $\widetilde{\mathcal{P}}^s(A) \leq \widetilde{\mathcal{P}}^s(\overline{A})$. On the other hand, if $\overline{B}_r(x)$ is a ball with $x \in \overline{A}$, then there is a point $x' \in A$ as close as we like to x and a radius $r' = r - \rho(x', x)$ such that $\overline{B}_{r'}(x') \subseteq \overline{B}_r(x)$ and $2r'$ is as close as we like to $2r$. Thus, given a centered-ball packing $\{\overline{B}_{r_i}(x_i)\}$ of $\overline{A}$, we may choose a centered-ball packing $\{\overline{B}_{r'_i}(x'_i)\}$ of A with $\sum (2r'_i)^s$ as close as we like to $\sum (2r_i)^s$. So $\widetilde{\mathcal{P}}^s(A) = \widetilde{\mathcal{P}}^s(\overline{A})$.

Since $\widetilde{\mathcal{P}}^s(A) = \widetilde{\mathcal{P}}^s(\overline{A})$, in the definition

$$\overline{\mathcal{P}}^s(A) = \inf\left\{ \sum_{D \in \mathcal{D}} \widetilde{\mathcal{P}}^s(D) : \mathcal{D} \text{ is a countable cover of } A \right\}$$

of $\overline{\mathcal{P}}^s$, it is enough to use covers $\mathcal{D}$ consisting of closed sets.

(1.2.11) Exercise. Let E be any set, possibly not measurable. There is a Borel set A with $A \supseteq E$ and $\mathcal{P}^s(A) = \overline{\mathcal{P}}^s(E)$. So $\mathcal{P}^s$ is a regular outer measure.

(1.2.12) Exercise. Let s and ε be positive numbers. Let E_n be an increasing sequence of (possibly not measurable) sets. Prove or disprove:

(a) $\lim_n \widetilde{\mathcal{P}}^s_\varepsilon(E_n) = \widetilde{\mathcal{P}}^s_\varepsilon(E)$.
(b) $\lim_n \widetilde{\mathcal{P}}^s(E_n) = \widetilde{\mathcal{P}}^s(E)$.
(c) $\lim_n \overline{\mathcal{P}}^s(E_n) = \overline{\mathcal{P}}^s(E)$.

Given a set A, there is a critical value $s_0 \in [0, \infty]$ such that

$$\overline{\mathcal{P}}^s(A) = \begin{cases} \infty, & \text{for all } s \text{ with } s < s_0, \\ 0, & \text{for all } s \text{ with } s > s_0. \end{cases}$$

This critical value is the **packing dimension** of the set A, which will be written $\operatorname{Dim} A = s_0$ in this book.

The packing and Hausdorff measures are related:

$$\overline{\mathcal{H}}^s(A) \le \overline{\mathcal{P}}^s(A)$$

for all s. This is proved below, Corollary 1.3.5. Therefore, $\dim A \le \operatorname{Dim} A$.

(1.2.13) Exercise. Consider some variant definitions for "packing": Let (S, ρ) be a metric space, and let π be a collection of constituents.

(a) π is an **(a)-packing** iff $\rho(x, x') > r \vee r'$ for all $(x, r) \ne (x', r')$ in π;
(b) π is a **(b)-packing** iff $\overline{B}_r(x) \cap \overline{B}_{r'}(x') = \varnothing$ for all $(x, r) \ne (x', r')$ in π;
(c) π is a **(c)-packing** iff $\rho(x, x') > r + r'$ for all $(x, r) \ne (x', r')$ in π.

Define packing measures $^{(a)}\mathcal{P}^s$, $^{(b)}\mathcal{P}^s$, and $^{(c)}\mathcal{P}^s$ using these three definitions. Show that $^{(c)}\mathcal{P}^s(E) \le {}^{(b)}\mathcal{P}^s(E) \le {}^{(a)}\mathcal{P}^s(E) \le 2^s \, {}^{(c)}\mathcal{P}^s(E)$. Conclude that the packing dimension $\operatorname{Dim} E$ is the same for each of the definitions.

(1.2.14) Exercise. Let S and T be metric spaces. Let p be a positive number. Suppose $f : S \to T$ satisfies a Hölder condition of order p. What can be said about the relation between the packing dimensions $\operatorname{Dim} A$ and $\operatorname{Dim} f[A]$ for $A \subseteq S$?

1.3 Vitali Theorems

The relation between the Hausdorff measure and the corresponding packing measure is clarified using the next result. It will be used again for other purposes.

Recall that a **fine cover** of a set E is a family β of closed balls $\overline{B}_r(x)$ with $x \in E, r > 0$, such that for every $x \in E$ and every $\varepsilon > 0$, there is $r > 0$ such that $r < \varepsilon$ and $\overline{B}_r(x) \in \beta$. In a general metric space, it is possible that $\overline{B}_r(x) = \overline{B}_{r'}(x')$, where $x \neq x'$ and/or $r' \neq r$. So it is more correct to think of a fine cover as a collection of pairs (x, r). Sometimes we will write $\overline{B}_r(x) \in \beta$, and sometimes $(x, r) \in \beta$.

Here is a version of the Vitali Covering Theorem. The idea in the proof is due to S. Banach.

(1.3.1) Theorem. *Let S be a metric space, let $E \subseteq S$ be a subset, and let β be a fine cover of E. Then there exists either* (a) *an infinite disjoint sequence $\overline{B}_{r_i}(x_i) \in \beta$ with $\inf r_i > 0$ or* (b) *a countable (possibly finite) disjoint sequence of balls $\overline{B}_{r_i}(x_i) \in \beta$ such that for all $j \in \mathbb{N}$,*

$$E \setminus \bigcup_{i=1}^{j} \overline{B}_{r_i}(x_i) \subseteq \bigcup_{i=j+1}^{\infty} \overline{B}_{3r_i}(x_i).$$

Proof. We define recursively a disjoint sequence of sets $\overline{B}_{r_n}(x_n) \in \beta$ and a decreasing sequence of fine covers $\beta_n \subseteq \beta$.

Let $\beta_1 = \left\{ \overline{B}_r(x) \in \beta : r \leq 1 \right\}$. Then β_1 is again a Vitali cover of E. Define

$$t_1 = \sup \left\{ r : \overline{B}_r(x) \in \beta_1 \right\},$$

and then choose a ball $\overline{B}_{r_1}(x_1) \in \beta_1$ with $r_1 \geq t_1/2$. Now suppose $\overline{B}_{r_1}(x_1)$, $\overline{B}_{r_2}(x_2), \cdots, \overline{B}_{r_n}(x_n)$ and $\beta_1, \beta_2, \cdots, \beta_n$ have been chosen. Let

$$\beta_{n+1} = \left\{ \overline{B}_r(x) \in \beta_n : \overline{B}_r(x) \cap \overline{B}_{r_n}(x_n) = \varnothing \right\}.$$

If β_{n+1} is empty, the construction terminates. If it is not empty, define

$$t_{n+1} = \sup \left\{ r : \overline{B}_r(x) \in \beta_{n+1} \right\},$$

and choose $\overline{B}_{r_{n+1}}(x_{n+1}) \in \beta_{n+1}$ with $r_{n+1} \geq t_{n+1}/2$. This completes the recursive construction.

If the construction terminates, say $\beta_{n+1} = \varnothing$, this means that $E \subseteq \bigcup_{i=1}^{n} \overline{B}_{r_i}(x_i)$, since $\bigcup_{i=1}^{n} \overline{B}_{r_i}(x_i)$ is closed. In this case, conclusion (b) holds.

So suppose the construction does not terminate, and (a) is false. I must prove (b). Fix j, and let $x \in E \setminus \bigcup_{i=1}^{j} \overline{B}_{r_i}(x_i)$. I must prove that $x \in \bigcup_{i=j+1}^{\infty} \overline{B}_{3r_i}(x_i)$. Now there is a positive distance between x and the closed set $\bigcup_{i=1}^{j} \overline{B}_{r_i}(x_i)$. Since β_1 is a Vitali cover of E, there is $r_0 > 0$ with $\overline{B}_{r_0}(x) \in \beta_1$ and $\overline{B}_{r_0}(x) \cap \bigcup_{i=1}^{j} \overline{B}_{r_i}(x_i) = \varnothing$.

Now $r_n \to 0$, so there exists a least n with $r_n < (1/2)r_0$. For this n, I claim that $\overline{B}_{r_0}(x) \cap \bigcup_{i=1}^{n-1} \overline{B}_{r_i}(x_i) \neq \varnothing$. Indeed, if $\overline{B}_{r_0}(x) \cap \bigcup_{i=1}^{n-1} \overline{B}_{r_i}(x_i) = \varnothing$, then $\overline{B}_{r_0}(x) \in \beta_n$, so $t_n \geq r_0 > 2r_n \geq t_n$, a contradiction. Now let k be the least integer with $\overline{B}_{r_0}(x) \cap \overline{B}_{r_k}(x_k) \neq \varnothing$. (Certainly $j < k < n$.) So $r_k \geq (1/2)r_0$. There is a point $z \in \overline{B}_{r_0}(x) \cap \overline{B}_{r_k}(x_k)$. So

$$\rho(x, x_k) \leq \rho(x, z) + \rho(z, x_k) \leq r_0 + r_k \leq 3r_k.$$

Therefore, $x \in \overline{B}_{3r_k}(x_k) \subseteq \bigcup_{i=j+1}^{\infty} \overline{B}_{3r_i}(x_i)$ as required. ☺

(1.3.2) Exercise. If S is a compact metric space, then conclusion (a) is impossible.

(1.3.3) Exercise. Suppose S is a countable union of compact sets. (This allows, for example, Euclidean space $\mathbb{R}^d$.) Let $E \subseteq S$ be a subset, and let β be a fine cover of E. Then there exists a countable disjoint collection of sets $\overline{B}_{r_i}(x_i) \in \beta$ such that for all $j \in \mathbb{N}$,

$$E \setminus \bigcup_{i=1}^{j} \overline{B}_{r_i}(x_i) \subseteq \bigcup_{i=j+1}^{\infty} \overline{B}_{3r_i}(x_i).$$

(1.3.4) Exercise. Suppose S is a compact ultrametric space. (See the definition in [MTFG, p. 43].) Let $E \subseteq S$ be a subset, and let β be a fine cover of E. Then there exists a countable disjoint collection of sets $\overline{B}_{r_i}(x_i) \in \beta$ such that

$$E \subseteq \bigcup_{i=1}^{\infty} \overline{B}_{r_i}(x_i).$$

Comparison of Hausdorff and Packing Measures.

(1.3.5) Corollary. *Let S be a metric space, $E \subseteq S$ a subset, and $s > 0$. Then $\overline{\mathcal{H}}^s(E) \leq \overline{\mathcal{P}}^s(E)$.*

Proof. First I claim that $\overline{\mathcal{H}}^s(E) \leq \widetilde{\mathcal{P}}^s(E)$. If $\widetilde{\mathcal{P}}^s(E) = \infty$, there is nothing to prove. So assume that $\widetilde{\mathcal{P}}^s(E) < \infty$. Then let ε be small enough that $\widetilde{\mathcal{P}}^s_\varepsilon(E) < \infty$. Now

$$\beta = \{(x, r) : x \in E, 0 < r < \varepsilon\}$$

is a fine cover of E. Apply Theorem 1.3.1 to this fine cover. Any disjoint sequence $\{\overline{B}_{r_i}(x_i)\}$ of sets from β is a packing of E by centered balls with radius $< \varepsilon$, so $\sum r_i^s \leq 2^{-s}\widetilde{\mathcal{P}}^s_\varepsilon(E) < \infty$; so conclusion (a) is impossible. Thus there is a disjoint family $\overline{B}_{r_i}(x_i) \in \beta$ with

$$E \setminus \bigcup_{i=1}^{j} \overline{B}_{r_i}(x_i) \subseteq \bigcup_{i=j+1}^{\infty} \overline{B}_{3r_i}(x_i)$$

for all j.

Now, $\{\overline{B}_{r_1}(x_1), \overline{B}_{r_2}(x_2), \cdots\}$ is a packing of E, so

$$\widetilde{\mathcal{P}}^s_\varepsilon(E) \geq \sum_i (2r_i)^s.$$

The series $\sum (2r_i)^s$ is convergent. For each j,

$$\left\{ \overline{B}_{r_i}(x_i) : 1 \leq i \leq j \right\} \cup \left\{ \overline{B}_{3r_i}(x_i) : i \geq j+1 \right\}$$

is a cover of E, so

$$\overline{\mathcal{H}}^s_{6\varepsilon}(E) \leq \sum_{i=1}^{j} (2r_i)^s + \sum_{i=j+1}^{\infty} (6r_i)^s.$$

Let $j \to \infty$ to obtain $\overline{\mathcal{H}}^s_{6\varepsilon}(E) \leq \sum (2r_i)^s$. Therefore, $\widetilde{\mathcal{P}}^s_\varepsilon(E) \geq \overline{\mathcal{H}}^s_{6\varepsilon}(E)$. Let $\varepsilon \to 0$ to obtain $\widetilde{\mathcal{P}}^s(E) \geq \overline{\mathcal{H}}^s(E)$.

Next suppose E is covered, $E \subseteq \bigcup_n E_n$. Now, $\overline{\mathcal{H}}^s$ is countably subadditive, so $\sum_n \widetilde{\mathcal{P}}^s(E_n) \geq \sum_n \overline{\mathcal{H}}^s(E_n) \geq \overline{\mathcal{H}}^s(E)$. This is true for all covers, so $\overline{\mathcal{P}}^s(E) \geq \overline{\mathcal{H}}^s(E)$. ☺

Other Vitali Theorems. A **Vitali cover** of a set E is a family β of sets such that for every $x \in E$ and every $\varepsilon > 0$, there is $B \in \beta$ with $x \in B$ and $0 < \operatorname{diam} B \leq \varepsilon$.

(1.3.6) Exercise. Imitate the proof for 1.3.1 above to prove the following Vitali Covering Theorem: Let S be a metric space with no isolated points, and let $E \subseteq S$. Let β be a Vitali cover of a set E by closed sets. Then there is a disjoint sequence of sets $B_i \in \beta$ such that either (a) $\inf \operatorname{diam} B_i > 0$ or (b) for every j,

$$E \setminus \bigcup_{i=1}^{j} B_i \subseteq \bigcup_{i=j+1}^{\infty} U_i,$$

where U_i is a closed ball with center in B_i and radius $3 \operatorname{diam} B_i$.

The most common version of Vitali's theorem is for Euclidean space with Lebesgue measure. If a set E is covered (in the sense of Vitali) by balls, then it is "almost covered" by a disjoint union of them.

(1.3.7) Vitali Covering Theorem. *Let $E \subseteq \mathbb{R}^d$ be a Borel subset, and let β be a Vitali cover of E by balls (open or closed). Then there exists a countable disjoint collection of sets $B_i \in \beta$ such that*

$$\mathcal{L}^d \left(E \setminus \bigcup_{i=1}^{\infty} B_i \right) = 0.$$

Proof. Since the boundary of a ball in $\mathbb{R}^d$ has Lebesgue measure 0, we may replace any open balls in β by their closures, then find a sequence B_i of closed balls, and finally replace some of these balls with the corresponding open balls again. So it is enough to consider the case when all balls in β are closed.

First, suppose the set E is bounded, say $E \subseteq B_M(0)$. Proceed as in Theorem 1.3.1 to get disjoint balls $\overline{B}_{r_i}(x_i) \in \beta_1$. Now all of the balls $\overline{B}_{r_i}(x_i)$ are contained inside $B_{M+1}(0)$, which has finite Lebesgue measure, so $\sum_{i=1}^{\infty} \mathcal{L}^d(\overline{B}_{r_i}(x_i)) < \infty$. Conclusion (a) fails, so conclusion (b) holds:

$$E \setminus \bigcup_{i=1}^{j} \overline{B}_{r_i}(x_i) \subseteq \bigcup_{i=j+1}^{\infty} \overline{B}_{3r_i}(x_i).$$

But $\mathcal{L}^d(\overline{B}_{3r_i}(x_i)) = 3^d \mathcal{L}^d(\overline{B}_{r_i}(x_i))$, so $\sum \mathcal{L}^d(\overline{B}_{3r_i}(x_i)) < \infty$. Now, for each j,

$$E \setminus \bigcup_{i=1}^{\infty} \overline{B}_{r_i}(x_i) \subseteq E \setminus \bigcup_{i=1}^{j} \overline{B}_{r_i}(x_i) \subseteq \bigcup_{i=j+1}^{\infty} \overline{B}_{3r_i}(x_i),$$

and

$$\mathcal{L}^d \left(\bigcup_{i=j+1}^{\infty} \overline{B}_{3r_i}(x_i) \right) \leq \sum_{i=j+1}^{\infty} \mathcal{L}^d(\overline{B}_{3r_i}(x_i)),$$

which approaches 0 as $j \to \infty$. Therefore, $\mathcal{L}^d(E \setminus \bigcup_{i=1}^{\infty} \overline{B}_{r_i}(x_i)) = 0$.

Now suppose E is unbounded. A **lattice cube** is a cube of side one and corners with integer coordinates:

$$C = (n_1, n_1 + 1) \times (n_2, n_2 + 1) \times \cdots \times (n_d, n_d + 1),$$

where the n_i are integers. The set $E \cap C$ is Vitali covered by

$$\{ B \in \beta : B \subseteq C \},$$

since C is an open set. By the first part of the proof, there is a countable disjoint family of sets $B \in \beta$ with $B \subseteq C$ that almost covers $E \cap C$. Combining these families, one for each lattice cube C, we obtain a countable disjoint family that almost covers E, since the part of $\mathbb{R}^d$ not contained in any lattice cube is a countable union of hyperplanes, and thus has Lebesgue measure zero.   ☺

(1.3.8) Exercise. Let S be a metric space, let $E \subseteq S$ be a Borel subset, let $s > 0$, and and let β be a Vitali cover of E. Then there exists a countable disjoint collection of sets $B_i \in \beta$ such that either $\sum (\operatorname{diam} B_i)^s = \infty$ or

$$\mathcal{H}^s \left(E \setminus \bigcup_{i=1}^{\infty} B_i \right) = 0.$$

The Strong Vitali Property. Let $\mathcal{M}$ be a Borel measure on S. We say that $\mathcal{M}$ has the **strong Vitali property** iff for any Borel set $E \subseteq S$ with

$\mathcal{M}(E) < \infty$ and any fine cover β of E, there exists a countable disjoint family $\{\overline{B}_{r_1}(x_1), \overline{B}_{r_2}(x_2), \cdots\} \subseteq \beta$ such that

$$\mathcal{M}\left(E \setminus \bigcup_{i=1}^{\infty} \overline{B}_{r_i}(x_i)\right) = 0.$$

Note that by Theorem 1.1.12, if $\mathcal{M}$ is a finite measure, we may also arrange that $\mathcal{M}\left(\bigcup \overline{B}_{r_i}(x_i) \setminus E\right)$ is as small as we like.

In many of the common metric spaces, every finite Borel measure has the strong Vitali property. For example, by Exercise 1.3.4, this is true for compact ultrametric spaces. We see next that it is also true for Euclidean space. Begin with an exercise in trigonometry:

(1.3.9) Exercise. Let x, x', x'' be three vertices of a triangle in the plane; let r, r', r'' be three positive numbers. Suppose $r' \leq |x - x'| \leq r' + r$, $r'' \leq |x - x''| \leq r'' + r$, $r' \leq |x' - x''|$, $r' \geq 2r$, $r'' \geq 2r$, $r'' \leq (4/3)r'$. Then the angle of the triangle at x measures at least $10°$.

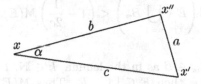

Figure 1.3.10. Triangle.

(1.3.11) Lemma. *Let d be a positive integer. There is an integer c (for example, $16^d + 1$ will do) such that for any fine cover β of a bounded set $E \subseteq \mathbb{R}^d$ there exist c sequences $\{B_{ik} : k \in \mathbb{N}\} \subseteq \beta$, $1 \leq i \leq c$, such that $B_{ik} \cap B_{il} = \varnothing$ for $k \neq l$ and $E \subseteq \bigcup_{i,k} B_{ik}$.*

Proof. We define a sequence $B_n = \overline{B}_{r_n}(x_n)$ recursively. Let

$$\beta_1 = \left\{\overline{B}_r(x) \in \beta : r \leq 1\right\},$$

$t_1 = \sup\left\{r : \overline{B}_r(x) \in \beta_1\right\}$ and choose $B_1 = \overline{B}_{r_1}(x_1) \in \beta_1$ with $r_1 > (3/4)t_1$. Let $\beta_2 = \left\{\overline{B}_r(x) \in \beta_1 : x \notin B_1\right\}$, $t_2 = \sup\left\{r : \overline{B}_r(x) \in \beta_2\right\}$ and choose $B_2 = \overline{B}_{r_2}(x_2) \in \beta_2$ with $r_2 > (3/4)t_2$. Continue in this way.

I claim that $E \subseteq \bigcup_n B_n$. If the construction ends at some point, it does so because E is covered by finitely many B_n. So assume that the construction does not end. The balls $\overline{B}_{r_n/3}(x_n)$ are disjoint and contained in a bounded set of $\mathbb{R}^d$, so by a volume computation, $r_n \to 0$. If there is some $x \in E \setminus \bigcup B_n$, there is $r > 0$ with $\overline{B}_r(x) \in \beta_1$, which contradicts the choice of B_n when r_n is small enough.

Let $c = 16^d + 1$. The next step involves distributing the sets B_n into c classes such that the balls in each class are disjoint. This will be possible to do inductively if for each n, at most $c-1$ of the balls $\{\, B_k : k < n \,\}$ meet B_n. Fix n. Let $\mathcal{A} = \{\, B_k : k < n, B_k \cap B_n \neq \varnothing \,\}$. Write $\mathcal{A} = \mathcal{A}_1 \cup \mathcal{A}_2$:

$$\mathcal{A}_1 = \left\{\, \overline{B}_{r_k}(x_k) \in \mathcal{A} : k < n, |x_k - x_n| \leq 3r_n \,\right\},$$
$$\mathcal{A}_2 = \left\{\, \overline{B}_{r_k}(x_k) \in \mathcal{A} : k < n, |x_k - x_n| > 3r_n \,\right\}.$$

For $B_k = \overline{B}_{r_k}(x_k) \in \mathcal{A}_1$, write $\widetilde{B}_k = \overline{B}_{r_n/4}(x_k)$. For $B_k = \overline{B}_{r_k}(x_k) \in \mathcal{A}_2$, let $\tilde{x}_k$ be the point on the line segment from x_k to x_n with distance $3r_n$ from x_n, and let $\widetilde{B}_k = \overline{B}_{r_n/4}(\tilde{x}_k)$. Now the sets $\widetilde{B}_k$ are disjoint (for two sets derived from $\mathcal{A}_2$, use the exercise; for the other cases use the triangle inequality). They are all contained in the set $\overline{B}_{4r_n}(x_n)$. Therefore, the number of sets in $\mathcal{A}$ is at most $(4r_n)^d/(r_n/4)^d = 16^d$.   ☺

(1.3.12) Corollary. *Let d be a positive integer, and let c be the constant of the lemma. Let $\mathcal{M}$ be a Borel measure on $\mathbb{R}^d$, let $E \subseteq \mathbb{R}^d$ be a subset, and let β be a fine cover of E. If $0 < \mathcal{M}(E) < \infty$, then there exist finitely many disjoint balls $B_1, B_2, \cdots, B_N \in \beta$ such that*

$$\mathcal{M}\left(E \setminus \bigcup_{n=1}^{N} B_n\right) < \left(1 - \frac{1}{2c}\right)\mathcal{M}(E).$$

Proof. Choose balls $B_{ik} \in \beta$ as in the lemma, $k \in \mathbb{N}$, $1 \leq i \leq c$. Now, if we write $E_i = E \cap \bigcup_k B_{ik}$, then $E \subseteq \bigcup_{i=1}^c E_i$. Thus $\mathcal{M}(E) \leq \sum_{i=1}^c \mathcal{M}(E_i)$, so $\mathcal{M}(E_i) \geq (1/c)\mathcal{M}(E)$ for some i. Then

$$\mathcal{M}\left(E \setminus \bigcup_{k=1}^{\infty} B_{ik}\right) = \mathcal{M}(E) - \mathcal{M}(E_i) \leq \left(1 - \frac{1}{c}\right)\mathcal{M}(E).$$

Therefore,

$$\mathcal{M}\left(E \setminus \bigcup_{k=1}^{N} B_{ik}\right) < \left(1 - \frac{1}{2c}\right)\mathcal{M}(E)$$

if N is large enough.   ☺

(1.3.13) Theorem. *Let $\mathcal{M}$ be a Borel measure on $\mathbb{R}^d$. Then $\mathcal{M}$ has the strong Vitali property.*

Proof. Let $E \subseteq \mathbb{R}^d$ with $\mathcal{M}(E) < \infty$. Let β be a fine cover of E. Use the lemma to choose disjoint balls $B_1, \cdots, B_{n_1} \in \beta$ with $\mathcal{M}(E \setminus \bigcup B_i) < \left(1 - 1/(2c)\right)\mathcal{M}(E)$. Now, $\bigcup B_i$ is closed, so

$$\left\{ B \in \beta : B \cap \bigcup_{i=1}^{n_1} B_i = \varnothing \right\}$$

is a fine cover of $E \setminus \bigcup_{i=1}^{n_1} B_i$. Repeat the process: obtain balls $B_{n_1+1}, \cdots, B_{n_2}$, disjoint from each other and from the previous B_i, such that

$$\mathcal{M}\left(E \setminus \bigcup_{i=1}^{n_1} B_i \setminus \bigcup_{i=n_1+1}^{n_2} B_i\right) < \left(1 - \frac{1}{2c}\right) \mathcal{M}\left(E \setminus \bigcup_{i=1}^{n_1} B_i\right)$$

$$< \left(1 - \frac{1}{2c}\right)^2 \mathcal{M}(E).$$

Continue in this way. This constructs an infinite disjoint sequence B_i with $\mathcal{M}(E \setminus \bigcup B_i) = 0$. ☺

(1.3.14) Corollary. *Let $E \subseteq \mathbb{R}^d$ be a Borel set, let $s > 0$, and let β be a fine cover of E. If $\mathcal{P}^s(E) < \infty$, then there exists a countable disjoint collection of sets $B_i \in \beta$ such that*

$$\mathcal{P}^s\left(E \setminus \bigcup_{i=1}^{\infty} B_i\right) = 0.$$

A Borel measure $\mathcal{M}$ is called a **Federer measure** iff there exist $K < \infty$ and $\delta > 0$ such that

$$\mathcal{M}(\overline{B}_{2r}(x)) \leq K\mathcal{M}(\overline{B}_r(x))$$

for all $r < \delta$ and all x.

(1.3.15) Proposition. *Let $\mathcal{M}$ be a finite Federer measure in a compact metric space S. Then $\mathcal{M}$ has the strong Vitali property.*

Proof. Let E be a Borel set with $\mathcal{M}(E) < \infty$ and let β be a fine cover of E. Let δ and K be the constants from the Federer property. If we apply the Federer property twice, we have $\mathcal{M}(\overline{B}_{4r}(x)) \leq K^2\mathcal{M}(\overline{B}_r(x))$ for all $r < \delta/2$. Now,

$$\beta' = \left\{(x,r) \in \beta : r < \delta/2\right\}$$

is also a fine cover of E. By Theorem 1.3.1 (and Exercise 1.3.2) there is a packing $\left\{(x_i, r_i)\right\} \subseteq \beta'$ such that for all $j \in \mathbb{N}$,

$$E \setminus \bigcup_{i=1}^{j} \overline{B}_{r_i}(x_i) \subseteq \bigcup_{i=j+1}^{\infty} \overline{B}_{3r_i}(x_i).$$

But since the balls $\overline{B}_{r_i}(x_i)$ are disjoint and $\mathcal{M}$ is a finite measure, the series

$$\sum_i \mathcal{M}(\overline{B}_{r_i}(x_i))$$

converges, so that $\sum_{i=j+1}^{\infty} \mathcal{M}(\overline{B}_{r_i}(x_i))$ goes to 0 as $j \to \infty$. Now for every j,

$$\mathcal{M}\left(E \setminus \bigcup_{i=1}^{\infty} \mathcal{B}_{r_i}(x_i)\right) \leq \mathcal{M}\left(E \setminus \bigcup_{i=1}^{j} \mathcal{B}_r(x_i)\right)$$

$$\leq \sum_{i=j+1}^{\infty} \mathcal{M}(\overline{B}_{3r_i}(x_i)) \leq K^2 \sum_{i=j+1}^{\infty} \mathcal{M}(\overline{B}_{r_i}(x_i)).$$

Therefore, $\mathcal{M}\left(E \setminus \bigcup_{i=1}^{\infty} \mathcal{B}_{r_i}(x_i)\right) = 0$, as required.   ☺

(1.3.16) Exercise. Show that a finite Federer measure on a σ-compact metric space has the strong Vitali property.

Vitali and Variation. For measures with the strong Vitali property (such as measures in Euclidean space), there is a result for the fine variation that may be considered a counterpart to Exercise 1.1.18 for the full variation.

(1.3.17) Lemma. *Let S be a metric space, and let $\mathcal{N}$ be a finite Borel measure on S with the strong Vitali property. Let a constituent function C be defined by $C(x, r) = \mathcal{N}(\overline{B}_r(x))$. Then $v^C(F) \geq \mathcal{N}(F)$ for all Borel sets F.*

Proof. Let $\varepsilon > 0$ be given. Let β be a fine cover of F. By the strong Vitali property, there is a finite packing $\pi \subseteq \beta$ such that $\mathcal{N}\left(F \setminus \bigcup_\pi \overline{B}_r(x)\right) < \varepsilon$. Then

$$\sum_{(x,r) \in \pi} \mathcal{N}(\overline{B}_r(x)) \geq \mathcal{N}(F) - \varepsilon.$$

Therefore, $v^C_\beta \geq \mathcal{N}(F) - \varepsilon$. Take the infimum over β to get $v^C(F) \geq \mathcal{N}(F) - \varepsilon$. Then take the infimum over ε to get $v^C(F) \geq \mathcal{N}(F)$.   ☺

(1.3.18) Proposition. *Let S be a metric space, and let $\mathcal{N}$ be a finite Borel measure on S with the strong Vitali property. Let a constituent function C be defined by $C(x, r) = \mathcal{N}(\overline{B}_r(x))$. Then $v^C(F) = V^C(F) = \mathcal{N}(F)$ for all Borel sets F.*

Proof. Combine (1.1.18), (1.1.20), and (1.3.17): $\mathcal{N}(F) \geq V^C(F) \geq v^C(F) \geq \mathcal{N}(F)$.   ☺

1.4 Other Local Fractal Measures

Next we will consider, more briefly, a few other local fractal measures, as well as set-functions that are not measures.

Covering Measure. A variant of the Hausdorff measure is obtained if we use only covers by centered balls. The resulting measure is useful because it is close to Hausdorff measure.

Let S be a metric space. Let $E \subseteq S$ be a subset. A centered-ball cover of E is a collection β of closed balls with centers in E such that $E \subseteq \bigcup_{B \in \beta} B$. Let $\varepsilon > 0$. Define

$$\widetilde{\mathcal{C}}^s_\varepsilon(E) = \inf \sum_i (2r_i)^s,$$

where the infimum is over all countable covers $\beta = \left\{ \overline{B}_{r_i}(x_i) : i = 1, 2, \cdots \right\}$ of E by centered closed balls with $r_i < \varepsilon$. As ε decreases, $\widetilde{\mathfrak{C}}^s_\varepsilon(E)$ increases, and we define

$$\widetilde{\mathfrak{C}}^s(E) = \lim_{\varepsilon \to 0} \widetilde{\mathfrak{C}}^s_\varepsilon(E).$$

Note that $\widetilde{\mathfrak{C}}^s$ need not be an outer measure.

(1.4.1) Exercise. An example with $A \subseteq B$ but $\widetilde{\mathfrak{C}}^s(A) > \widetilde{\mathfrak{C}}^s(B)$. Let $A \subseteq \mathbb{R}$ be the triadic Cantor dust [MTFG, p. 1]; let $s = \log 2 / \log 3$ be its similarity dimension. Show that $\widetilde{\mathfrak{C}}^s(A) \geq (4/3)^s > 1$. Let B be the set A together with the center points of each of the complementary intervals ("tremas"). Show that $\widetilde{\mathfrak{C}}^s(B) = 1$.

Finally, define the **s-dimensional covering outer measure** of E by

$$\overline{\mathfrak{C}}^s(E) = \sup \left\{ \widetilde{\mathfrak{C}}^s(A) : A \subseteq E \right\},$$

and write $\mathfrak{C}^s$ for the restriction of $\overline{\mathfrak{C}}^s$ to the measurable sets. We will call $\mathfrak{C}^s$ the **s-dimensional covering measure**.

(1.4.2) Proposition. *Let S be a metric space and $s > 0$. The set-function $\overline{\mathfrak{C}}^s$ is a metric outer measure.*

Proof. The only centered-ball cover of the empty set is the empty cover, so $\widetilde{\mathfrak{C}}^s(\varnothing) = 0$. The only subset of $\varnothing$ is $\varnothing$, so $\overline{\mathfrak{C}}^s(\varnothing) = 0$.

Suppose $E \subseteq F$. I claim that $\overline{\mathfrak{C}}^s(E) \leq \overline{\mathfrak{C}}^s(F)$. Indeed, if $A \subseteq E$, then $A \subseteq F$, so by the definition of $\mathfrak{C}^s(F)$ we have

$$\widetilde{\mathfrak{C}}^s(A) \leq \overline{\mathfrak{C}}^s(F).$$

Now take the supremum over all $A \subseteq E$ to obtain $\overline{\mathfrak{C}}^s(E) \leq \overline{\mathfrak{C}}^s(F)$, as required.

Fix $\varepsilon > 0$. I claim that $\widetilde{\mathfrak{C}}^s_\varepsilon$ is countably subadditive. Suppose $E = \bigcup_{n=1}^\infty E_n$. For each n, let $\beta_n = \left\{ \overline{B}_{r_{in}}(x_{in}) : i = 1, 2, \cdots \right\}$ be a centered-ball cover of E_n by closed balls with diameter $< \varepsilon$. The union $\beta = \bigcup_n \beta_n$ is then a centered-ball cover of E. And

$$\widetilde{\mathfrak{C}}^s_\varepsilon(E) \leq \sum_{i,n} (2r_{in})^s = \sum_{n=1}^\infty \sum_i (2r_{in})^s.$$

Let $\delta > 0$, and choose β_n such that $\sum_i (2r_{in})^s < \widetilde{\mathfrak{C}}^s_\varepsilon(E_n) + \delta/2^n$. Then

$$\widetilde{\mathfrak{C}}^s_\varepsilon(E) \leq \sum_{n=1}^\infty \widetilde{\mathfrak{C}}^s_\varepsilon(E_n) + \delta.$$

Now let $\delta \to 0$ to obtain $\widetilde{\mathcal{C}}^s_\varepsilon(E) \leq \sum_{n=1}^\infty \widetilde{\mathcal{C}}^s_\varepsilon(E_n)$. Thus $\widetilde{\mathcal{C}}^s_\varepsilon$ is countably subadditive.

Next, I claim that $\widetilde{\mathcal{C}}^s$ is countably subadditive. Suppose $E = \bigcup_{n=1}^\infty E_n$. For each $\varepsilon > 0$,

$$\widetilde{\mathcal{C}}^s_\varepsilon(E) \leq \sum_{n=1}^\infty \widetilde{\mathcal{C}}^s_\varepsilon(E_n) \leq \sum_{n=1}^\infty \widetilde{\mathcal{C}}^s(E_n).$$

Then let $\varepsilon \to 0$ to obtain $\widetilde{\mathcal{C}}^s(E) \leq \sum_{n=1}^\infty \widetilde{\mathcal{C}}^s(E_n)$.

Next, $\overline{\mathcal{C}}^s$ is countably subadditive. Suppose $E = \bigcup_{n=1}^\infty E_n$. If $A \subseteq E$, then $A \cap E_n \subseteq E_n$ for each n. So

$$\widetilde{\mathcal{C}}^s(A) \leq \sum_{n=1}^\infty \widetilde{\mathcal{C}}^s(A \cap E_n) \leq \sum_{n=1}^\infty \overline{\mathcal{C}}^s(E_n).$$

Now take the supremum over all $A \subseteq E$ to obtain $\overline{\mathcal{C}}^s(E) \leq \sum_{n=1}^\infty \overline{\mathcal{C}}^s(E_n)$.

Finally, we prove the metric property. Suppose E and F are sets with $\mathrm{dist}(E, F) = \alpha > 0$. If $A \subseteq E$ and $B \subseteq F$, then also $\mathrm{dist}(A, B) \geq \alpha$. Now, if $\varepsilon < \alpha/2$, then the sets of any centered covering of A by balls with radius $< \varepsilon$ are disjoint from the sets of any centered covering of B by balls with radius $< \varepsilon$. So $\widetilde{\mathcal{C}}^s_\varepsilon(A \cup B) = \widetilde{\mathcal{C}}^s_\varepsilon(A) + \widetilde{\mathcal{C}}^s_\varepsilon(B)$. Let $\varepsilon \to 0$ to obtain $\widetilde{\mathcal{C}}^s(A \cup B) = \widetilde{\mathcal{C}}^s(A) + \widetilde{\mathcal{C}}^s(B)$. Then take the supremum over all $A \subseteq E$ and $B \subseteq F$ to obtain $\overline{\mathcal{C}}^s(E \cup F) = \overline{\mathcal{C}}^s(E) + \overline{\mathcal{C}}^s(F)$. ☺

The covering measure $\mathcal{C}^s$ and the Hausdorff measure $\mathcal{H}^s$ are within a constant factor of each other.

(1.4.3) Proposition. *Let S be a metric space, let $s > 0$, and let $E \subseteq S$. Then*

$$2^{-s}\,\overline{\mathcal{C}}^s(E) \leq \overline{\mathcal{H}}^s(E) \leq \overline{\mathcal{C}}^s(E).$$

Proof. If $\overline{B}_r(x)$ is a closed ball, then $\mathrm{diam}\,\overline{B}_r(x) \leq 2r$. Any centered-ball covering of E is in particular a covering of E, so for any $\varepsilon > 0$, we have $\overline{\mathcal{H}}^s_{2\varepsilon}(E) \leq \widetilde{\mathcal{C}}^s_\varepsilon(E)$. Let $\varepsilon \to 0$ to obtain $\overline{\mathcal{H}}^s(E) \leq \widetilde{\mathcal{C}}^s(E) \leq \overline{\mathcal{C}}^s(E)$.

Now let $A \subseteq E$. Suppose $A \subseteq \bigcup_i A_i$ is a cover of A and $\mathrm{diam}\,A_i < \varepsilon$ for all i. If some A_i is disjoint from A, remove it from the union; the remaining sets still cover A. For each A_i, choose $x_i \in A_i \cap A$. Then

$$B_i = \overline{B}_{\mathrm{diam}\,A_i}(x_i) \supseteq A_i.$$

So the sets B_i are a centered-ball cover of A. Thus

$$\widetilde{\mathcal{C}}^s_\varepsilon(A) \leq \sum_i (2\,\mathrm{diam}\,A_i)^s = 2^s \sum_i (\mathrm{diam}\,A_i)^s.$$

This is true for all such covers A_i, so $\widetilde{\mathcal{C}}^s_\varepsilon(A) \leq 2^s \overline{\mathcal{H}}^s_\varepsilon(A)$. Let $\varepsilon \to 0$ to obtain $\widetilde{\mathcal{C}}^s(A) \leq 2^s \overline{\mathcal{H}}^s(A) \leq 2^s \overline{\mathcal{H}}^s(E)$. Finally, take the supremum over all $A \subseteq E$ to obtain $\overline{\mathcal{C}}^s(E) \leq 2^s \overline{\mathcal{H}}^s(E)$. ☺

Because the measures $\mathcal{C}^s$ and $\mathcal{H}^s$ are related in this way, if one of the values $\mathcal{C}^s(E)$ and $\mathcal{H}^s(E)$ is 0, then so is the other. If one is ∞, then so is the other. So the measures $\mathcal{C}^s$ may be used in the computation of the Hausdorff dimension.

The covering measure $\mathcal{C}^s$ is also related to the packing measure $\mathcal{P}^s$.

(1.4.4) Exercise. Let S be a metric space, let $s > 0$, and let $E \subseteq S$. Then

$$\widetilde{\mathcal{C}}^s(E) \leq \overline{\mathcal{C}}^s(E) \leq \overline{\mathcal{P}}^s(E) \leq \widetilde{\mathcal{P}}^s(E).$$

Note that the equality $\overline{\mathcal{C}}^s(E) = \overline{\mathcal{P}}^s(E)$ holds only rarely unless they are both 0 or both ∞ ([232, 184]). In [257], there is an example of a compact set $E \subseteq \mathbb{R}^2$ with $\overline{\mathcal{C}}^1(E) = 0$ and $\overline{\mathcal{P}}^1(E) = \infty$.

Diameter-Based Packing Measure. In the definition of the packing measure, we use $(2r)^s$ for centered balls $\overline{B}_r(x)$. In Euclidean space (and many other metric spaces), diam $\overline{B}_r(x) = 2r$. If we use $(\operatorname{diam} \overline{B}_r(x))^s$ in place of $(2r)^s$, we obtain a variant of the packing measure.

Begin with

$$\widetilde{\mathcal{R}}_c^s(E) = \sup \sum_{i=1}^{\infty} \left(\operatorname{diam} \overline{B}_{r_i}(x_i) \right)^s,$$

where the supremum is over all centered packings $\{\overline{B}_{r_1}(x_1), \overline{B}_{r_2}(x_2), \cdots \}$ of E. Then define $\widetilde{\mathcal{R}}^s$ and $\overline{\mathcal{R}}^s$ in the same way as the packing measure $\mathcal{P}^s$ is defined.

Now in general, diam $\overline{B}_r(x) \leq 2r$, so $\widetilde{\mathcal{P}}_\varepsilon^s(E) \geq \widetilde{\mathcal{R}}_\varepsilon^s(E)$; $\widetilde{\mathcal{P}}^s(E) \geq \widetilde{\mathcal{R}}^s(E)$; $\overline{\mathcal{P}}^s(E) \geq \overline{\mathcal{R}}^s(E)$.

For $S = \mathbb{R}^d$, we have diam $\overline{B}_r(x) = 2r$, so for $E \subseteq \mathbb{R}^d$ we have $\widetilde{\mathcal{P}}_\varepsilon^s(E) = \widetilde{\mathcal{R}}_\varepsilon^s(E)$; $\widetilde{\mathcal{P}}^s(E) = \widetilde{\mathcal{R}}^s(E)$; $\overline{\mathcal{P}}^s(E) = \overline{\mathcal{R}}^s(E)$.

Suppose the metric space S satisfies a lower diameter inequality: there is a constant $c > 0$ such that diam $\overline{B}_r(x) \geq cr$ for all $x \in S$ and all $r > 0$. Then $(c/2)^s \widetilde{\mathcal{P}}_\varepsilon^s(E) \leq \widetilde{\mathcal{R}}_\varepsilon^s(E)$; $(c/2)^s \widetilde{\mathcal{P}}^s(E) \leq \widetilde{\mathcal{R}}^s(E)$; $(c/2)^s \overline{\mathcal{P}}^s(E) \leq \overline{\mathcal{R}}^s(E)$. Thus, in such a space the measures $\overline{\mathcal{R}}^s$ may be used for the computation of the packing dimension.

(In [MTFG, §6.5] I used the diameter definition for packing measure. So in fact, the measures in that book are $\mathcal{R}^s$. If radius-packing measure $\mathcal{P}^s$ had been used instead, the only difference would have been in [MTFG, Theorem 6.5.9]. With radius-packing measure, we get an inequality $\mathcal{P}^s(E) \leq C\mathcal{M}(E)$ for a certain constant C. This inequality is enough for the proof of [MTFG, Theorem 6.5.10]. Similar considerations apply to [MTFG, Exercise 6.5.11].)

(1.4.5) Exercise. In a general metric space, the critical value for $\overline{\mathcal{P}}^s$ need not be the same as for $\overline{\mathcal{R}}^s$. As an example, use a tree: fix an integer $m > 1$, and let level k of the tree have 2^{m^k} nodes in it, each of diameter 2^{-m^k}. Each node in level k has $2^{m^{k+1}}/2^{m^k}$ children. Then the Hausdorff dimension for the space of branches is 1, and the critical value for the diameter-packing measures $\overline{\mathcal{R}}^s$ is also 1. But the critical value for the radius-packing measures $\overline{\mathcal{P}}^s$ is $s = m$.

Balls of the Same Size. For the set-functions defined above, we have allowed sets of varying sizes in the packings or coverings. Now we will consider what happens when we use balls all of the same size. In a general metric space, it is conceivable that no two balls have exactly the same diameter, so requiring covers or packings by balls with the same diameter is not useful. We can consider coverings or packings by balls $\overline{B}_\varepsilon(x)$ for a fixed value of ε: these balls may have diameter much smaller than ε. If the balls are to be disjoint, then certainly the center of one must be outside the others, so the distances between the centers must be $> \varepsilon$.

Let S be a metric space, and let $U \subseteq S$ be a subset. We say that U is ε-**separated** iff $\rho(x, y) > \varepsilon$ for all pairs $x, y \in U$, $x \neq y$. Let $E \subseteq S$ be a set. Define $M_\varepsilon(E)$ to be the maximum size of an ε-separated subset of E.

Define $N_\varepsilon(E)$ to be the minimum size of a cover of E by sets with diameter $\leq 2\varepsilon$. It is easy to see that N_ε and M_ε are closely related. Because of this relation, we will use only M_ε most of the time.

(1.4.6) Proposition. $M_{2\varepsilon}(E) \leq N_\varepsilon(E) \leq M_\varepsilon(E)$.

Proof. Let $\{x_1, x_2, \cdots x_M\}$ be an ε-separated set of maximum size, $M = M_\varepsilon(E)$. For every point y of E, there is x_i with $\rho(y, x_i) \leq \varepsilon$, since otherwise $\{x_1, x_2, \cdots, x_M, y\}$ would be ε-separated and larger than M. Thus, the balls $\overline{B}_\varepsilon(x_i)$ cover E and have diameter $\leq 2\varepsilon$. Therefore, $N_\varepsilon(E) \leq M = M_\varepsilon(E)$.

Suppose $\{U_1, U_2, \cdots, U_N\}$ is a cover of E and diam $U_i \leq 2\varepsilon$ for all i. Each U_i can contain at most one point of a 2ε-separated set. So $M_{2\varepsilon}(E) \leq N$. ☺

The number $\log_2 M_\varepsilon(E)$ is sometimes called the ε-**capacity** of E, and the number $\log_2 N_\varepsilon(E)$ is sometimes called the ε-**entropy** of E. Kolmogorov and Tikhomirov [155] provide this information on the terminology:

> These names are connected with designations from the theory of information. To explain them, the following approximation remarks suffice: a) in the theory of information, the unit of a "collection of information" is the amount of information in one binary sign (that is, designating whether it is 0 or 1); b) the "entropy" of a collection of possible "communications," undergoing preservation or transmission with a specified accuracy, is defined as the number of binary signs necessary to transmit an arbitrary one of these communications with the given accuracy (that is, an h such that to every communication x, one can assign a sequence $s(x)$ of h binary signs from which the communication x can be reconstructed with the needed accuracy); c) the "capacity" of a transmitting or remembering apparatus is defined as the number of binary signs that it can reliably transmit or remember.

If we regard A as a set of possible communications and suppose that finding x' with distance $\rho(x, x') \leq \varepsilon$ permits us to reconstruct the communication x with sufficient accuracy, then it is evident that it suffices to demand $N_\varepsilon(A)$ different signals to transmit any of the communications $x \in A$; as these signals, we may take sequences of binary signs of length not greater than $\log_2 N_\varepsilon(A) + 1$. To succeed with sequences of length less than $\log_2 N_\varepsilon(A)$ is obviously impossible.

On the other hand, if we consider A as a set of possible signals and suppose that two signals $x_1 \in A$ and $x_2 \in A$ are reliably different in the case $\rho(x_1, x_2) > \varepsilon$, then it is evident that one can choose in A $M_\varepsilon(A)$ reliably different signals and with their help fix for storage or transmission any binary sequence of length $\log_2 M_\varepsilon(A) - 1$. It is obviously impossible to find a system of reliably different signals in A for the transmission of arbitrary binary sequences of length $\log_2 M_\varepsilon(A) + 1$.

We will be interested in sets E where $M_\varepsilon(E)$ is finite. Recall that a set E is called **totally bounded** iff $M_\varepsilon(E)$ is finite for every $\varepsilon > 0$; that is, E has no infinite ε-separated subset.

If E has n points, then $M_\varepsilon(E) = n$ for small enough ε. If E is an infinite set, then $M_\varepsilon(E) \to \infty$ as $\varepsilon \to 0$. The speed at which $M_\varepsilon(E)$ grows is related to the fractal dimension of the set E. If we examine this carefully, we will be able to define some fractal measures.

Let S be a metric space, let $s > 0$, and let $E \subseteq S$ be a totally bounded set. Define
$$\widetilde{\mathcal{J}}_\varepsilon^s(E) = M_\varepsilon(E) \cdot (2\varepsilon)^s.$$
In general, this does not converge as $\varepsilon \to 0$. But define
$$\widetilde{\mathcal{J}}^s(E) = \liminf_{\varepsilon \to 0} \widetilde{\mathcal{J}}_\varepsilon^s(E).$$

Then apply method I to obtain an outer measure $\overline{\mathcal{J}}^s$ on S. It is a metric outer measure. Write $\mathcal{J}^s$ for its restriction to the measurable sets.

The value $\mathcal{J}^s(E)$ will be ∞ except possibly for sets E covered by countably many totally bounded sets. For example, Euclidean space $\mathbb{R}^d$ is not totally bounded, but each ball $B_M(0)$ is, so $\mathcal{J}^s(\mathbb{R}^d) < \infty$ is possible. In fact, we will see that $\mathcal{J}^s(\mathbb{R}^d) = 0$ for $s > d$.

(1.4.7) Proposition. $\overline{\mathcal{J}}^s(E) \geq \overline{\mathcal{C}}^s(E) \geq \overline{\mathcal{H}}^s(E)$.

Proof. Let $\varepsilon > 0$ be given. Write $M = M_\varepsilon(E)$. If $\{x_1, x_2, \cdots, x_M\}$ is an ε-separated set, then the balls $\overline{B}_\varepsilon(x_i)$ cover E. Now, diam $\overline{B}_\varepsilon(x_i) \leq 2\varepsilon$, so
$$M \cdot (2\varepsilon)^s = \sum_i (2\varepsilon)^s \geq \widetilde{\mathcal{C}}_\varepsilon^s(E).$$

That is, $\widetilde{\mathcal{J}}_\varepsilon^s(E) \geq \widetilde{\mathcal{C}}_\varepsilon^s(E)$. Let $\varepsilon \to 0$ to obtain $\widetilde{\mathcal{J}}^s(E) \geq \widetilde{\mathcal{C}}^s(E)$. Apply this to an arbitrary subset of E to obtain $\widetilde{\mathcal{J}}^s(E) \geq \overline{\mathcal{C}}^s(E)$. Apply the countable subadditivity of $\overline{\mathcal{C}}^s$ to obtain $\overline{\mathcal{J}}^s(E) \geq \overline{\mathcal{C}}^s(E)$. ☺

The critical value for the set-functions $\overline{\mathcal{J}}^s$ will be called the **lower entropy dimension** and written led E. Thus,

$$\overline{\mathcal{J}}^s(E) = \begin{cases} \infty & \text{for } s < \text{led } E, \\ 0 & \text{for } s > \text{led } E. \end{cases}$$

The inequalities above show that dim $E \leq$ led E.

Sometimes it is useful to use the set-functions $\widetilde{\mathcal{J}}^s$, even though they are not measures. The critical value for $\widetilde{\mathcal{J}}^s$ will be called the **lower entropy index** and written lei E. Thus,

$$\widetilde{\mathcal{J}}^s(E) = \begin{cases} \infty & \text{for } s < \text{lei } E, \\ 0 & \text{for } s > \text{lei } E. \end{cases}$$

Note that led $E \leq$ lei E. There is a "formula" in this case:

(1.4.8) Exercise. Let E be a totally bounded set. Then

$$\text{lei } E = \liminf_{\varepsilon \to 0} \frac{\log M_\varepsilon(E)}{-\log \varepsilon} = \liminf_{\varepsilon \to 0} \frac{\log N_\varepsilon(E)}{-\log \varepsilon}.$$

Next, we replace "lim inf" by "lim sup" in the definitions. Let S be a metric space, let $s > 0$, and let $E \subseteq S$ be a totally bounded set. Define

$$\widetilde{\mathcal{K}}^s_\varepsilon(E) = M_\varepsilon(E) \cdot (2\varepsilon)^s.$$

Then define

$$\widetilde{\mathcal{K}}^s(E) = \limsup_{\varepsilon \to 0} \widetilde{\mathcal{K}}^s_\varepsilon(E).$$

Then apply method I to obtain an outer measure $\overline{\mathcal{K}}^s$. It is a metric outer measure. Write $\mathcal{K}^s$ for its restriction to the measurable sets.

(1.4.9) Exercise. Compute $\mathcal{J}^s(E)$ and $\mathcal{K}^s(E)$, where E is the space of Exercise 1.4.5.

(1.4.10) Theorem. *Let S be a metric space, let $s > 0$, and let $E \subseteq S$ be a subset.*

(a) $\overline{\mathcal{J}}^s(E) \leq \overline{\mathcal{K}}^s(E)$.

(b) $\overline{\mathcal{K}}^s(E) \leq 2^s \overline{\mathcal{P}}^s(E)$.

(c) *If S satisfies a lower diameter estimate* diam $B_r(x) \geq cr$, *then* $\overline{\mathcal{K}}^s(E) \leq (4/c)^s \overline{\mathcal{R}}^s(E)$.

(d) *If $\overline{\mathcal{K}}^s(E) < \infty$ and $s < t$, then $\overline{\mathcal{P}}^t(E) = 0$.*

Proof. (a) First, $\widetilde{\mathcal{J}}^s_\varepsilon(E) = \widetilde{\mathcal{K}}^s_\varepsilon(E)$, so $\overline{\mathcal{J}}^s(E) \leq \widetilde{\mathcal{K}}^s(E)$. Apply method I to obtain $\overline{\mathcal{J}}^s(E) \leq \overline{\mathcal{K}}^s(E)$.

(b) Let $\varepsilon > 0$ be given, and write $M = M_\varepsilon(E)$. Suppose $\{x_1, x_2, \cdots, x_M\}$ is an ε-separated set. Then the balls $\overline{B}_{\varepsilon/2}(x_i)$ are disjoint. So

$$M \cdot (2\varepsilon)^s = \sum_{i=1}^{M} (2\varepsilon)^s \le 2^s \, \widetilde{\mathcal{P}}_\varepsilon^s(E).$$

That is, $\widetilde{\mathcal{K}}_\varepsilon^s(E) \le 2^s \widetilde{\mathcal{P}}_\varepsilon^s(E)$. Let $\varepsilon \to 0$; then apply method I to conclude that $\overline{\mathcal{K}}^s(E) \le 2^s \overline{\mathcal{P}}^s(E)$.

(c) follows from (b).

(d) If $\overline{\mathcal{K}}^s(E) < \infty$, then E is covered by a sequence of sets E_n with $\widetilde{\mathcal{K}}^s(E_n) < \infty$. So it is enough to prove that if $\widetilde{\mathcal{K}}^s(E) < \infty$ and $s < t$, then $\widetilde{\mathcal{P}}^t(E) = 0$. Let u satisfy $s < u < t$; it is enough to prove that $\widetilde{\mathcal{P}}^u(E) < \infty$.

Let $C = \widetilde{\mathcal{K}}^s(E) + 1 < \infty$. Then there is ε with $0 < \varepsilon < 1$ such that $M_r(E) \cdot (2r)^s < C$ for all $r < \varepsilon$. Now suppose $\{ \overline{B}_{r_i}(x_i) : i \in \mathbb{N} \}$ is a centered-ball packing of E with $r_i < \varepsilon$ for all i. Now, if $2^{-k} \le r_i < 2^{-k+1}$, then $x_j \notin \overline{B}_{r_i}(x_i)$ for $j \ne i$, so $\rho(x_j, x_i) > r_i \ge 2^{-k}$. Thus, for each k, the set

$$\left\{ x_i : 2^{-k} \le r_i < 2^{-k+1} \right\}$$

is a 2^{-k}-separated set in E. So it has at most $M_{2^{-k}}(E)$ elements. Therefore,

$$\sum_{i=1}^{\infty} (2r_i)^u \le \sum_{k=1}^{\infty} M_{2^{-k}}(E) \cdot \left(2^{-k+2}\right)^u \le 2^u \cdot C \cdot \sum_{k=1}^{\infty} \left(2^{-k+1}\right)^{u-s}.$$

This bound is finite and independent of the packing, so $\widetilde{\mathcal{P}}_\varepsilon^u(E) < \infty$. Thus $\widetilde{\mathcal{P}}^u(E) \le \widetilde{\mathcal{P}}_\varepsilon^u(E) < \infty$, as required. ☺

The critical value for the set-functions $\overline{\mathcal{K}}^s$ will be called the **upper entropy dimension**, and written ued E. Thus,

$$\overline{\mathcal{K}}^s(E) = \begin{cases} \infty & \text{for } s < \text{ued } E, \\ 0 & \text{for } s > \text{ued } E. \end{cases}$$

According to the theorem, parts (b) and (d), ued is also the critical value for the family $\overline{\mathcal{P}}^s$:

$$\overline{\mathcal{P}}^s(E) = \begin{cases} \infty & \text{for } s < \text{ued } E, \\ 0 & \text{for } s > \text{ued } E. \end{cases}$$

So in fact, Dim $E = $ ued E.

The critical value for $\widetilde{\mathcal{K}}^s$ will be called the **upper entropy index** and written uei E:

$$\widetilde{\mathcal{K}}^s(E) = \begin{cases} \infty & \text{for } s < \text{uei } E, \\ 0 & \text{for } s > \text{uei } E. \end{cases}$$

This time,

$$\text{uei } E = \limsup_{\varepsilon \to 0} \frac{M_\varepsilon(E)}{-\log \varepsilon}.$$

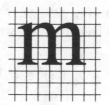

Figure 1.4.11. Count the squares.

Box Dimension. A computation adapted for Euclidean space is often used. For a fixed value of $\varepsilon > 0$, the **lattice cubes** of side ε are sets in $\mathbb{R}^d$ of the form

$$[n_1\varepsilon, (n_1 + 1)\varepsilon) \times [n_2\varepsilon, (n_2 + 1)\varepsilon) \times \cdots \times [n_d\varepsilon, (n_d + 1)\varepsilon),$$

where the n_i are integers. Given a set $E \subseteq \mathbb{R}^d$, let $K_\varepsilon(E)$ be the number of lattice cubes of side ε that intersect E.

(1.4.12) Exercise. $M_{\varepsilon\sqrt{d}}(E) \le K_\varepsilon(E) \le 3^d M_\varepsilon(E)$. Can you improve the factor $\sqrt{d}$ or 3^d?

The **box dimension** of E is defined to be

$$\limsup_{\varepsilon \to 0} \frac{\log K_\varepsilon(E)}{-\log \varepsilon}.$$

By the inequalities above, clearly the box dimension is equal to uei E. In some books, the box dimension is defined with lim inf, which yields lei E. If necessary, we can proceed to define set-functions corresponding to the box dimension. The total volume of all the boxes that meet E is $K_\varepsilon(E) \cdot \varepsilon^d$, so perhaps the s-dimensional set-function should be

$$K_\varepsilon(E) \cdot \varepsilon^s.$$

It is not used in this book, so I have not given it a notation.

Bouligand Dimension. First Cantor and then Minkowski used the idea of the "ε-neighborhood" of a set to determine its volume or area. Bouligand proposed its use for non-integer values of the dimension.

Let (S, ρ) be a metric space. Let $A \subseteq S$ be a subset and $\varepsilon > 0$. The **ε-neighborhood** (or **parallel body** or **Minkowski sausage**) of A is

$$S_\varepsilon(A) = \{\, x \in S : \rho(x, y) < \varepsilon \text{ for some } y \in A \,\} = \bigcup_{y \in S} B_\varepsilon(y).$$

The ε-neighborhood contains all points within distance ε of the set. The closure $\overline{A}$ of A consists of all points with distance 0 from A, so

$$\overline{A} = \bigcap_{\varepsilon > 0} S_\varepsilon(A).$$

Figure 1.4.13. A set and ε-neighborhoods for two values of ε.

We now specialize to the case of Euclidean space $S = \mathbb{R}^d$. The ε-neighborhood of a set A is an open set, so it is Lebesgue measurable. Except in the case where $\overline{A}$ has positive measure or A is unbounded, the d-dimensional volume $\mathcal{L}^d(S_\varepsilon(A))$ decreases to 0 as $\varepsilon \to 0$. The rate at which it goes to 0 tells us about the fractal dimension of A. The fractal dimension should be s if the volume behaves like ε^{d-s}.

(1.4.14) Proposition. *Let d be a positive integer, and let $0 < s \le d$. There are positive constants C and c such that for every $E \subseteq \mathbb{R}^d$ and every $\varepsilon > 0$,*

$$c\widetilde{\mathcal{J}}^s_{2\varepsilon}(E) \le \frac{\mathcal{L}^d(S_\varepsilon(E))}{\varepsilon^{d-s}} \le C\widetilde{\mathcal{J}}^s_\varepsilon(E).$$

Therefore,

$$\text{uei } E = d - \liminf_{\varepsilon \to 0} \frac{\log \mathcal{L}^d(S_\varepsilon(E))}{\log \varepsilon},$$

$$\text{lei } E = d - \limsup_{\varepsilon \to 0} \frac{\log \mathcal{L}^d(S_\varepsilon(E))}{\log \varepsilon}.$$

Proof. Write $M = M_\varepsilon(E)$. Let $\{x_1, x_2, \cdots, x_M\}$ be a maximal ε-separated set in E. Then $S_\varepsilon(E) \subseteq \bigcup_{i=1}^M B_{2\varepsilon}(x_i)$, so that

$$\mathcal{L}^d(S_\varepsilon(E)) \le M_\varepsilon(E)\mathcal{L}^d(B_1(0))(2\varepsilon)^d.$$

Therefore,

$$\frac{\mathcal{L}^d(S_\varepsilon(E))}{\varepsilon^{d-s}} \le C\widetilde{\mathcal{J}}^s_\varepsilon(E),$$

where $C = 2^{d-s}\mathcal{L}^d(B_1(0))$.

Now write $M = M_{2\varepsilon}(E)$. Let $\{x_1, x_2, \cdots, x_M\}$ be a maximal 2ε-separated set in E. The balls $B_\varepsilon(x_i)$ are disjoint and contained in $S_\varepsilon(E)$, so

$$\mathcal{L}^d(S_\varepsilon(E)) \ge M_{2\varepsilon}(E)\mathcal{L}^d(B_1(0))\varepsilon^d.$$

Therefore,

$$\frac{\mathcal{L}^d(S_\varepsilon(E))}{\varepsilon^{d-s}} \ge c\widetilde{\mathcal{J}}^s_{2\varepsilon}(E),$$

where $c = 4^{-s}\mathcal{L}^d(B_1(0))$. ☺

Divider Dimension. In his essay "How Long Is the Coast of Britain?" [175], Mandelbrot proposed a method for computing a (fractional) dimension of a curve. We call it the "divider dimension" after the tool sometimes used for measuring lengths of curved lines on a map.

Let S be a metric space, and let $f\colon [a,b] \to S$ be a continuous function. Fix a (small) positive number ε. Now define $t_1 = 0$ and recursively, if t_k has been defined, then

$$t_{k+1} = \inf \left\{ t > t_k : \rho\big(f(t), f(t_k)\big) \geq \varepsilon \right\}.$$

When no further continuation is possible (when there are no $t > t_k$ with $\rho\big(f(t), f(t_k)\big) \geq \varepsilon$), we stop. If t_n is the last point defined, then let $D_\varepsilon(f) = n$.

Figure 1.4.15. Divider.

Now of course, $D_\varepsilon(f)$ increases as ε decreases. The rate at which it increases is related to the fractal dimension of the curve f. The **divider dimension** of the curve f is

$$\liminf_{\varepsilon \to 0} \frac{\log D_\varepsilon(f)}{-\log \varepsilon}.$$

(1.4.16) Exercise. Let $f\colon [a,b] \to S$ be a continuous function, and let $C = \{ f(t) : a \leq t \leq b \}$ be its range. Show that $N_{\varepsilon/2}(C) \leq D_\varepsilon(f)$. Conclude that the divider dimension of f is at least as large as the lower entropy index of C.

(1.4.17) Exercise. Investigate the possible equality of the divider dimension of a continuous function f and lower entropy index of its range C. Certainly, you will want to assume that f is one-to-one; are other conditions on f needed as well?

Complementary Intervals. Let K be a compact set in $\mathbb{R}$ with Lebesgue measure 0. Write $a = \inf K$ and $b = \sup K$. The complement $\mathbb{R} \setminus K$ then consists of two unbounded intervals, $(-\infty, a), (b, \infty)$, and countably many bounded intervals. Since the total length of these bounded complementary intervals is $b - a$, their lengths go to 0. So let us number them in decreasing order of size:

$$[a, b] \setminus K = \bigcup_k (u_k, v_k), \qquad v_1 - u_1 \geq v_2 - u_2 \geq v_3 - u_3 \geq \cdots .$$

For example, let K be the countable compact set consisting of $1, 1/2, 1/3, 1/4, \cdots$ and the limit point 0. The bounded complementary intervals have lengths

$$\frac{1}{2}, \frac{1}{6}, \frac{1}{12}, \frac{1}{20}, \frac{1}{30}, \frac{1}{42}, \frac{1}{56}, \cdots .$$

For another example, let K be the ternary Cantor dust. The bounded complementary intervals have lengths

$$\frac{1}{3}, \frac{1}{9}, \frac{1}{9}, \frac{1}{27}, \frac{1}{27}, \frac{1}{27}, \frac{1}{27}, \frac{1}{81}, \cdots .$$

Now let us consider the ε-neighborhood $S_\varepsilon(K)$ of the set K. Of course, it contains $(a - \varepsilon, a)$ and $(b, b + \varepsilon)$. It contains the complete interval (u_k, v_k) when k is so large that $v_k - u_k < 2\varepsilon$. And for the smaller values of k, where $v_k - u_k \geq 2\varepsilon$, the ε-neighborhood of K contains $(u_k, u_k + \varepsilon)$ and $(v_k - \varepsilon, v_k)$. Thus the total length of $S_\varepsilon(K)$ is

$$2\varepsilon m + \sum_{k=m}^{\infty} (v_k - u_k),$$

where m is the least integer with $v_m - u_m < 2\varepsilon$.

In the example $K = \{0, 1, 1/2, 1/3, \cdots\}$, the complementary intervals have lengths $1/k - 1/(k+1) = 1/(k(k+1))$. For given $\varepsilon > 0$, the solution k of the equation

$$\frac{1}{k(k+1)} = 2\varepsilon$$

is asymptotic to $1/\sqrt{2\varepsilon}$. The sum of a tail of the series

$$\sum_{k=m}^{\infty} \frac{1}{k(k+1)}$$

is $1/m$. So $\mathcal{L}(S_\varepsilon(K))$ is asymptotic to

$$2\varepsilon \frac{1}{\sqrt{2\varepsilon}} + \sqrt{2\varepsilon} = 2\sqrt{2}\sqrt{\varepsilon}.$$

We may compute

$$1 - \lim_{\varepsilon \to 0} \left(\frac{\log \mathcal{L}(S_\varepsilon(K))}{\log \varepsilon} \right) = 1 - \lim_{\varepsilon} \frac{(1/2) \log \varepsilon}{\log \varepsilon} = 1 - \frac{1}{2} = \frac{1}{2}.$$

So the Bouligand dimension (= uei K = lei K) is $1/2$.

(1.4.18) Exercise. Use the lengths of the complementary intervals to compute the Bouligand dimension of the ternary Cantor dust.

(1.4.19) Exercise. Construct a countable compact set K such that the complementary intervals have exactly the same lengths as the complementary intervals of the ternary Cantor dust.

Full and Fine Variations. The constructions of the packing measures $\mathcal{P}^s$ are a bit awkward because of the extra "method I" step required at the end to create an actual outer measure. There is an alternative construction as a full variation that avoids this last step.

Recall: a **gauge** on a set E is a function with positive values, $\Delta \colon E \to (0, \infty)$. If $\pi = \{\, (x_i, r_i) : i \in \mathbb{N} \,\}$ is a packing of E by centered balls $\overline{B}_{r_i}(x_i)$, we say that π is Δ-**fine** iff $r_i < \Delta(x_i)$ for all i.

For $s > 0$, consider the constituent function $C(x, r) = (2r)^s$. Then define

$$V_\Delta^C(E) = \sup \sum_i (2r_i)^s,$$

where the supremum is over all Δ-fine packings π of E. Of course, if the gauge Δ is the constant ε, then $V_\Delta^C = \widetilde{\mathcal{P}}_\varepsilon^s$. Then the full variation is the infimum over all gauges,

$$V^C(E) = \inf_\Delta V_\Delta^C(E).$$

As we know (1.1.16), this is a metric outer measure.

(1.4.20) Proposition. *Let S be a metric space, and let $s > 0$. Define the constituent function C by $C(x, r) = (2r)^s$. Then $V^C(E) = \overline{\mathcal{P}}^s(E)$ for all $E \subseteq S$.*

Proof. Constants are among the gauges, so $V^C(E) \le \widetilde{\mathcal{P}}^s(E)$.

Suppose $E \subseteq \bigcup_n E_n$. Then $V^C(E) \le V^C\bigl(\bigcup_n E_n\bigr) \le \sum_n V^C(E_n) \le \sum \widetilde{\mathcal{P}}^s(E_n)$. Therefore, $V^C(E) \le \overline{\mathcal{P}}^s(E)$.

On the other hand, suppose a gauge Δ is given. For each $n \in \mathbb{N}$, let

$$E_n = \left\{\, x \in E : \Delta(x) \ge \frac{1}{n} \,\right\}.$$

So $V_\Delta^C(E) \ge V_\Delta^C(E_n) \ge V_{1/n}^C(E_n) = \widetilde{\mathcal{P}}_{1/n}^s(E_n) \ge \widetilde{\mathcal{P}}^s(E_n) \ge \overline{\mathcal{P}}^s(E_n)$. Now E_n increases to E as $n \to \infty$, so $\lim_n \overline{\mathcal{P}}^s(E_n) = \overline{\mathcal{P}}^s(E)$, and thus $V_\Delta^C(E) \ge \overline{\mathcal{P}}^s(E)$. This is true for all gauges Δ, so $V^C(E) \ge \overline{\mathcal{P}}^s(E)$.   ☺

We will next consider the covering measure $\mathcal{C}^s$. When it was defined, there was an awkward step added on the end to ensure that the resulting set-function is monotone. This step is not needed when we realize this measure as a fine variation.

As before, consider the constituent function C defined by $C(x, r) = (2r)^s$. Recall the definition: Let E be a subset of S. If β is a (fine) cover of E, write

$$v_\beta^C = \sup \sum_{(x,r) \in \pi} (2r)^s,$$

where the supremum is over all packings $\pi \subseteq \beta$. The fine variation of E is

$$v^C(E) = \inf v_\beta^C,$$

where the infimum is over all fine covers β of E.

(1.4.21) Exercise. Let E be the Cantor dust (obtained by repeatedly removing middle thirds). Using the definitions above, estimate as closely as possible (above and below) the value of $v^C(E)$.

The next result asserts that this fine variation coincides with $\overline{\mathcal{C}}^s$. This will be proved in (1.5.10), after our discussion of densities.

(1.4.22) Theorem. *Let S be a separable metric space, and let $s > 0$. Define the constituent function C by $C(x, r) = (2r)^s$. Then $v^C(E) = \overline{\mathcal{C}}^s(E)$ for all $E \subseteq S$.*

1.5 Geometry of Fractals

Next we will consider some of the "geometric" properties of a set that are related to the fractal dimension of the set. We use the term "geometric" in a broad sense to include some miscellaneous topics: semifiniteness, Cartesian products, densities.

Semifinite Measure. Let S be a metric space, and let $\mathcal{M}$ be a Borel measure on S. The measure $\mathcal{M}$ is called **semifinite** iff for every Borel set $E \subseteq S$,

$$\mathcal{M}(E) = \sup \{ \mathcal{M}(A) : A \subseteq E, A \text{ Borel}, \mathcal{M}(A) < \infty \}.$$

Of course, if $\mathcal{M}(E) < \infty$, then this equation is true by Theorem 1.1.12. So the question becomes, If $\mathcal{M}(E) = \infty$, must there be a subset $A \subseteq E$ with $0 < \mathcal{M}(A) < \infty$?

(1.5.1) Exercise. Let S be a metric space, and let $\mathcal{M}$ be a Borel measure on S. Are the following equivalent?

(a) For every Borel set $E \subseteq S$,

$$\mathcal{M}(E) = \sup \{ \mathcal{M}(A) : A \subseteq E, A \text{ Borel}, \mathcal{M}(A) < \infty \}.$$

(b) For every Borel set $E \subseteq S$ with $\mathcal{M}(E) = \infty$, there exists a Borel set $A \subseteq E$ with $0 < \mathcal{M}(A) < \infty$.

The question of semifiniteness for Hausdorff and packing measures has been settled only recently.

(1.5.2) Theorem. *Let S be a complete separable metric space, and let $s > 0$. The Hausdorff measure $\mathcal{H}^s$ is semifinite on S.*

This result is due to Howroyd (1995) [138]. We will not include the general proof here. We will prove it in certain special cases in Sections 1.6 and 1.7.

(1.5.3) Theorem. *Let S be a complete separable metric space, and let $s > 0$. The packing measure $\mathcal{P}^s$ is semifinite on S.*

This result is due to Joyce and Preiss (1995) [146]. We will not include the proof here.

Product Sets. Let X_1 and X_2 be two sets. The **Cartesian product** of X_1 and X_2 is the set of ordered pairs

$$X_1 \times X_2 = \{ (x, y) : x \in X_1, y \in X_2 \}.$$

Sometimes we will use a single letter to represent an element of a product: if $a \in X_1 \times X_2$, then a is of the form $a = (x, y)$ with $x \in X_1$ and $y \in X_2$.

There are two natural "projection" maps,

$$\pi_1 \colon X_1 \times X_2 \to X_1 \quad \text{and} \quad \pi_2 \colon X_1 \times X_2 \to X_2,$$

defined by $\pi_1(x, y) = x$, $\pi_2(x, y) = y$. The Cartesian product satisfies a **universal mapping property**: If Z is any set, and $g_1 \colon Z \to X_1$, $g_2 \colon Z \to X_2$ are two functions, then there is a unique function $g \colon Z \to X_1 \times X_2$ such that $g_1 = \pi_1 \circ g$ and $g_2 = \pi_2 \circ g$. This is shown in the diagram

$$
\begin{array}{ccc}
 & Z & \\
{}^{g_1}\swarrow \ \ \downarrow{\scriptstyle g} & & \searrow{}^{g_2} \\
X_1 \xleftarrow{\ \pi_1\ } & X_1 \times X_2 & \xrightarrow{\ \pi_2\ } X_2
\end{array}
$$

This function g is defined by $g(z) = \big(g_1(z), g_2(z)\big)$.

Product Metric Spaces. Suppose (S_1, ρ_1) and (S_2, ρ_2) are metric spaces. The Cartesian product set $S_1 \times S_2$ can be given the structure of a metric space in many ways. The **maximum metric** on $S_1 \times S_2$ is the metric ρ defined by

$$\rho\big((a_1, a_2), (b_1, b_2)\big) = \rho_1(a_1, b_1) \vee \rho_2(a_2, b_2).$$

Unless otherwise indicated, this is the metric that will be used on Cartesian products.

(1.5.4) Exercise.

(a) The maximum metric ρ is, indeed, a metric. The projections π_1 and π_2 satisfy a Lipschitz condition:

$$\rho_i\big(\pi_i(a), \pi_i(b)\big) \leq \rho(a, b) \qquad \text{for all } a, b \in S_1 \times S_2.$$

(b) Suppose (T, θ) is a metric space, and $g_1 \colon T \to S_1$, $g_2 \colon T \to S_2$ are two functions satisfying

$$\rho_1\big(g_1(z), g_1(w)\big) \leq \theta(z, w), \qquad \rho_2\big(g_2(z), g_2(w)\big) \leq \theta(z, w),$$

for all $z, w \in T$. Then the natural map g in

$$
\begin{array}{ccc}
 & T & \\
\mathllap{g_1}\swarrow & \downarrow{\scriptstyle g} & \searrow\mathrlap{g_2} \\
S_1 \xleftarrow{\pi_1} & S_1 \times S_2 & \xrightarrow{\pi_2} S_2
\end{array}
$$

also satisfies a Lipschitz condition

$$\rho\big(g(z), g(w)\big) \leq \theta(z, w).$$

(c) g is continuous if and only if g_1 and g_2 are both continuous.

(d) Is the maximum metric ρ the only metric on the product $S_1 \times S_2$ that satisfies (a) and (b)? Is it the only metric that satisfies (a) and (c)?

(1.5.5) Exercise. Show that

$$\theta\big((a_1, a_2), (b_1, b_2)\big) = \sqrt{\rho_1(a_1, b_1)^2 + \rho_2(a_2, b_2)^2}$$

defines a metric θ on $S_1 \times S_2$. The formal identity function $(S_1 \times S_2, \rho) \to (S_1 \times S_2, \theta)$ has "bounded distortion" in the sense that there are constants $C, D > 0$ with

$$C\,\theta\big((a_1, a_2), (b_1, b_2)\big) \leq \rho\big((a_1, a_2), (b_1, b_2)\big) \leq D\,\theta\big((a_1, a_2), (b_1, b_2)\big).$$

(Thus θ has property (c) of the preceding exercise.)

Fractal Dimension of a Product. If $E \subseteq S_1$ has fractal dimension s and $F \subseteq S_2$ has fractal dimension t, then what is the fractal dimension of the product $E \times F \subseteq S_1 \times S_2$? In many cases, the natural answer, $s + t$, is correct. But not in all cases.

The following example is copied from Falconer [83]. It is a construction of two compact sets $E, F \subseteq \mathbb{R}$ with $\dim E = \dim F = 0$ but $\dim(E \times F) \geq 1$. Let $0 = m_0 < m_1 < m_2 < \cdots$ be a rapidly increasing sequence of integers (satisfying a condition to be given later). Let E consist of those real numbers in $[0, 1]$ whose decimal expansion has 0 in the rth place for all r with $m_k + 1 \leq r \leq m_{k+1}$ and k even; let F consist of those real numbers in $[0, 1]$ whose decimal expansion has 0 in the rth place for all r with $m_k + 1 \leq r \leq m_{k+1}$

and k odd. Now, for k even, looking at the first m_{k+1} decimal places, we see that there is a cover of E by 10^{j_k} intervals of length $\varepsilon_k = 10^{-m_{k+1}}$, where

$$j_k = (m_2 - m_1) + (m_4 - m_3) + \cdots + (m_k - m_{k-1}).$$

For computation of the lower entropy index,

$$(*) \qquad\qquad \frac{\log 10^{j_k}}{-\log 10^{-m_{k+1}}}$$

tends to 0 as $k \to \infty$ provided that the m_k are chosen to increase sufficiently rapidly. (Once m_k is chosen, j_k is defined, so then choose m_{k+1} large enough that $(*)$ is $< 1/k$.) Therefore, we have lei $E = 0$, so dim $E = 0$. A similar calculation shows that dim $F = $ lei $F = 0$. Now, the sum $E + F$ clearly contains the whole interval $[0,1]$. The addition function $\mathbb{R} \times \mathbb{R} \to \mathbb{R}$ is a Lipschitz function, so

$$\dim(E \times F) \geq \dim(E + F) = \dim[0,1] = 1.$$

We will see below (3.2.11) that in general, $\dim(E \times F) \geq \dim E + \dim F$. On the other hand, for the packing dimension we have the opposite inequality, $\mathrm{Dim}(E \times F) \leq \mathrm{Dim}\, E + \mathrm{Dim}\, F$.

(1.5.6) Proposition.

(a) *Let $M_\varepsilon(E)$ denote the maximum size of an ε-separated set in E. Then*
$$M_\varepsilon(E \times F) \geq M_\varepsilon(E) M_\varepsilon(F).$$
(b) *Let $N_\varepsilon(E)$ denote the minimum size of a cover by sets of diameter $\leq 2\varepsilon$. Then $N_\varepsilon(E \times F) \leq N_\varepsilon(E) N_\varepsilon(F)$.*
(c) uei $(E \times F) \leq$ uei $E +$ uei F, lei $(E \times F) \geq$ lei $E +$ lei F, ued $(E \times F) \leq$ ued $E +$ ued F, $\mathrm{Dim}(E \times F) \leq \mathrm{Dim}\, E + \mathrm{Dim}\, F$.

Proof. (a) Simply note that if $\{ x_i : i = 1, \cdots, n \}$ is an ε-separated set in E and $\{ y_j : j = 1, \cdots, m \}$ is an ε-separated set in F, then

$$\{ (x_i, y_j) : i = 1, \cdots, n; j = 1, \cdots, m \}$$

is an ε-separated set in $E \times F$ with the maximum metric.

(b) If $\{ U_i : i = 1, \cdots, n \}$ is a cover of E with diam $U_i \leq 2\varepsilon$ and $\{ V_j : j = 1, \cdots, m \}$ is a cover of F with diam $V_j \leq 2\varepsilon$, then

$$\{ U_i \times V_j : i = 1, \cdots, n; j = 1, \cdots, m \}$$

is a cover of $E \times F$ and diam $U_i \times V_j \leq 2\varepsilon$.

(c) From (a), we have

$$\begin{aligned}
\widetilde{\mathfrak{J}}^{s+t}_\varepsilon(E \times F) &= M_\varepsilon(E \times F) \cdot (2\varepsilon)^{s+t} \\
&\geq M_\varepsilon(E) \cdot (2\varepsilon)^s \cdot M_\varepsilon(F) \cdot (2\varepsilon)^t \\
&= \widetilde{\mathfrak{J}}^s_\varepsilon(E) \widetilde{\mathfrak{J}}^t_\varepsilon(F).
\end{aligned}$$

In general, $M_{2\varepsilon} \leq N_\varepsilon \leq M_\varepsilon$, so from (b) we have

$$\widetilde{\mathcal{K}}_{2\varepsilon}^{s+t}(E \times F) = M_{2\varepsilon}(E \times F) \cdot (4\varepsilon)^{s+t}$$
$$\leq 2^{s+t} \cdot N_\varepsilon(E \times F) \cdot (2\varepsilon)^{s+t}$$
$$\leq 2^{s+t} \cdot N_\varepsilon(E) \cdot (2\varepsilon)^s \cdot N_\varepsilon(F) \cdot (2\varepsilon)^t$$
$$\leq 2^{s+t} \cdot M_\varepsilon(E) \cdot (2\varepsilon)^s \cdot M_\varepsilon(F) \cdot (2\varepsilon)^t$$
$$= 2^{s+t} \widetilde{\mathcal{K}}_\varepsilon^s(E) \widetilde{\mathcal{K}}_\varepsilon^t(F).$$

Now, when $\varepsilon \to 0$, we obtain

$$\widetilde{\mathcal{J}}^{s+t}(E \times F) \geq \widetilde{\mathcal{J}}^s(E)\widetilde{\mathcal{J}}^t(F),$$
$$\widetilde{\mathcal{K}}^{s+t}(E \times F) \leq 2^{s+t}\widetilde{\mathcal{K}}^s(E)\widetilde{\mathcal{K}}^t(F).$$

Therefore, uei $(E \times F) \leq$ uei $E +$ uei F, and lei $(E \times F) \geq$ lei $E +$ lei F.

Now, if $\{A_i\}$ is a cover of E and $\{B_j\}$ is a cover of F, then $\{A_i \times B_j\}$ is a cover of $E \times F$. So

$$\overline{\mathcal{K}}^{s+t}(E \times F) \leq \sum_{i,j} \widetilde{\mathcal{K}}^{s+t}(A_i \times B_j) \leq 2^{s+t} \sum_i \widetilde{\mathcal{K}}^s(A_i) \sum_j \widetilde{\mathcal{K}}^t(B_j).$$

Take the infimum over all covers $\{A_i\}$ of E and $\{B_j\}$ of F:

$$\overline{\mathcal{K}}^{s+t}(E \times F) \leq 2^{s+t}\overline{\mathcal{K}}^s(E)\overline{\mathcal{K}}^t(F),$$

and therefore ued $(E \times F) \leq$ ued $E +$ ued F. Finally, Dim $=$ ued .   ☺

Now, we have inequalities dim $(E \times F) \geq$ dim $E +$ dim F and Dim $(E \times F) \leq$ Dim $E +$ Dim F. So if E and F are fractals in the sense of Taylor, that is,

$$\dim E = \text{Dim } E, \qquad \dim F = \text{Dim } F,$$

then

$$\dim (E \times F) \leq \text{Dim } (E \times F) \leq \text{Dim } E + \text{Dim } F$$
$$= \dim E + \dim F \leq \dim (E \times F),$$

so $E \times F$ is also a fractal in the sense of Taylor, and dim $(E \times F) =$ dim $E +$ dim F.

(1.5.7) Exercise. The small inductive dimension ind is defined in [MTFG, §3.1]. (a) Prove (by induction, of course) that ind $(E \times F) \leq$ ind $E +$ ind F. (b) If E and F are fractals in the sense of Mandelbrot, that is,

$$\text{ind } E < \dim E, \qquad \text{ind } F < \dim F,$$

then so is the Cartesian product $E \times F$.

Density. If A is a set in $\mathbb{R}^d$, and $B_r(x)$ is a ball, then the **average density** of A in $B_r(x)$ is the fraction of the ball that is occupied by the set A; in symbols, $\mathcal{L}^d(A \cap B_r(x))/\mathcal{L}^d(B_r(x))$. The "fraction" is measured by d-dimensional volume. The **density** of A at a point x is the limit $\lim_{r\to 0} \mathcal{L}^d(A \cap B_r(x))/\mathcal{L}^d(B_r(x))$, if it exists. Lebesgue's Density Theorem (for example [48, p. 184]) asserts that if A is a measurable set, then the density is 1 at almost every point of A and 0 at almost every point of the complement.

If $\mathcal{M}$ is a finite measure on $\mathbb{R}^d$, then the average density of $\mathcal{M}$ in a ball $B_r(x)$ is $\mathcal{M}(B_r(x))/\mathcal{L}^d(B_r(x))$, and the density of $\mathcal{M}$ at the point x is $\lim_{r\to 0} \mathcal{M}(B_r(x))/\mathcal{L}^d(B_r(x))$.

When we consider a set A with Lebesgue measure zero (for example, a fractal with dimension $< d$ or a measure with dimension $< d$) then we get density zero almost everywhere. But it may still be relevant to compute a density using a different denominator. Now, the denominator $\mathcal{L}^d(B_r(x))$ is simply r^d (up to a constant factor). So for a measure or set of fractal dimension $s < d$, it is useful to do these computations with denominator r^s. The diameter of $B_r(x)$ is $2r$ when we use Euclidean space, but it may not be in other metric spaces.

With these preliminary explanations, we now turn to the formal mathematical definitions.

(1.5.8) Definition. Let (S, ρ) be a metric space, let $\mathcal{M}$ be a finite Borel measure on S, let $s > 0$, and let $x \in S$. The **upper s-density** of $\mathcal{M}$ at x is

$$\overline{D}^s_{\mathcal{M}}(x) = \limsup_{r\to 0} \frac{\mathcal{M}(\overline{B}_r(x))}{(2r)^s}.$$

The **lower s-density** of $\mathcal{M}$ at x is

$$\underline{D}^s_{\mathcal{M}}(x) = \liminf_{r\to 0} \frac{\mathcal{M}(\overline{B}_r(x))}{(2r)^s}.$$

In general, $\underline{D}^s_{\mathcal{M}}(x) \leq \overline{D}^s_{\mathcal{M}}(x)$. If $\underline{D}^s_{\mathcal{M}}(x) = \overline{D}^s_{\mathcal{M}}(x)$, then we say x is a **regular point** for $\mathcal{M}$, and the common value $\underline{D}^s_{\mathcal{M}}(x) = \overline{D}^s_{\mathcal{M}}(x)$ is called the **s-density** of $\mathcal{M}$ at x and written $D^s_{\mathcal{M}}(x)$.

The same values are obtained if we use open balls $B_r(x)$ rather than closed balls $\overline{B}_r(x)$:

$$\limsup_{r\to 0} \frac{\mathcal{M}(\overline{B}_r(x))}{(2r)^s} = \limsup_{r\to 0} \frac{\mathcal{M}(B_r(x))}{(2r)^s},$$

$$\liminf_{r\to 0} \frac{\mathcal{M}(\overline{B}_r(x))}{(2r)^s} = \liminf_{r\to 0} \frac{\mathcal{M}(B_r(x))}{(2r)^s}.$$

Indeed, for any open ball $B_r(x)$, there exist closed balls $\overline{B}_{r-1/n}(x)$ with radius and measure as close as we like to $B_r(x)$. And for any any closed ball $\overline{B}_r(x)$, there exist open balls $B_{r+1/n}(x)$ with radius and measure as close as we like to $\overline{B}_r(x)$.

Proof of measurability of densities uses some information about Borel measurable functions (Chapter 2 of this book).

(1.5.9) Proposition.   *The functions $\overline{D}^s_{\mathcal{M}}$ and $\underline{D}^s_{\mathcal{M}}$ are Borel measurable functions from S to $[0, \infty]$.*

Proof. First, for a fixed $r > 0$, I claim that the function $x \mapsto \mathcal{M}(B_r(x))$ is a Borel function. For $t \in \mathbb{R}$, I must show that $V = \{ x \in S : \mathcal{M}(B_r(x)) > t \}$ is a Borel set. In fact, I will show that V is an open set.[3] Let $x_0 \in V$. That is, $\mathcal{M}(B_r(x_0)) > t$. The balls $B_{r-1/n}(x_0)$ increase to $B_r(x_0)$, so the measures $\mathcal{M}(B_{r-1/n}(x_0))$ converge to $\mathcal{M}(B_r(x_0))$. There is n such that $\mathcal{M}(B_{r-1/n}(x_0)) > t$. I will show that $B_{1/n}(x_0) \subseteq V$. Suppose $\rho(x, x_0) < 1/n$. Then $B_r(x) \supseteq B_{r-1/n}(x_0)$, so $\mathcal{M}(B_r(x)) \geq \mathcal{M}(B_{r-1/n}(x_0)) > t$, and thus $x \in V$. This completes the proof that V is open and that $x \mapsto \mathcal{M}(B_r(x))$ is Borel measurable.

Next, I claim that in the definition

$$\overline{D}^s_{\mathcal{M}}(x) = \limsup_{r \to 0} \frac{\mathcal{M}(B_r(x))}{(2r)^s}$$

the same value is obtained if the lim sup is taken using only rational numbers r. If r_n is a sequence of rational numbers that increases to the real number r, then

$$\lim_{n \to \infty} \mathcal{M}(B_{r_n}(x)) = \mathcal{M}(B_r(x)), \text{ and } \lim_{n \to \infty} (2r_n)^s = (2r)^s,$$

so that

$$\lim_{n \to \infty} \frac{\mathcal{M}(B_{r_n}(x))}{(2r_n)^s} = \frac{\mathcal{M}(B_r(x))}{(2r)^s}.$$

Therefore, for $\varepsilon > 0$,

$$\sup \left\{ \frac{\mathcal{M}(B_r(x))}{(2r)^s} : 0 < r < \varepsilon, r \text{ real} \right\}$$

$$= \sup \left\{ \frac{\mathcal{M}(B_r(x))}{(2r)^s} : 0 < r < \varepsilon, r \text{ rational} \right\}.$$

If this value is called $Y(\varepsilon)$, then $\overline{D}^s_{\mathcal{M}}(x) = \inf_{\varepsilon > 0} Y(\varepsilon) = \lim_{n \to \infty} Y(1/n)$.

So we conclude that $\overline{D}^s_{\mathcal{M}}(x)$ is a Borel function of x. And $\underline{D}^s_{\mathcal{M}}(x)$ may be done in the same way.   ☺

Now we will prove that the fine variation v^C defined by the constituent function $C(x, r) = (2r)^s$ coincides with the covering measure $\overline{\mathcal{C}}^s$, as promised above.

---

[3] The function is thus "lower semicontinuous."

(1.5.10) Theorem. *Let S be a separable metric space, and let $s > 0$. Define the constituent function C by $C(x,r) = (2r)^s$. Then $v^C(E) = \overline{\mathbb{C}}^s(E)$ for all $E \subseteq S$.*

Proof. (a) We first prove $\overline{\mathbb{C}}^s(E) \leq v^C(E)$. If $v^C(E) = \infty$, there is nothing to prove, so suppose $v^C(E) < \infty$. Let β be a fine cover of E with $v^C_\beta < \infty$. Let $\varepsilon > 0$. By the Vitali Theorem (1.3.1, see Exercise 1.3.8), there is a packing $\{(x_i, r_i)\} \subseteq \beta$ with $r_i < \varepsilon$ and either $\sum (2r_i)^s = \infty$ or $\overline{\mathbb{C}}^s\big(E \setminus \bigcup \overline{B}_{r_i}(x_i)\big) = 0$. But $\sum (2r_i)^s \leq v^C_\beta < \infty$, so in fact $\overline{\mathbb{C}}^s\big(E \setminus \bigcup \overline{B}_{r_i}(x_i)\big) = 0$. Then $\widetilde{\mathbb{C}}^s_\varepsilon\big(E \setminus \bigcup \overline{B}_{r_i}(x_i)\big) = 0$ for all $\varepsilon > 0$. Thus

$$\widetilde{\mathbb{C}}^s_\varepsilon(E) \leq \widetilde{\mathbb{C}}^s_\varepsilon\left(\bigcup_{i=1}^\infty \overline{B}_{r_i}(x_i)\right) \leq \sum_{i=1}^\infty (2r_i)^s \leq v^C_\beta.$$

Now let $\varepsilon \to 0$ to obtain $\widetilde{\mathbb{C}}^s(E) \leq v^C_\beta$. Then take the infimum over β to obtain $\widetilde{\mathbb{C}}^s(E) \leq v^C(E)$. Finally, take the supremum of this inequality over all subsets to obtain $\overline{\mathbb{C}}^s(E) \leq v^C(E)$.

(b) Next we prove: if $\overline{\mathbb{C}}^s(E) = 0$, then $v^C(E) = 0$. Since S is separable and $v^C(\{x\}) = 0$ for any single point x, we may reduce to the case where E contains no isolated points.

Let $\varepsilon > 0$. For each positive integer n, since $\widetilde{\mathbb{C}}^s_{1/n}(E) = 0$, there is a centered cover

$$\big\{ (x_{in}, r_{in}) : i \in \mathbb{N} \big\}$$

of E with $r_{in} < 1/n$, $x_{in} \in E$, and

$$\sum_i (2r_{in})^s < \frac{\varepsilon}{2^{n+1}}.$$

Now for each i and n let

$$\beta_{in} = \{ (y, r_{in}) : y \in E, \rho(y, x_{in}) \leq r_{in} \}.$$

Then

$$\beta = \bigcup_{i,n} \beta_{in}$$

is a fine cover of E. Let $\pi \subseteq \beta$ be a packing. For each i, n, there is at most one element of β_{in} in π (because all elements of β_{in} contain the point x_{in}). Thus

$$\sum_{(x,r) \in \pi} (2r)^s \leq \sum_{i,n} (2r_{in})^s \leq \sum_n \frac{\varepsilon}{2^{n+1}} \leq \varepsilon.$$

Thus $v^C_\beta \leq \varepsilon$. So $v^C(E) \leq \varepsilon$. But $\varepsilon > 0$ was arbitrary, so $v^C(E) = 0$.

(c) Finally, we prove that $v^C(E) \leq \overline{\mathbb{C}}^s(E)$. Again, we may reduce to the case where E contains no isolated points of S. If $\overline{\mathbb{C}}^s(E) = \infty$, there is nothing to prove, so assume $\overline{\mathbb{C}}^s(E) < \infty$. Let $\mathcal{M}$ be the restriction of $\overline{\mathbb{C}}^s$ to E; that is, $\mathcal{M}(A) = \overline{\mathbb{C}}^s(A \cap E)$ for all A. Then $\mathcal{M}$ is a finite metric outer measure.

We will decompose E using the upper s-density $\overline{D}_M^s$. Fix a number $\alpha > 1$. Write

$$E_1 = \left\{ x \in E : \overline{D}_M^s(x) \leq \alpha^{-3} \right\},$$
$$E_2 = \left\{ x \in E : \overline{D}_M^s(x) > \alpha^{-3} \right\}.$$

Consider first E_1. For $n \in \mathbb{N}$, write

$$F_n = \left\{ x \in E_1 : \frac{M(\overline{B}_r(x))}{(2r)^s} < \alpha^{-2} \text{ for all } r < \frac{1}{n} \right\}.$$

Then F_n increases to E_1 as $n \to \infty$, since $\alpha^{-2} > \alpha^{-3}$.

I claim that $\overline{\mathcal{C}}^s(F_n) = 0$. If $\varepsilon < 1/n$, then when F_n is covered by $\{\overline{B}_{r_i}(x_i)\}$ with $r_i < \varepsilon$, we have

$$\sum (2r_i)^s \geq \alpha^2 \sum M(\overline{B}_{r_i}(x_i)) \geq \alpha^2 M \left(\bigcup \overline{B}_{r_i}(x_i) \right)$$
$$\geq \alpha^2 M(F_n) = \alpha^2 \overline{\mathcal{C}}^s(F_n).$$

Therefore, $\widetilde{\mathcal{C}}_\varepsilon^s(F_n) \geq \alpha^2 \overline{\mathcal{C}}^s(F_n)$. Let $\varepsilon \to 0$ to obtain $\widetilde{\mathcal{C}}^s(F_n) \geq \alpha^2 \overline{\mathcal{C}}^s(F_n)$. Therefore, $\overline{\mathcal{C}}^s(F_n) \geq \alpha^2 \overline{\mathcal{C}}^s(F_n)$. Now, $\overline{\mathcal{C}}^s(F_n) < \infty$ and $\alpha^2 > 1$, so $\overline{\mathcal{C}}^s(F_n) = 0$.

Thus $\overline{\mathcal{C}}^s(F_n) = 0$ for all n. By countable subadditivity, we conclude that $\overline{\mathcal{C}}^s(E_1) = 0$. By part (b), $v^C(E_1) = 0$ as well.

Next consider the set E_2. Since $\alpha^{-4} < \alpha^{-3}$, the set

$$\beta = \left\{ (x, r) \text{ constituent} : x \in E_2, \frac{M(\overline{B}_r(x))}{(2r)^s} > \alpha^{-4} \right\}$$

is a fine cover of E_2. Now, if $\pi \subseteq \beta$ is a packing, then

$$\sum_{(x,r) \in \pi} (2r)^s < \alpha^4 \sum_\pi \overline{\mathcal{C}}^s(\overline{B}_r(x) \cap E) \leq \alpha^4 \overline{\mathcal{C}}^s(E).$$

This is true for all $\pi \subseteq \beta$, so $v_\beta^C \leq \alpha^4 \overline{\mathcal{C}}^s(E)$, and thus $v^C(E_2) \leq \alpha^4 \overline{\mathcal{C}}^s(E)$.

Combining the two parts, we have

$$v^C(E) \leq v^C(E_1) + v^C(E_2) \leq 0 + \alpha^4 \overline{\mathcal{C}}^s(E).$$

Take the infimum over all $\alpha > 1$ to obtain $v^C(E) \leq \overline{\mathcal{C}}^s(E)$.   ☺

Density and Fractal Measure. In many cases, when a fractal set K is constructed, a measure M concentrated on K is also implicitly constructed. Generally speaking, if a set has fractal dimension s, then we expect that the amount of the set inside a ball $B_r(x)$ decreases to 0 as $r \to 0$ at a rate approximately like r^s. For example, on a smooth surface K, the area of $K \cap B_r(x)$ is roughly equal to the area of a circular disk of radius r, so it decreases to 0 at a rate proportional to r^2. It should be easy to believe that the general idea here can be connected with general densities.

We will see that the lower density $\underline{D}_{\mathcal{M}}^s$ is connected with the packing measure $\mathcal{P}^s$, and the upper density $\overline{D}_{\mathcal{M}}^s$ is connected with the Hausdorff measure $\mathcal{H}^s$, or (more correctly) with the covering measure $\mathcal{C}^s$.

Recall that in many metric spaces (such as Euclidean space $\mathbb{R}^d$), every finite Borel measure has the strong Vitali property.

(1.5.11) Theorem. *Let S be a metric space, and let $\mathcal{M}$ be a finite Borel measure with the strong Vitali property. Let $E \subseteq S$ be a Borel set, and let $s > 0$. Then*

$$\mathcal{P}^s(E) \inf_{x \in E} \underline{D}_{\mathcal{M}}^s(x) \leq \mathcal{M}(E) \leq \mathcal{P}^s(E) \sup_{x \in E} \underline{D}_{\mathcal{M}}^s(x).$$

Proof. (a) We begin with the proof of $\mathcal{P}^s(E) \inf_{x \in E} \underline{D}_{\mathcal{M}}^s(x) \leq \mathcal{M}(E)$. Let $h > 0$ such that $\underline{D}_{\mathcal{M}}^s(x) > h$ for all $x \in E$. I must show that $h\mathcal{P}^s(E) \leq \mathcal{M}(E)$. Let $\varepsilon > 0$ be given. By (1.1.12) there is an open set $V \supseteq E$ such that $\mathcal{M}(V) < \mathcal{M}(E) + \varepsilon$. For $x \in E$, let $\Delta(x) > 0$ be so small that

$$\text{(a)} \quad \frac{\mathcal{M}(\overline{B}_r(x))}{(2r)^s} > h \text{ for all } r < \Delta(x),$$

$$\text{(b)} \quad \Delta(x) < \text{dist}(x, S \setminus V).$$

Then Δ is a gauge for E. Let $T_i = \overline{B}_{r_i}(x_i)$ be a Δ-fine centered-ball packing of E. Then $\bigcup T_i$ is contained in V, and

$$\sum_i (2r_i)^s < \frac{1}{h} \sum \mathcal{M}(T_i) = \frac{1}{h}\mathcal{M}(\bigcup T_i) \leq \frac{1}{h}\mathcal{M}(V).$$

This shows that

$$\overline{\mathcal{P}}^s(E) \leq \widetilde{\mathcal{P}}_\Delta^s(E) \leq \frac{1}{h}\mathcal{M}(V) \leq \frac{1}{h}(\mathcal{M}(E) + \varepsilon).$$

Let $\varepsilon \to 0$ to obtain $\overline{\mathcal{P}}^s(E) \leq (1/h)\mathcal{M}(E)$, as required.

(b) Next we prove $\mathcal{M}(E) \leq \mathcal{P}^s(E) \sup_{x \in E} \underline{D}_{\mathcal{M}}^s(x)$. Suppose $h < \infty$ satisfies $\underline{D}_{\mathcal{M}}^s(x) < h$ for all $x \in E$. I must show that $\mathcal{M}(E) \leq h\mathcal{P}^s(E)$. Let Δ be a gauge on E. Then

$$\left\{ B_r(x) : x \in E, r < \frac{1}{2}\Delta(x), \frac{\mathcal{M}(\overline{B}_r(x))}{(2r)^s} \leq h \right\}$$

is a centered Vitali cover of E. So by the strong Vitali property we may select from it a centered-ball packing $T_i = \overline{B}_{r_i}(x_i)$ of E such that $\mathcal{M}(E) = \mathcal{M}(E \cap \bigcup T_i)$. Thus,

$$\mathcal{M}(E) = \mathcal{M}(E \cap \bigcup T_i) \leq \sum_i \mathcal{M}(T_i) \leq h \sum_i (2r_i)^s.$$

So $\mathcal{M}(E) \le h\widetilde{\mathcal{P}}^s_\Delta(E)$. But Δ was arbitrary, so $\mathcal{M}(E) \le h\mathcal{P}^s(E)$, as required.   ☺

An "integral" form of the preceding result can be found below, Exercise 2.3.11.

(1.5.12) Exercise. Is the strong Vitali property needed in Theorem 1.5.11?

(1.5.13) Theorem. *Let $E \subseteq S$ be a Borel set, let $\mathcal{M}$ be a finite Borel measure on S, and let $s > 0$. Then*

$$\mathcal{C}^s(E) \inf_{x \in E} \overline{D}^s_\mathcal{M}(x) \le \mathcal{M}(E) \le \mathcal{C}^s(E) \sup_{x \in E} \overline{D}^s_\mathcal{M}(x).$$

Proof. (a) First I prove $\mathcal{C}^s(E) \inf_{x \in E} \overline{D}^s_\mathcal{M}(x) \le \mathcal{M}(E)$. Let $h > 0$ satisfy $\overline{D}^s_\mathcal{M}(x) > h$ for all $x \in E$. I must show that $h\mathcal{C}^s(E) \le \mathcal{M}(E)$. It is enough to show that $h\widetilde{\mathcal{C}}^s(F) \le \mathcal{M}(E)$ for any subset $F \subseteq E$.

Let $\varepsilon > 0$, and let $V \supseteq E$ be an open set. Define a gauge Δ on F such that $\Delta(x) < \varepsilon$ and $\Delta(x) < \operatorname{dist}(x, S \setminus V)$. Since $\overline{D}^s(x) > h$ on F,

$$\beta = \left\{ (x, r) : x \in F, r < \frac{\Delta(x)}{10}, \frac{\mathcal{M}(\overline{B}_r(x))}{(2r)^s} \ge h \right\}$$

is a fine cover of F. So by Theorem 1.3.1 we may choose from it a centered-ball packing $\{\overline{B}_{r_i}(x_i)\}$ for F such that

$$F \setminus \left(\bigcup_{i=1}^{j} \overline{B}_{r_i}(x_i) \right) \subseteq \bigcup_{i=j+1}^{\infty} \overline{B}_{3r_i}(x_i)$$

for all j.[4] Now, $\bigcup \overline{B}_{r_i}(x_i)$ is contained in V, so

$$\sum (2r_i)^s \le \frac{1}{h} \sum \mathcal{M}(\overline{B}_{r_i}(x_i)) = \frac{1}{h}\mathcal{M}(\bigcup \overline{B}_{r_i}(x_i)) \le \frac{1}{h}\mathcal{M}(V).$$

In particular, $\sum(2r_i)^s < \infty$. But then also, $\sum(6r_i)^s < \infty$. Now,

$$\widetilde{\mathcal{C}}^s_\varepsilon(F) \le \sum_{i=1}^{j} (2r_i)^s + \sum_{i=j+1}^{\infty} (6r_i)^s$$

for all j, so

$$\widetilde{\mathcal{C}}^s_\varepsilon(F) \le \sum_{i=1}^{\infty} (2r_i)^s \le \frac{1}{h}\mathcal{M}(V).$$

Take the infimum over V to obtain $\mathcal{C}^s_\varepsilon(F) \le (1/h)\mathcal{M}(E)$. Now let $\varepsilon \to 0$ to obtain $\mathcal{C}^s(F) \le (1/h)\mathcal{M}(E)$, as required.

---

[4] Case (a) of 1.3.1 is impossible by the same reasoning as below.

(b) Now we prove $\mathcal{M}(E) \leq \mathcal{C}^s(E) \sup_{x \in E} \overline{D}^s_{\mathcal{M}}(x)$. Suppose $\overline{D}^s_{\mathcal{M}}(x) < h < \infty$ for all $x \in E$. I must prove that $\mathcal{M}(E) \leq h\mathcal{C}^s(E)$. For $n \in \mathbb{N}$, let

$$E_n = \left\{ x \in E : \frac{\mathcal{M}(\overline{B}_r(x))}{(2r)^s} < h \text{ for all } r < \frac{1}{n} \right\}.$$

Now $\overline{D}^s_{\mathcal{M}}(x) < h$ for all $x \in E$, so the sets E_n increase to E. Let $\varepsilon < 1/n$ and let $B_i = \overline{B}_{r_i}(x_i)$ be a centered-ball cover of E_n with $r_i < \varepsilon$. Then

$$\sum (2r_i)^s \geq \frac{1}{h} \sum \mathcal{M}(B_i) \geq \frac{1}{h}\mathcal{M}(\bigcup B_i) \geq \frac{1}{h}\mathcal{M}(E_n).$$

Therefore, $\widetilde{\mathcal{C}}^s_\varepsilon(E_n) \geq (1/h)\mathcal{M}(E_n)$. Now if we let $\varepsilon \to 0$, we have $\mathcal{C}^s(E) \geq \mathcal{C}^s(E_n) \geq \widetilde{\mathcal{C}}^s(E_n) \geq (1/h)\mathcal{M}(E_n)$. Finally, let $n \to \infty$ to obtain $\mathcal{C}^s(E) \geq (1/h)\mathcal{M}(E)$, as required.   ☺

Note that the proof in part (a) is valid even if $\overline{D}^s_{\mathcal{M}}(x) = \infty$ on E. The conclusion is, of course, that $\mathcal{C}^s(E) \cdot h \leq \mathcal{M}(E)$ for any positive h, and thus $\mathcal{C}^s(E) = 0$. And part (b) is valid even if $\overline{D}^s_{\mathcal{M}}(x) = 0$ on E. The conclusion is that $\mathcal{M}(E) \leq \mathcal{C}^s(E) \cdot h$ for every positive h, and thus $\mathcal{C}^s(E) = \infty$. It is important that $0 \cdot \infty$ is not interpreted as 0 here. For example, if $\mathcal{M}$ is Lebesgue measure on $E = [0, 1]$ and $0 < s < 1$, then $\overline{D}^s_{\mathcal{M}}(x) = 0$ for all x even though $\mathcal{M}(E) > 0$.

(1.5.14) Theorem. *Let $E \subseteq S$ be a Borel set, let $\mathcal{M}$ be a finite Borel measure on S, and let $s > 0$. Then*

$$\mathcal{H}^s(E) \inf_{x \in E} \overline{D}^s_{\mathcal{M}}(x) \leq \mathcal{M}(E) \leq 2^s \mathcal{H}^s(E) \sup_{x \in E} \overline{D}^s_{\mathcal{M}}(x).$$

Proof. Apply Proposition 1.4.3 to the theorem.   ☺

These results can be used in the following way.

(1.5.15) Proposition. *Let $E \subseteq S$ be a Borel set, let $\mathcal{M}$ be a finite Borel measure on S, and let $s > 0$.*

(a) *If $\overline{D}^s_{\mathcal{M}}(x) = 0$ for all $x \in E$ and $\mathcal{M}(E) > 0$, then $\mathcal{H}^s(E) = \infty$.*
(b) *If $\overline{D}^s_{\mathcal{M}}(x) < \infty$ for all $x \in E$ and $\mathcal{M}(E) > 0$, then $\mathcal{H}^s(E) > 0$.*
(c) *If $\overline{D}^s_{\mathcal{M}}(x) > 0$ for all $x \in E$, then E is σ-finite for $\mathcal{H}^s$.*
(d) *If $\overline{D}^s_{\mathcal{M}}(x) = \infty$ for all $x \in E$, then $\mathcal{H}^s(E) = 0$.*

Proof. (a) If $\mathcal{H}^s(E) < \infty$, apply the theorem: $\mathcal{M}(E) \leq \mathcal{C}^s(E) \sup \overline{D}^s_{\mathcal{M}}(x) \leq 2^s \mathcal{H}^s(E) \sup \overline{D}^s_{\mathcal{M}}(x) = 0$.

(b) Assume $\mathcal{H}^s(E) = 0$. For positive m, let $E_m = \left\{ x \in E : \overline{D}^s_{\mathcal{M}}(x) \leq m \right\}$. Then $\mathcal{M}(E_m) \leq 2^s \mathcal{H}^s(E_m) \cdot m = 0$. So $\mathcal{M}(E) = \lim_m \mathcal{M}(E_m) = 0$.

(c) For positive δ, let $E_\delta = \left\{ x \in E : \overline{D}^s_{\mathcal{M}}(x) \geq \delta \right\}$. Then $\inf_{x \in E_\delta} \overline{D}^s_{\mathcal{M}}(x) \geq \delta$, so $\mathcal{H}^s(E_\delta) \leq \mathcal{M}(E)/\delta$. Thus $\mathcal{H}^s(E_\delta) < \infty$. Thus E is σ-finite for $\mathcal{H}^s$.

(d) Now, $\sup \overline{D}^s_{\mathcal{M}}(x) = \infty$ and $\mathcal{M}(E) < \infty$, so by Theorem 1.5.14, $\mathcal{H}^s(E) = 0$. ☺

Now suppose $E \subseteq S$ is a Borel set, $\mathcal{M}$ a finite Borel measure on S, and $s > 0$. For $\delta > 0$, write $E_\delta = \left\{ x \in E : \overline{D}^s_{\mathcal{M}}(x) \geq \delta \right\}$; then E is partitioned into three parts:

$$E_0 = \left\{ x \in E : \overline{D}^s_{\mathcal{M}}(x) = 0 \right\},$$

$$E_* = \left\{ x \in E : 0 < \overline{D}^s_{\mathcal{M}}(x) < \infty \right\},$$

$$E_\infty = \left\{ x \in E : \overline{D}^s_{\mathcal{M}}(x) = \infty \right\}.$$

The previous result can be applied to estimate the Hausdorff dimension of the parts. By (d), $\dim E_\infty \leq s$; in fact, $\mathcal{H}^s(E_\infty) = 0$. By (c), $\dim E_* \leq s$; in fact, $\mathcal{H}^s(E_\delta) < \infty$ for all $\delta > 0$, so E_* is σ-finite. By (b), either $\mathcal{M}(E_*) = 0$ or $\mathcal{H}^s(E_*) > 0$ and $\dim E_* \geq s$. So if $\mathcal{M}(E_*) > 0$, then $\dim E_* = s$. By (a), either $\mathcal{M}(E_0) = 0$ or $\mathcal{H}^s(E_0) = \infty$ and $\dim E_0 \geq s$.

The same facts can be used to estimate the Hausdorff dimension of these parts:

(1.5.16) Corollary. *Let S be a metric space, let $\mathcal{M}$ be a finite Borel measure on S, and let $E \subseteq S$ be a Borel set with $\mathcal{M}(E) > 0$. (α) Suppose $\overline{D}^s_{\mathcal{M}}(x) = 0$ almost everywhere on E. Then $\mathcal{H}^s(E_0) = \infty$ and $\dim E \geq \dim E_0 \geq s$. ($\beta$) Suppose $\overline{D}^s_{\mathcal{M}}(x) < \infty$ almost everywhere on E. Then $\mathcal{H}^s(E_0 \cup E_*) > 0$ and $\dim E \geq \dim (E_0 \cup E_*) \geq s$. ($\gamma$) Suppose S is complete. Suppose $\dim E > s$. Then there is a finite Borel measure $\mathcal{M}$ with $\mathcal{M}(E) > 0$ such that $\overline{D}^s_{\mathcal{M}}(x) = 0$ almost everywhere on E.*

Proof. (α) and (β) are easy. For (γ), choose t with $s < t < \dim E$. Because S is complete, the measure $\mathcal{H}^t$ is semifinite (Theorem 1.5.2). Then $\mathcal{H}^t(E) = \infty$, so by semifiniteness there is a set $A \subseteq E$ with $0 < \mathcal{H}^t(A) < \infty$. Let $\mathcal{M} = \mathcal{H}^t \upharpoonright A$. Now, $\mathcal{M}(E) = \mathcal{H}^t(A) > 0$, but $\dim (E_* \cup E_\infty) \leq s$, so

$$\mathcal{M}(E_* \cup E_\infty) \leq \mathcal{H}^t(E_* \cup E_\infty) = 0.$$

That is, $\overline{D}^s_{\mathcal{M}}(x) = 0$ almost everywhere on E. ☺

Let $s > 0$. We say that a set $E \subseteq S$ is an *s-set* iff $0 < \mathcal{H}^s(E) < \infty$. This implies $\dim E = s$ but tells us much more. The **Hausdorff upper s-density** of E at a point $x \in S$ is

$$\limsup_{r \to 0} \frac{\mathcal{H}^s\left(E \cap \overline{B}_r(x)\right)}{(2r)^s} = \overline{D}^s_{\mathcal{H}^s \upharpoonright E}(x).$$

Similarly, the **covering upper s-density** of E at a point $x \in S$ is $\overline{D}^s_{\mathcal{C}^s \upharpoonright E}(x)$.

(1.5.17) Corollary. *Let S be a metric space and let $E \subseteq S$ be an s-set. Then (a) $\overline{D}^s_{\mathcal{C}^s \restriction E}(x) = 1$ for $\mathcal{C}^s$-almost all $x \in E$; (b) $2^{-s} \leq \overline{D}^s_{\mathcal{H}^s \restriction E}(x) \leq 1$ for $\mathcal{H}^s$-almost all $x \in E$; (c) $\overline{D}^s_{\mathcal{C}^s \restriction E}(x) = \overline{D}^s_{\mathcal{H}^s \restriction E}(x) = 0$ for $\mathcal{H}^s$-almost all $x \notin E$.*

Proof. Let $\mathcal{M} = \mathcal{C}^s \restriction E$. Then $\mathcal{M}$ is a finite Borel measure, since E is an s-set.

First, I claim that $\overline{D}^s_{\mathcal{M}}(x) \leq 1$ almost everywhere on E. Let $c > 1$. Then the set $E_c = \left\{ x \in E : \overline{D}^s_{\mathcal{M}}(x) \geq c \right\}$ is a Borel set, and $\mathcal{C}^s(E_c) < \infty$. So by Theorem 1.5.13 we have $\mathcal{C}^s(E_c) \cdot c \leq \mathcal{M}(E_c) = \mathcal{C}^s(E_c)$. But $c > 1$, so we must have $\mathcal{C}^s(E_c) = 0$. This is true for all $c > 1$, so $\overline{D}^s_{\mathcal{C}^s \restriction E}(x) \leq 1$ almost everywhere on E.

Next, I claim that $\overline{D}^s_{\mathcal{M}}(x) \geq 1$ almost everywhere on E. Let $c < 1$. Then $E_c = \left\{ x \in E : \overline{D}^s_{\mathcal{M}}(x) \leq c \right\}$ is a Borel set, and $\mathcal{C}^s(E_c) < \infty$. So $\mathcal{M}(E_c) \leq \mathcal{C}^s(E_c) \cdot c = \mathcal{M}(E_c) \cdot c$, and therefore $\mathcal{M}(E_c) = 0$. This is true for all $c < 1$, so $\overline{D}^s_{\mathcal{M}}(x) \geq 1$ almost everywhere on E.

Now we turn to the complement of E. We must show that

$$\mathcal{C}^s \left(\left\{ x \notin E : \overline{D}^s_{\mathcal{M}}(x) > 0 \right\} \right) = 0.$$

For $c > 0$, let $F_c = \left\{ x \notin E : \overline{D}^s_{\mathcal{M}}(x) \geq c \right\}$. Then $\mathcal{C}^s(F_c) \cdot c \leq \mathcal{M}(F_c) = 0$, so $\mathcal{C}^s(F_c) = 0$. This is true for any $c > 0$, so $\mathcal{C}^s \left(\left\{ x \notin E : \overline{D}^s_{\mathcal{M}}(x) > 0 \right\} \right) = 0$. Therefore, $\overline{D}^s_{\mathcal{C}^s \restriction E}(x) = 0$ for $\mathcal{C}^s$-almost all $x \notin E$.

The results for Hausdorff measure $\mathcal{H}^s$ follow from those for covering measure $\mathcal{C}^s$ using the inequalities (1.4.3).   ☺

With $s = d$ in Euclidean space $\mathbb{R}^d$, we have $\mathcal{C}^d(\overline{B}_r(x)) = \mathcal{H}^d(\overline{B}_r(x)) = (2r)^d$, so in this case both densities are 1 at an interior point of E.

(1.5.18) Exercise. Let E be a Borel set with $0 < \mathcal{P}^s(E) < \infty$. Then $\underline{D}^s_{\mathcal{P}^s \restriction E}(x) = 1$ for $\mathcal{P}^s$-almost all $x \in E$, and $\underline{D}^s_{\mathcal{P}^s \restriction E}(x) = 0$ for $\mathcal{P}^s$-almost all $x \notin E$.

*1.6 Ultrametric Spaces

This section and the next one may be considered optional sections. In this section we will prove the semifiniteness of Hausdorff measures on compact ultrametric spaces. Although the proof is fairly long, it is elementary. The reader may skip this section (and the next one) and rely on the assertion (Theorem 1.5.2) that Hausdorff measures $\mathcal{H}^s$ are semifinite in all complete separable metric spaces.

---

* An optional section.

As a preparation for the proof of semifiniteness for Hausdorff measures, we prove semifiniteness for certain method I outer measures. Note that $\overline{\mathbb{M}}$ is an outer measure, but usually not a metric outer measure, so compact sets need not be measurable.

(1.6.1) Proposition. *Let S be a compact ultrametric space. Let $\mathcal{A}$ be a Vitali cover of S by balls, let $s > 0$, and let $\overline{\mathbb{M}}$ be the method I measure defined by $\mathcal{A}$ and the set-function $(\mathrm{diam}\ A)^s$. If α is any number satisfying $0 < \alpha < \overline{\mathbb{M}}(S)$, then there is a compact set $K \subseteq S$ such that $\overline{\mathbb{M}}(K) = \alpha$.*

Proof. (a) We prove first that if α and $\delta > 0$ satisfy $0 \leq \alpha - \delta < \alpha < \alpha + \delta < \overline{\mathbb{M}}(S)$, then there are clopen sets U and V with $U \subseteq V \subseteq S$ and $\alpha - \delta < \overline{\mathbb{M}}(U) \leq \alpha \leq \overline{\mathbb{M}}(V) < \alpha + \delta$.

Let ε be so small that $\varepsilon^s < \delta$. Cover S by balls A_i from $\mathcal{A}$ with diam $A_i < \varepsilon$; by compactness we may assume that this is a finite cover:

$$S \subseteq \bigcup_{i=1}^{n} A_i.$$

Now let k be chosen such that

$$\overline{\mathbb{M}}\left(\bigcup_{i=1}^{k-1} A_i\right) \leq \alpha < \overline{\mathbb{M}}\left(\bigcup_{i=1}^{k} A_i\right).$$

This is possible since $\overline{\mathbb{M}}(\bigcup_{i=1}^{0} A_i) = 0$ and $\overline{\mathbb{M}}(\bigcup_{i=1}^{n} A_i) = \overline{\mathbb{M}}(S) > \alpha$. Then $U = \bigcup_{i=1}^{k-1} A_i$ and $V = \bigcup_{i=1}^{k} A_i$ satisfy the required conclusion, since

$$\alpha \leq \overline{\mathbb{M}}\left(\bigcup_{i=1}^{k} A_i\right) = \overline{\mathbb{M}}(V)$$

$$\leq \overline{\mathbb{M}}\left(\bigcup_{i=1}^{k-1} A_i\right) + \overline{\mathbb{M}}(A_k) \leq \alpha + (\mathrm{diam}\ A_k)^s < \alpha + \delta,$$

and similarly for U.

(b) Now we turn to the statement of the proposition. By part (a), there exist clopen sets $U_1 \subseteq V_1$ with $\alpha - 2^{-1} < \overline{\mathbb{M}}(U_1) \leq \alpha \leq \overline{\mathbb{M}}(V_1) < \alpha + 2^{-1}$. Applying (a) again, starting with the space $S_1 = V_1 \setminus U_1$ and the restriction of $\mathcal{A}$ to S_1, we obtain clopen sets U_2 and V_2 with $U_1 \subseteq U_2 \subseteq V_2 \subseteq V_1$ and $\alpha - 2^{-2} < \overline{\mathbb{M}}(U_2) \leq \alpha \leq \overline{\mathbb{M}}(V_2) < \alpha + 2^{-2}$. Continuing recursively, we obtain a decreasing sequence (V_n) of clopen sets and an increasing sequence (U_n) of clopen sets with $\alpha - 2^{-n} < \overline{\mathbb{M}}(U_n) \leq \alpha \leq \overline{\mathbb{M}}(V_n) < \alpha + 2^{-n}$ for all n. Let $K = \bigcap_{n \in \mathbb{N}} V_n$. Then K is compact, and $\overline{\mathbb{M}}(U_n) \leq \overline{\mathbb{M}}(K) \leq \overline{\mathbb{M}}(V_n)$ for all n, so that $\overline{\mathbb{M}}(K) = \alpha$. ☺

Semifiniteness Theorem. Now we turn to the proof that Hausdorff measures $\mathcal{H}^s$ are semifinite in compact ultrametric spaces. Let S be a compact

ultrametric space. (We may assume that it has no isolated points.) List all of
the closed balls of S,

$$\{B_1, B_2, B_3, \cdots\},$$

with diam $B_1 \geq$ diam $B_2 \geq$ diam $B_3 \geq \cdots$. This is possible, since for any
given $\varepsilon > 0$, there are only finitely many distinct balls with diameter $> \varepsilon$. Fix
a number $s > 0$. For $n \in \mathbb{N}$, let

$$\mathcal{B}_n = \{B_{n+1}, B_{n+2}, \cdots\},$$

and let $\overline{\mathcal{M}}_n$ be the method I outer measure defined using the Vitali cover $\mathcal{B}_n$
and the set-function $(\text{diam } A)^s$. Then of course, $\lim_{n \to \infty} \overline{\mathcal{M}}_n(A) = \overline{\mathcal{H}}^s(A)$ for
all $A \subseteq S$.

(1.6.2) Lemma. *If $m \leq n$, then B_m is $\overline{\mathcal{M}}_n$-measurable.*

Proof. Let E be any set. I must show that $\overline{\mathcal{M}}_n(E) = \overline{\mathcal{M}}_n(E \cap B_m) + \overline{\mathcal{M}}_n(E \setminus B_m)$.
The inequality $\leq$ is true in general, so it suffices to prove $\geq$. Let $\mathcal{A} \subseteq \mathcal{B}_n$ be a
cover of E. Each $A \in \mathcal{A}$ satisfies either $A \subseteq B_m$ or $A \cap B_m = \varnothing$. Thus $\mathcal{A}$ may
be split into two parts:

$$\mathcal{A}_1 = \{A \in \mathcal{A} : A \subseteq B_m\},$$
$$\mathcal{A}_2 = \{A \in \mathcal{A} : A \cap B_m = \varnothing\}.$$

Now, $\mathcal{A}_1$ covers $E \cap B_m$, and $\mathcal{A}_2$ covers $E \setminus B_m$. Thus

$$\sum_{A \in \mathcal{A}_1} (\text{diam } A)^s \geq \overline{\mathcal{M}}_n(E \cap B_m),$$

$$\sum_{A \in \mathcal{A}_2} (\text{diam } A)^s \geq \overline{\mathcal{M}}_n(E \setminus B_m).$$

So

$$\sum_{A \in \mathcal{A}} (\text{diam } A)^s \geq \overline{\mathcal{M}}_n(E \cap B_m) + \overline{\mathcal{M}}_n(E \setminus B_m).$$

This is true for all covers, so we have $\overline{\mathcal{M}}_n(E) \geq \overline{\mathcal{M}}_n(E \cap B_m) + \overline{\mathcal{M}}_n(E \setminus B_m)$,
as required.   ☺

(1.6.3) Semifiniteness Theorem (Ultrametric Case). *Let S be a com-
pact ultrametric space, and let $s > 0$. Suppose $\mathcal{H}^s(S) > \alpha \geq 0$. Then there is
a compact set $K \subseteq S$ with $\mathcal{H}^s(K) = \alpha$.*

Proof. We will define recursively a decreasing sequence $(K_n)_{n=m}^{\infty}$ of compact
sets. Since $\mathcal{H}^s(S) > \alpha$, there is $m \in \mathbb{N}$ with $\overline{\mathcal{M}}_m(S) > \alpha$. By Proposition
1.6.1, there is a compact set $K_m \subseteq S$ with $\overline{\mathcal{M}}_m(K_m) = \alpha$.

Now let $n \geq m$, and suppose that K_n has been defined. To define K_{n+1}
we will specify the two parts $K_{n+1} \cap B_n$ and $K_{n+1} \setminus B_n$. The part outside B_n

is unchanged: $K_{n+1} \setminus B_n = K_n \setminus B_n$. To define the part inside B_n, note that $\overline{\mathcal{M}}_{n+1}(K_n \cap B_n) \geq \overline{\mathcal{M}}_n(K_n \cap B_n)$. There are two cases:

(i) If $\overline{\mathcal{M}}_{n+1}(K_n \cap B_n) = \overline{\mathcal{M}}_n(K_n \cap B_n)$, let $K_{n+1} \cap B_n = K_n \cap B_n$, so that $K_{n+1} = K_n$.

(ii) $\overline{\mathcal{M}}_{n+1}(K_n \cap B_n) > \overline{\mathcal{M}}_n(K_n \cap B_n)$: This happens only if the set B_n is needed for a minimal cover in the computation of $\overline{\mathcal{M}}_n(K_n \cap B_n)$, so that $\overline{\mathcal{M}}_n(K_n \cap B_n) = (\text{diam } B_n)^s$. In this case, use Proposition 1.6.1 to choose $K_{n+1} \cap B_n \subseteq K_n \cap B_n$ with $\overline{\mathcal{M}}_{n+1}(K_{n+1} \cap B_n) = \overline{\mathcal{M}}_n(K_n \cap B_n) = (\text{diam } B_n)^s$.

This completes the recursive construction.

I claim that $\overline{\mathcal{M}}_n(K_n) = \alpha$ for $m \leq n$. This will be proved by induction on n. It is true for $n = m$ by the construction of K_m. Suppose it is true for a given value of n, and consider $n + 1$. The set B_n is not needed in the computation of $\overline{\mathcal{M}}_n(K_{n+1} \setminus B_n)$, and $K_n \setminus B_n = K_{n+1} \setminus B_n$, so

$$\overline{\mathcal{M}}_{n+1}(K_{n+1} \setminus B_n) = \overline{\mathcal{M}}_n(K_n \setminus B_n).$$

Next, $\overline{\mathcal{M}}_{n+1}(K_{n+1} \cap B_n) = \overline{\mathcal{M}}_n(K_n \cap B_n)$ by construction, so

$$\overline{\mathcal{M}}_{n+1}(K_{n+1} \cap B_n) = \overline{\mathcal{M}}_n(K_n \cap B_n).$$

Finally, since B_n is both $\overline{\mathcal{M}}_n$ and $\overline{\mathcal{M}}_{n+1}$ measurable, we may add to obtain

$$\overline{\mathcal{M}}_{n+1}(K_{n+1}) = \overline{\mathcal{M}}_n(K_n).$$

This completes the induction.

Next, I claim that $\overline{\mathcal{M}}_n(K_p) = \alpha$ for $m \leq n \leq p$. But $\overline{\mathcal{M}}_m(K_p) \leq \overline{\mathcal{M}}_n(K_p) \leq \overline{\mathcal{M}}_p(K_p) = \alpha$, so it is enough to prove $\overline{\mathcal{M}}_m(K_p) \geq \alpha$. This will be proved by induction on p. For $p = m$, it is already known. Suppose it is true for $p = n$, and consider $p = n+1$. In the case $K_{n+1} = K_n$, we are finished. So suppose we have the other case, $\overline{\mathcal{M}}_{n+1}(B_n \cap K_n) > \overline{\mathcal{M}}_{n+1}(B_n \cap K_{n+1}) = \overline{\mathcal{M}}_n(B_n \cap K_n) = (\text{diam } B_n)^s$. Now let $\mathcal{A} \subseteq \mathcal{B}_m$ be a cover of K_{n+1},

$$K_{n+1} \subseteq \bigcup_{A \in \mathcal{A}} A.$$

There are two cases:

(1) $B_n \subseteq A$ for some $A \in \mathcal{A}$. Then $\mathcal{A}$ also covers K_n, so $\sum_{A \in \mathcal{A}} (\text{diam } A)^s \geq \overline{\mathcal{M}}_m(K_n) = \alpha$ by the induction hypothesis.

(2) For all $A \in \mathcal{A}$, either $A \cap B_n = \emptyset$ or $A \subset B_n$. Divide $\mathcal{A}$ into two parts:

$$\mathcal{A}_1 = \{ A \in \mathcal{A} : A \subset B_n \},$$
$$\mathcal{A}_2 = \{ A \in \mathcal{A} : A \cap B_n = \emptyset \}.$$

The sets in $\mathcal{A}_1$ are proper subsets of B_n, so $\mathcal{A}_1 \subseteq \mathcal{B}_{n+1}$. But $\mathcal{A}_1$ covers $K_{n+1} \cap B_n$, so

$$\sum_{A \in \mathcal{A}_1} (\text{diam } A)^s \geq \overline{\mathcal{M}}_{n+1}(K_{n+1} \cap B_n) = (\text{diam } B_n)^s.$$

But $\mathcal{A}_2 \cup \{B_n\}$ covers K_n, so

$$\sum_{A \in \mathcal{A}} (\text{diam } A)^s = \sum_{A \in \mathcal{A}_1} (\text{diam } A)^s + \sum_{A \in \mathcal{A}_2} (\text{diam } A)^s$$

$$\geq (\text{diam } B_n)^s + \sum_{A \in \mathcal{A}_2} (\text{diam } A)^s$$

$$\geq \overline{\mathcal{M}}_m(K_n) = \alpha.$$

So in both cases $\sum_{A \in \mathcal{A}} (\text{diam } A)^s \geq \alpha$. This is true for any such cover $\mathcal{A}$, so $\overline{\mathcal{M}}_m(K_{n+1}) \geq \alpha$. This completes the proof by induction.

Now define $K = \bigcap K_n$. Since each K_n is compact, so is K. I claim that $\overline{\mathcal{M}}_n(K) = \alpha$ for all $n \geq m$. Indeed, if $\mathcal{A} \subseteq \mathcal{B}_n$ is a cover of K, by compactness we may assume it is finite, so it covers some K_p, and we may assume $p \geq n$. So $\sum_{A \in \mathcal{A}} (\text{diam } A)^s \geq \overline{\mathcal{M}}_m(K_p) = \alpha$. Thus $\overline{\mathcal{M}}_n(K) \geq \alpha$. But also $K \subseteq K_n$, so $\overline{\mathcal{M}}_n(K) \leq \overline{\mathcal{M}}_n(K_n) = \alpha$. Thus $\overline{\mathcal{M}}_n(K) = \alpha$.

Finally, we have $\mathcal{H}^s(K) = \lim_n \overline{\mathcal{M}}_n(K) = \alpha$.   ☺

*1.7 Comparable Net Measures

In this section we turn to a proof of the semifiniteness theorem for Euclidean space.

If $\mathcal{A} \subseteq \mathcal{B}$ are two Vitali covers of the space S, then it is easy to see that

$$\mathcal{H}^s_{\mathcal{A}}(F) \geq \mathcal{H}^s_{\mathcal{B}}(F)$$

for all Borel sets F. We may consider this to be the case $p = q = 1$ of the next comparison theorem:

(1.7.1) Theorem. *Let $\mathcal{A}$ and $\mathcal{B}$ be Vitali covers of the space S. Suppose there are positive constants p and q such that for every $A \in \mathcal{A}$ there exist at most p sets $B_i \in \mathcal{B}$ with diam $B_i \leq q$ diam A and $A \subseteq \bigcup_{i=1}^p B_i$. Then*

$$\mathcal{H}^s_{\mathcal{B}}(F) \leq pq^s \mathcal{H}^s_{\mathcal{A}}(F)$$

for all $s > 0$ and all $F \subseteq S$.

Proof. Let $F \subseteq \bigcup_j A_j$ be a countable cover of F by sets from $\mathcal{A}$ with diameter $< \varepsilon$. For each A_j there exist at most p sets B_{ij} by assumption. The sets B_{ij} have diameter $< q\varepsilon$ and cover F, so

---

* An optional section.

$$\overline{\mathcal{H}}^s_{\mathcal{B},q\varepsilon}(F) \le \sum_{i,j}(\text{diam } B_{ij})^s \le p\sum_j (q\,\text{diam }A_j)^s = pq^s\sum_j(\text{diam }A_j)^s.$$

Taking the infimum over all such covers, we obtain

$$\overline{\mathcal{H}}^s_{\mathcal{B},q\varepsilon}(F) \le pq^s\overline{\mathcal{H}}^s_{\mathcal{A},\varepsilon}(F).$$

Now let $\varepsilon \to 0$ to obtain

$$\mathcal{H}^s_{\mathcal{B}}(F) \le pq^s\mathcal{H}^s_{\mathcal{A}}(F),$$

as required.   ☺

Variants for two spaces related by a transformation have similar proofs:

(1.7.2) Exercise. Let S_1 and S_2 be metric spaces, let $h\colon S_1 \to S_2$ be a function, let $\mathcal{A}_1$ be a Vitali cover of S_1, and $\mathcal{A}_2$ a Vitali cover of S_2.

(a) Suppose that there are positive constants p, q such that for every $A \in \mathcal{A}_1$, there exist at most p sets $B_i \in \mathcal{A}_2$ with diam $B_i \le q\,$diam A and $h[A] \subseteq \bigcup_{i=1}^p B_i$. Then $\mathcal{H}^s_{\mathcal{A}_2}(h[F]) \le pq^s\mathcal{H}^s_{\mathcal{A}_1}(F)$ for all $s > 0$ and all $F \subseteq S_1$.

(b) Suppose that there are positive constants p, q such that for every $A \in \mathcal{A}_2$, there exist at most p sets $B_i \in \mathcal{A}_1$ with diam $B_i \le q\,$diam A and $h^{-1}[A] \subseteq \bigcup_{i=1}^p B_i$. Then $\mathcal{H}^s_{\mathcal{A}_1}(h^{-1}[E]) \le pq^s\mathcal{H}^s_{\mathcal{A}_2}(E)$ for all $s > 0$ and all $E \subseteq S_2$.

The Besicovitch Net Measure in $\mathbb{R}^d$. Although Euclidean space $\mathbb{R}^d$ is not an ultrametric space, we will show next that it is "comparable" to one. A **dyadic cube** in $\mathbb{R}^d$ is a set of the form

$$\left[\frac{i_1-1}{2^k}, \frac{i_1}{2^k}\right) \times \left[\frac{i_2-1}{2^k}, \frac{i_2}{2^k}\right) \times \cdots \times \left[\frac{i_d-1}{2^k}, \frac{i_d}{2^k}\right),$$

where k and $i_1, \cdots, i_d$ are integers.

Let $\mathcal{U}^d_k$ be the set of all dyadic cubes with a fixed value of k. Note that these cubes are disjoint and cover all of $\mathbb{R}^d$. The **Besicovitch net** on $\mathbb{R}^d$ is the set $\mathcal{U}^d = \bigcup_{k=0}^{\infty}\mathcal{U}^d_k$. If $A, B \in \mathcal{U}^d$, then we must have one of

$$A \subseteq B, \qquad A \supseteq B, \qquad A \cap B = \varnothing.$$

Also, because we have allowed only $k \ge 0$, if $A \in \mathcal{U}^d$, then there are only finitely many $B \in \mathcal{U}^d$ with $A \subseteq B$.

(1.7.3) Proposition. *Let $A \subseteq \mathbb{R}^d$ be a set with diameter r, $0 < r < 1$. Then A is covered by at most 3^d sets $B_i \in \mathcal{U}^d$ with diameter at most $2\sqrt{d}\,r$.*

Proof. Choose $k \in \mathbb{N}$ such that $2^{-k-1} \le r < 2^{-k}$. Let x be a point of A. Let $B_1 \in \mathcal{U}^d_k$ be the dyadic cube with side 2^{-k} such that $x \in B_1$. The cube B_1 and all its neighbors, taken together, form a group of $3 \times 3 \times \cdots \times 3$ cubes, whose union contains A. The diameter of each of these cubes is $2^{-k}\sqrt{d}$, which is at most $2r\sqrt{d}$.   ☺

(1.7.4) Corollary. *Let $p = 3^d$ and $q = 2\sqrt{d}$. For all $F \subseteq \mathbb{R}^d$ and all $s > 0$,*

$$\mathcal{H}^s(F) \leq \mathcal{H}^s_{\mathcal{U}^d}(F) \leq pq^s\mathcal{H}^s(F).$$

The Besicovitch net $\mathcal{U}^d$ is called a **comparable net** because of this fact. The Hausdorff measure $\mathcal{H}^s$ is comparable to the net measure $\mathcal{H}^s_{\mathcal{U}^d}$ in the sense that they differ by at most a constant factor (depending on s). So, in particular, $\mathcal{H}^s(F) = 0$ if and only if $\mathcal{H}^s_{\mathcal{U}^d}(F) = 0$, and similarly for ∞. Thus, the Hausdorff dimension of a set $F \subseteq \mathbb{R}^d$ may be computed using the Besicovitch net measures $\mathcal{H}^s_{\mathcal{U}^d}$.

(1.7.5) Exercise. Show that the Besicovitch net measure is comparable to Hausdorff measure with a constant factor not depending on s. That is, find a number p such that each set $A \subseteq \mathbb{R}^d$ (with small enough diameter) is contained in the union of at most p sets $B_i \in \mathcal{U}^d$ with diam $B_i \leq$ diam A.

The dyadic cubes $B \in \mathcal{U}^d$ contained in the unit cube $[0, 1)^d$ form a tree in the usual way: each cube corresponds to a node; a cube A is an ancestor of B iff $A \supseteq B$. A space of strings corresponds to this tree. Use an alphabet E of 2^d letters; the finite strings $E^{(*)}$ correspond to the nodes of the tree; the infinite strings $E^{(\omega)}$ are the points of a compact ultrametric space; the metric $\rho_{1/2}$ is defined such that diam $[\alpha] = (1/2)^k$ for a string α of length k. The model map $h: E^{(\omega)} \to \mathbb{R}^d$, formed using the base 2 expansion in each coordinate of $\mathbb{R}^d$, maps onto the unit cube $[0, 1]^d$ and satisfies

$$|h(\sigma) - h(\tau)| \leq \sqrt{d}\,\rho_{1/2}(\sigma, \tau).$$

(Another way to think of this is to consider the closed cube $[0, 1]^d$ as the attractor of an iterated function system consisting of 2^d contractions with ratio $1/2$. The cube is "self-similar," since it is made up of smaller cubes obtained by bisecting each of the edges of the large cube $[0, 1]^d$.)

(1.7.6) Proposition. *Let $F \subseteq [0, 1]^d$. Then $h^{-1}[F] \subseteq E^{(\omega)}$, and*

$$d^{-s/2}\mathcal{H}^s(F) \leq \mathcal{H}^s\big(h^{-1}[F]\big) \leq 3^d2^s\mathcal{H}^s(F).$$

Proof. First, if $A \subseteq E^{(\omega)}$, then $h[A]$ is covered by one set (itself) with diam $h[A] \leq \sqrt{d}\,$diam A. Therefore, $\mathcal{H}^s(h[A]) \leq d^{s/2}\mathcal{H}^s(A)$ for any $A \subseteq E^{(\omega)}$. Now if $F \subseteq [0, 1]^d$, then $F = h[h^{-1}[F]]$, so $\mathcal{H}^s(F) \leq d^{s/2}\mathcal{H}^s(h^{-1}[F])$.

On the other hand, any set $A \subseteq [0, 1]^d$ is covered by at most 3^d sets $h[B_i]$, where B_i is a closed ball in $E^{(\omega)}$ such that $h[B_i]$ is a cube with side ≤ 2 diam A, so diam $B_i \leq 2$ diam A. So $\mathcal{H}^s(h^{-1}[F]) \leq 3^d2^s\mathcal{H}^s(F)$ for $F \subseteq [0, 1]^d$.   ☺

(1.7.7) Corollary. *Let $F \subseteq [0,1]^d$ be a Borel set. Let h be the model map defined above. Then* $\dim F = \dim h^{-1}[F]$.

(1.7.8) Semifiniteness Theorem (Euclidean Case). *Let $s > 0$, and let $F \subseteq \mathbb{R}^d$ be a closed set. Suppose $\mathcal{H}^s(F) = \infty$. Then there is a compact set $K \subseteq F$ with $0 < \mathcal{H}^s(K) < \infty$.*

Proof. If there is some cube $A \in \mathcal{U}_0^d$ of side one with $0 < \mathcal{H}^s(F \cap \overline{A}) < \infty$, let $K = F \cap \overline{A}$. If there is no such cube, then there is $A \in \mathcal{U}_0^d$ with $\mathcal{H}^s(F \cap \overline{A}) = \infty$. By translation, we may assume that this is true for the unit cube $\overline{A} = [0,1]^d$. Now consider the set $h^{-1}[F]$, where $h: E^{(\omega)} \to [0,1]^d$ is the model map discussed above. Then $\mathcal{H}^s(h^{-1}[F]) = \infty$. Now the string model $E^{(\omega)}$ is a compact ultrametric space, and $h^{-1}[F]$ is a closed subset, so $h^{-1}[F]$ is also a compact ultrametric space. By Theorem 1.6.3 there is a compact set $K_0 \subseteq h^{-1}[F]$ with $0 < \mathcal{H}^s(K_0) < \infty$. Now, $K = h[K_0]$ is compact, since h is continuous. But $\mathcal{H}^s(K) \leq d^{s/2} \mathcal{H}^s(K_0) < \infty$, and $\mathcal{H}^s(K) = \mathcal{H}^s(h[h^{-1}[K]]) \geq 3^{-d} 2^{-s} \mathcal{H}^s(h^{-1}[K]) \geq 3^{-d} 2^{-s} \mathcal{H}^s(K_0) > 0$. So $0 < \mathcal{H}^s(K) < \infty$. ☺

This version of semifiniteness will be enough for most of the applications we have in mind. But can you do better?

(1.7.9) Exercise. Improve the semifiniteness conclusion: Let $F \subseteq \mathbb{R}^d$ be a closed set. Suppose $\mathcal{H}^s(F) > \alpha \geq 0$. Then there is a compact set $K \subseteq S$ with $\mathcal{H}^s(K) = \alpha$.

Analytic Sets. The semifiniteness theorem is stated above only for closed sets in $\mathbb{R}^d$. In fact, it is true more generally; for example for Borel sets in $\mathbb{R}^d$. But the proof requires some knowledge of "analytic sets."

Let A_n be a decreasing sequence of sets, and let $A = \bigcap A_n$. We say that $A_n \searrow A$ **strongly** iff for any open set $U \supseteq A$, there is N such that for all $n \geq N$ we have $A_n \subseteq U$.

(1.7.10) Lemma. *Suppose $A_n \searrow A$ strongly, and $\mathcal{H}^s(A) = 0$. Then $\lim_{n \to \infty} \mathcal{H}_\delta^s(A_n) = 0$ for all $\delta > 0$.*

Proof. Let $\varepsilon > 0$. Since $\mathcal{H}^s(A) = 0$, we have $\mathcal{H}_\delta^s(A) = 0$. So there is a cover $A \subseteq \bigcup_{i=1}^\infty E_i$, where the E_i are open and $\operatorname{diam} E_i < \delta$, so that $\sum (\operatorname{diam} E_i)^s < \varepsilon$. Now, $\bigcup E_i \supseteq A$ is open, so by hypothesis, there is N such that for all $n \geq N$, we have $A_n \subseteq \bigcup E_i$, and thus $\mathcal{H}_\delta^s(A_n) < \varepsilon$. ☺

Note that if the sets A_n are compact, and $A_n \searrow A$, then $A_n \searrow A$ strongly. This is seen as follows: if $U \supseteq A$ is open, then the sets $A_n \setminus U$ are compact and decrease to $\varnothing$, so they are themselves empty for large n.

We write $\mathbb{N}^{\mathbb{N}}$ for Baire's space of sequences of natural numbers. An **analytic set** in $\mathbb{R}^d$ is a set E that may be written in the form

$$E = \bigcup_{(n_1, n_2, \cdots) \in \mathbb{N}^{\mathbb{N}}} \bigcap_{k \in \mathbb{N}} C(n_1, n_2, \cdots, n_k),$$

where the sets $C(n_1, n_2, \cdots, n_k)$ are compact, diam $C(n_1, n_2, \cdots, n_k) < 1/k$, and

$$C(n_1, \cdots, n_k) \supseteq \bigcup_{j=1}^{\infty} C(n_1, \cdots, n_k, j)$$

for all $n_1, \cdots, n_k$ [48, Proposition 8.2.7]. In particular, every Borel set is analytic [48, Prop. 8.2.3].

(1.7.11) Theorem. *Let $E \subseteq \mathbb{R}^d$ be an analytic set, and let $s > 0$. If $\mathcal{H}^s(E) > 0$, then there is a compact set $K \subseteq E$ with $\mathcal{H}^s(K) > 0$.*

Proof. First, write

$$E = \bigcup_{(n_1, n_2, \cdots) \in \mathbb{N}^{\mathbb{N}}} \bigcap_{k \in \mathbb{N}} C(n_1, n_2, \cdots, n_k),$$

as above. Now, $\mathcal{H}^s(E) > 0$. Let α be positive, real (and finite) such that $\mathcal{H}^s(E) > \alpha$. Choose $\delta > 0$ such that $\mathcal{H}^s_\delta(E) > \alpha$. Now, $E \subseteq \bigcup_{n_1=1}^{\infty} C(n_1)$, so there is k_1 with

$$\mathcal{H}^s_\delta \left(E \cap \bigcup_{n_1=1}^{k_1} C(n_1) \right) > \alpha.$$

Then each $E \cap C(n_1) = E \cap \bigcup_{n_2=1}^{\infty} C(n_1, n_2)$, so there is k_2 with

$$\mathcal{H}^s_\delta \left(E \cap \bigcup_{n_1=1}^{k_1} \bigcup_{n_2=1}^{k_2} C(n_1, n_2) \right) > \alpha.$$

Continuing in this way, we obtain a sequence $k_1, k_2, \cdots$ such that

$$\mathcal{H}^s_\delta \left(E \cap \bigcup_{n_1=1}^{k_1} \bigcup_{n_2=1}^{k_2} \cdots \bigcup_{n_j=1}^{k_j} C(n_1, n_2, \cdots, n_j) \right) > \alpha$$

for all j. But the sets

$$A_j = \bigcup_{n_1=1}^{k_1} \bigcup_{n_2=1}^{k_2} \cdots \bigcup_{n_j=1}^{k_j} C(n_1, n_2, \cdots, n_j)$$

form a decreasing sequence of compact sets. If $K = \bigcap A_j$, then $A_j \searrow K$ strongly, and certainly K is a compact subset of E. By the lemma, $\mathcal{H}^s(K) > 0$. ☺

(1.7.12) Corollary. *Let $E \subseteq \mathbb{R}^d$ be a Borel set, and let $s > 0$. If $\mathcal{H}^s(E) > 0$, then there is a compact set $K \subseteq E$ with $0 < \mathcal{H}^s(K) < \infty$.*

*1.8 Remarks

For the study of "fractal measures," one of the fundamental mathematical ideas is that of "measure." Further material on the topic can be found in many graduate-level texts; for example [48, 102, 111, 124, 135, 172, 200, 229, 230]. Material on Borel functions is in [130, 157]. The classical theory of the integral (in Euclidean space, not in abstract measure spaces) may be found in [233].

C. Carathéodory's theory of metric outer measures is found in [42]. This paper contains the constructions called here "Method I" and "Method II"; the terms "Method I" and "Method II" are attributed to M. Munroe [200]. Full and fine variations are due to Thomson [258, 259], who in turn credits R. Henstock for some of the ideas. Thomson originally called these constructions "Method III" and "Method IV"; they roughly correspond to Henstock's "variation" and "inner variation."

The Pi–Lambda Theorem (1.1.7) is probably due to E. B. Dynkin; it is similar to the "Monotone Class Theorem" from the older literature.

For Theorem 1.1.14 (approximation of measure from within by compact sets in complete separable metric space), see [207].

Hausdorff measure comes from F. Hausdorff [130]. Carathéodory [42] published a paper where arc length was generalized to a one-dimensional measure (today we would call it one-dimensional Hausdorff measure). Carathéodory concluded by saying that the same method could be used for an n-dimensional measure for any natural number n. When Hausdorff read that paper, he realized that in fact, the definition would work perfectly well even when n is not an integer. That was the beginning of his paper, but there are other significant ideas there as well.

The theory of the Hausdorff measure was elaborated by many papers of A. S. Besicovitch, for example [23–29]. Introductory material on the Hausdorff measures may be found in the mathematically oriented texts on fractal geometry, such as [13] or [87].

The proof of the "Increasing Sets Lemma" (1.2.8) is from [227].

Packing measure is from C. Tricot [264] and is elaborated in [232, 256]; independent definition of the premeasures $\widetilde{\mathcal{P}}^s$ is in [251]. The question of whether to use $2r$ or the diameter of $\overline{B}_r(x)$ in general metric space is discussed in [52, 120].

G. Vitali provided the prototype for the "Vitali theorems." A proof by S. Banach simplified the result and pointed the way to many generalizations. The history of the result, and its generalizations, is discussed in [233, p. 109]. The strong Vitali property for Euclidean space (1.3.13) goes back to A. S. Besicovitch [25]. Modern generalizations of Vitali covering, and its connection with derivation, may be found in [132]. Another source, including the Federer condition, is Federer's book [96]. Haase [122] studies Vitali theorems and packing measures (and includes a version of our Theorem 1.5.11).

---

* An optional section.

The covering measure $\mathcal{C}^s$ is found explicitly in [232]. But Hausdorff [130] contained many different variations of the definition, so one might attribute $\mathcal{C}^s$ to Hausdorff himself. (Some authors in fact call it the Hausdorff measure.)

This seems to be unknown:

(1.8.1) Question. Is $\mathcal{C}^s$ a regular measure?

The "packing measure" in most of the literature is defined in Euclidean space, so it could be either $\mathcal{P}^s$ or $\mathcal{R}^s$.

Use of balls the same size for packing and covering can be found in many works. Its use for fractal dimension goes back to G. Bouligand [36]; it was elaborated by A. N. Kolmogorov [155]; additional commentary on this is in [74]. Using balls the same size, H. Wegmann [268] defined a fractal dimension that is (by Theorem 1.4.10(d)) the same as the packing dimension; but note that Wegmann's paper preceeds the papers that define the packing dimension.

Divider dimension is due to Mandelbrot, for example [175].

(1.8.2) Exercise. Did you think that Figure 1.4.15 is drawn incorrectly? The parametrization starts at the right-hand endpoint—do you still think it is drawn incorrectly? We see in this example that $D_\varepsilon(f)$ may be different from $D_\varepsilon(\tilde{f})$, where $\tilde{f}$ parametrizes the same curve in the opposite direction. How much difference can there be between the two? Can opposite parametrizations yield different values for the divider dimension itself?

Bouligand dimension is from [36]; this method had previously been used by Minkowski for integer dimension, so this dimension is also known as "Minkowski dimension" (in addition to "box dimension"). Complementary intervals and their relation to fractal dimensions are discussed in [28]; twelve equivalent definitions are supplied by Tricot [263].

The gauge limit comes from the Henstock–Kurzweil theory of the integral; see [113, 18, 215, 158, 133, 134]. The packing measure as a gauge limit comes from [192, 76].

The fractal dimension of a product was studied by Besicovitch and Moran [27] and by Marstrand [183]. Other methods were introduced by Kelly [150] and Wegmann [268]. A modern summary is in [139].

Density results for Hausdorff measure date back practically to the beginnings of the subject; here, I have followed [232]. See [267].

Combining (1.5.17) and (1.5.18) might lead us to believe that the density $D_{\mathcal{M}}^s(x) = 1$ on a set E with $0 < \mathcal{C}^s(E) \le \mathcal{P}^s(E) < \infty$. But that is rarely the case. The density in (1.5.17) is computed using $\mathcal{C}^s$, and the density in (1.5.18) with respect to $\mathcal{P}^s$. So the upper $\mathcal{C}^s$-density is 1 and the lower $\mathcal{P}^s$-density is 1. In fact, let $\mathcal{M}$ be any finite measure (with the strong Vitali property) on our set E satisfying $0 < \mathcal{C}^s(E) < \infty$ and $0 < \mathcal{P}^s(E) < \infty$. Suppose

$$\overline{D}_{\mathcal{M}}^s(x) = \underline{D}_{\mathcal{M}}^s(x) = 1 \qquad \text{on } E.$$

From (1.5.11) we conclude that $\mathcal{M} = \mathcal{P}^s$ on E, and from (1.5.13) we conclude that $\mathcal{M} = \mathcal{C}^s$ on E. So in particular, $\mathcal{C}^s(E) = \mathcal{P}^s(E)$, which is positive and finite. This can happen when s is an integer and E is contained in a countable union of s-dimensional rectifiable manifolds. But that is the only case in which it happens; see [186, Ch. 14 and 17], and also [96, 184, 257]. If the fractal dimension s is not an integer, almost every point of E is *irregular* in the sense that $\underline{D}^s_{\mathcal{M}}(x) \neq \overline{D}^s_{\mathcal{M}}(x)$. This is also true for the "typical" set E even when s is an integer. More information on densities may be found in [83, 232, 184, 257].

Certain geometric questions become simpler in ultrametric spaces. Comparable net measures were introduced by Besicovitch. The proof of the semifiniteness theorem is taken from [227]; see also [186, p. 121].

The theory of analytic sets may be found in texts on descriptive set theory, or in some texts on measure theory or point-set topology. For example, [48, 38, 131].

Dimension and Category. Lebesgue's theory of measure and Baire's theory of category are two ways to determine "small" and "large" sets. They are independent of each other: sets of measure zero may be of second category and sets of first category may have large measure. The same is true if we replace Lebesgue measure with Hausdorff dimension.

(1.8.3) Exercise. Give an example of a dense G_δ set $E \subseteq \mathbb{R}$ with Hausdorff dimension 0.

However, this is not true for packing dimension.

(1.8.4) Exercise. A set $E \subseteq \mathbb{R}$ with $\mathrm{Dim}\, E < 1$ must be of the first category.

On the Exercises. *(1.1.10):* If $\mathrm{dist}(A, B) = r > 0$, then the $r/2$-neighborhood V of A is open, hence measurable. So

$$\overline{\mathcal{M}}(A \cup B) = \overline{\mathcal{M}}((A \cup B) \cap V) + \overline{\mathcal{M}}((A \cup B) \setminus V) = \overline{\mathcal{M}}(A) + \overline{\mathcal{M}}(B).$$

(1.2.4): $\dim A \geq p \dim f[A]$.

(1.2.11): For each $n \in \mathbb{N}$, choose a countable cover $\{D_{in}\}_{i=1}^{\infty}$ of E by closed sets D_{in} with

$$\sum_{i=1}^{\infty} \widetilde{\mathcal{P}}^s(D_{in}) \leq \overline{\mathcal{P}}^s(E) + \frac{1}{n}.$$

The Borel set $A = \bigcap_n \bigcup_i D_{in}$ will suffice.

(1.2.12): Parts (a) and (c) are correct. For (a): If $\mathcal{B}$ is a finite packing of E, then it is also a packing of E_n for large enough n. Conclude that $\lim_n \widetilde{\mathcal{P}}^s_\varepsilon(E_n) \geq \widetilde{\mathcal{P}}^s_\varepsilon(E)$. For (c), use Theorem 1.1.3 and the regularity.

(1.2.14): $\mathrm{Dim}\, A \geq p\, \mathrm{Dim}\, f[A]$.

(1.3.4): In an ultrametric space, either two balls are disjoint or else one is contained in the other. A closed ball is also an open set. So in a *compact*

ultrametric space, for any given $r > 0$ there are only finitely many distinct balls with radius $\geq r$.

The proof for (1.3.4) follows the proof given for Theorem 1.3.1 with a few minor refinements. Note that conclusion (a) of (1.3.1) is impossible in a compact metric space (by total boundedness). Because of the remark in the previous paragraph, we may make the choice of the balls $\overline{B}_{r_i}(x_i)$ such that $\overline{B}_{r_i}(x_i) = \overline{B}_{t_i}(x_i)$.

We must prove that $E \subseteq \bigcup_{i=1}^{\infty} \overline{B}_{r_i}(x_i)$. Let $x \in E$ and assume $x \notin \bigcup \overline{B}_{r_i}(x_i)$. There is $r_0 > 0$ such that $\overline{B}_{r_0}(x) \in \beta_1$. Let n be the least natural number with $t_n < r_0$. Now, $\overline{B}_{r_0}(x) \cap \bigcup_{i=1}^{n-1} \overline{B}_{r_i}(x_i) \neq \varnothing$, since if it were $\varnothing$, then $\overline{B}_{r_0}(x) \in \beta_n$, contradicting $t_n < r_0$. Let k be the least natural number with $\overline{B}_{r_0}(x) \cap \overline{B}_{r_k}(x_k) \neq \varnothing$. Then $t_k \geq r_0$. If $z \in \overline{B}_{r_0}(x) \cap \overline{B}_{r_k}(x_k)$, then $\rho(x, x_k) \leq r_0 \vee t_k = t_k$, so $x \in \overline{B}_{t_k}(x_k) = \overline{B}_{r_k}(x_k)$.

(1.3.9): Write $a = |x' - x''|$, $b = |x - x''|$, $c = |x - x'|$, and let α be the angle in question (opposite a). We will use the formula

$$\sin \frac{\alpha}{2} = \sqrt{\frac{(s-b)(s-c)}{bc}},$$

where s is the semiperimeter $(a + b + c)/2$. The given inequalities yield

$$b \leq r'' + r \leq r'' + \frac{1}{2} r'' = \frac{3}{2} r'',$$

$$c \leq r' + r \leq r' + \frac{1}{2} r' = \frac{3}{2} r',$$

$$2(s-b) = a - b + c \geq r' - (r'' + r) + r' \geq 2r' - \frac{4}{3} r' - \frac{1}{2} r' = \frac{1}{6} r',$$

$$2(s-c) = a + b - c \geq r' + r'' - (r' + r) \geq r'' - \frac{1}{2} r'' = \frac{1}{2} r'',$$

$$4 \sin^2 \frac{\alpha}{2} = \frac{4(s-b)(s-c)}{bc} \geq \frac{\frac{1}{6} r' \frac{1}{2} r''}{\frac{3}{2} r'' \frac{3}{2} r'} = \frac{1}{27}.$$

Thus $\sin^2(\alpha/2) \geq 1/108$, so $\sin(\alpha/2) \geq \sqrt{1/108} \approx 0.096$, and thus $\alpha/2 \geq 5.52°$, and $\alpha \geq 11.04°$.[5]

(1.4.4): Hint. Copy the proof of (1.3.5) with the fine cover

$$\beta = \left\{ \overline{B}_r(x) : x \in E, 0 < r \leq \operatorname{diam} \overline{B}_r(x) < \varepsilon \right\}$$

to show $\widetilde{\mathcal{C}}^s(E) \leq \widetilde{\mathcal{P}}^s(E)$. Use this to prove $\overline{\mathcal{C}}^s(E) \leq \widetilde{\mathcal{P}}^s(E)$ and then $\overline{\mathcal{C}}^s(E) \leq \overline{\mathcal{P}}^s(E)$.

---

[5] When the problem was published in *Mathematics Magazine* [2], the Anchorage Math Solutions Group found that the least possible measure for the angle is $\cos^{-1}(11/12) \approx 23.5°$.

(1.5.18): Copy the proof of Corollary 1.5.17; use Theorem 1.5.11 in place of Theorem 1.5.13. In order to apply Theorem 1.5.11, we must assume also that $\mathcal{P}^s \restriction E$ has the strong Vitali property.

Do we really need this extra assumption?

(1.8.5) Exercise. Give an example of a set E with $0 < \mathcal{P}^s(E) < \infty$ such that $\mathcal{P}^s \restriction E$ fails the strong Vitali property.

2. Integrals

We will discuss the modern (twentieth-century) theory of integration. Just as the Lebesgue measure generalizes the classical ideas on areas and volumes, the Lebesgue integral generalizes and clarifies the classical ideas on the integral.

First, we define the abstract integral, using our knowledge of abstract measure theory. Then we discuss integrals as linear functionals on spaces of functions. Finally, we talk about spaces of measures, and in particular narrow convergence for the set of tight Borel probability measures on a metric space.

2.1 Product Measures

Product σ-algebras. Let $(X_1, \mathcal{F}_1)$ and $(X_2, \mathcal{F}_2)$ be two Borel spaces. We will define a σ-algebra $\mathcal{F}$ for the Cartesian product $X_1 \times X_2$ in such a way that it satisfies the following properties:

(1) The projections π_1, π_2 are Borel maps ($\mathcal{F}$ to $\mathcal{F}_1$ and $\mathcal{F}$ to $\mathcal{F}_2$, respectively).
(2) If $(Z, \mathcal{G})$ is any Borel space, and $g_1 : Z \to X_1$, $g_2 : Z \to X_2$ are Borel maps, then the corresponding map $g : Z \to X_1 \times X_2$ is Borel ($\mathcal{G}$ to $\mathcal{F}$).

(2.1.1) Exercise. A **measurable rectangle** is a set of the form $A \times B$, where $A \in \mathcal{F}_1$ and $B \in \mathcal{F}_2$. Let $\mathcal{F}_1 \otimes \mathcal{F}_2$ denote the σ-algebra generated by the measurable rectangles. Prove:

(a) $\mathcal{F}_1 \otimes \mathcal{F}_2$ has properties (1) and (2).
(b) $\mathcal{F}_1 \otimes \mathcal{F}_2$ is the only σ-algebra on $X_1 \times X_2$ satisfying (1) and (2).

The σ-algebra $\mathcal{F}_1 \otimes \mathcal{F}_2$ will be called the **product σ-algebra** or **product Borel structure** defined by $\mathcal{F}_1$ and $\mathcal{F}_2$. A description of all the sets in $\mathcal{F}_1 \otimes \mathcal{F}_2$ is usually quite complicated. But some of the properties of these sets can be seen easily.

(2.1.2) Proposition. *Let $D \in \mathcal{F}_1 \otimes \mathcal{F}_2$. Then for each $x \in X_1$, the cross-section*

$$D_{[x]} = \{ y \in X_2 : (x, y) \in D \}$$

belongs to $\mathcal{F}_2$; *and for each* $y \in X_2$,

$$D^{[y]} = \{ x \in X_1 : (x, y) \in D \}$$

belongs to $\mathcal{F}_1$.

Proof. Fix $y \in X_2$. The collection of all sets $D \subseteq X_1 \times X_2$ such that $\{ y \in X_2 : (x, y) \in D \} \in \mathcal{F}_2$ is a σ-algebra and includes all measurable rectangles. So it includes the product σ-algebra $\mathcal{F}_1 \otimes \mathcal{F}_2$.   ☺

(2.1.3) Proposition. *Let* S_1 *and* S_2 *be separable metric spaces. Then the Borel sets of the product* $S_1 \times S_2$ *are given by*

$$\mathcal{B}(S_1 \times S_2) = \mathcal{B}(S_1) \otimes \mathcal{B}(S_2).$$

Proof. I shall show that the two σ-algebras are equal by showing that each is a subset of the other.

First, consider the family of subsets of S_1 defined by

$$\mathcal{A} = \{ A \subseteq S_1 : A \times S_2 \in \mathcal{B}(S_1 \times S_2) \}.$$

Now, if $A \subseteq S_1$ is an open set in S_1, then $A \times S_2$ is an open set in $S_1 \times S_2$, so all open sets belong to $\mathcal{A}$. But $\mathcal{A}$ is a σ-algebra. So $\mathcal{A} \supseteq \mathcal{B}(S_1)$. That is, if $A \in \mathcal{B}(S_1)$, then $A \times S_2 \in \mathcal{B}(S_1 \times S_2)$. Similarly, if $B \in \mathcal{B}(S_2)$, then $S_1 \times B \in \mathcal{B}(S_1 \times S_2)$. Now, $\mathcal{B}(S_1 \times S_2)$ is a σ-algebra, so if $A \in \mathcal{B}(S_1)$ and $B \in \mathcal{B}(S_2)$, then $A \times B = (A \times S_2) \cap (S_1 \times B)$ belongs to $\mathcal{B}(S_1 \times S_2)$. Since $\mathcal{B}(S_1 \times S_2)$ contains all measurable rectangles $A \times B$ and it is a σ-algebra, it must contain $\mathcal{B}(S_1) \otimes \mathcal{B}(S_2)$. Thus I have proved that $\mathcal{B}(S_1 \times S_2) \supseteq \mathcal{B}(S_1) \otimes \mathcal{B}(S_2)$.

On the other hand, suppose $V \subseteq S_1 \times S_2$ is an open set. Now, since $S_1 \times S_2$ is a separable metric space, by the Lindelöf property the set V is a countable union of balls (in the maximum metric):

$$V = \bigcup_{i=1}^{\infty} B_{r_i}(x_i, y_i) = \bigcup_{i=1}^{\infty} (B_{r_i}(x_i) \times B_{r_i}(y_i)).$$

Therefore, $V \in \mathcal{B}(S_1) \otimes \mathcal{B}(S_2)$. All open sets of $S_1 \times S_2$ belong to $\mathcal{B}(S_1) \otimes \mathcal{B}(S_2)$, and it is a σ-algebra, so $\mathcal{B}(S_1 \times S_2) \subseteq \mathcal{B}(S_1) \otimes \mathcal{B}(S_2)$.   ☺

(2.1.4) Exercise. Suppose S_1 and S_2 are nonseparable metric spaces. Then is it necessarily true that $\mathcal{B}(S_1 \times S_2) = \mathcal{B}(S_1) \otimes \mathcal{B}(S_2)$?

Now, for example, addition is a continuous function from $\mathbb{R} \times \mathbb{R} \to \mathbb{R}$, and therefore addition is a Borel function from $\mathcal{B}(\mathbb{R}) \otimes \mathcal{B}(\mathbb{R})$ to $\mathcal{B}(\mathbb{R})$. Similarly for subtraction and multiplication.

Measurable Functions. Let $(X, \mathcal{F}, \mathcal{M})$ be a measure space. A function $f : X \to \mathbb{R}$ is said to be $\mathcal{F}$-**measurable** iff it is a measurable function from

$\mathcal{F}$ to the σ-algebra $\mathcal{B}(\mathbb{R})$ of Borel sets of $\mathbb{R}$. Now, $\mathcal{B}(\mathbb{R})$ is generated by the intervals

$$(t, \infty), \qquad t \in \mathbb{R},$$

so a function $f \colon X \to \mathbb{R}$ is measurable if and only if $\{x \in X : f(x) > t\} \in \mathcal{F}$ for every $t \in \mathbb{R}$. The same is true for functions with values in $[-\infty, \infty]$.

The sum of two measurable functions is measurable. The general abstract reasoning for this fact is as follows. Suppose $f_1 \colon X \to \mathbb{R}$ and $f_2 \colon X \to \mathbb{R}$ are both measurable from $\mathcal{F}$ to $\mathcal{B}(\mathbb{R})$. Then the map $f \colon X \to \mathbb{R} \times \mathbb{R}$ defined by

$$f(x) = \big(f_1(x), f_2(x)\big)$$

is measurable from $\mathcal{F}$ to $\mathcal{B}(\mathbb{R}) \otimes \mathcal{B}(\mathbb{R})$. Addition $\alpha(x, y) = x + y$ is measurable from $\mathcal{B}(\mathbb{R}) \otimes \mathcal{B}(\mathbb{R})$ to $\mathcal{B}(\mathbb{R})$. So the composition

$$(\alpha \circ f)(x) = f_1(x) + f_2(x)$$

is measurable from $\mathcal{F}$ to $\mathcal{B}(\mathbb{R})$.

In the same way, the difference $f_1(x) - f_2(x)$ and the product $f_1(x)\, f_2(x)$ are measurable. The quotient $f_1(x)/f_2(x)$ is measurable, provided that for all x, $f_2(x) \neq 0$. The maximum of two measurable functions

$$g(x) = \max\{f_1(x), f_2(x)\} = f_1(x) \vee f_2(x)$$

is also measurable, since the "maximum" function $\mathbb{R} \times \mathbb{R} \to \mathbb{R}$ is continuous, and therefore Borel. Similarly, the minimum of two measurable functions is measurable.

The collection of all measurable functions from a measure space X to the real line $\mathbb{R}$ is a **linear space** over $\mathbb{R}$: the constant 0 is measurable; a scalar multiple of a measurable function is measurable; and the sum of two measurable functions is measurable (see Definition 2.4.2).

Measurable functions may also be combined in certain countable ways. If $f_n \colon X \to [-\infty, \infty]$ are countably many measurable functions, then their pointwise supremum

$$f(x) = \sup\{f_n(x) : n \in \mathbb{N}\}$$

is also measurable. To see this, note that

$$\{x \in X : f(x) > t\} = \bigcup_n \{x \in X : f_n(x) > t\}.$$

Similarly, the pointwise infimum of a sequence of measurable functions is measurable. Now we see that the lim sup and lim inf of a sequence of measurable functions is measurable, since

$$\limsup_{n \to \infty} f_n(x) = \inf_m \, \sup_{k \geq m} f_k(x).$$

If a sequence of measurable functions converges pointwise, then the limit is also measurable, since that limit is the lim sup of the sequence.

The measurability of a function $f\colon X \to \mathbb{R}$ is related to the product σ-algebra $\mathcal{F} \otimes \mathcal{B}(\mathbb{R})$:

(2.1.5) Proposition. *Let $f\colon X \to [-\infty, \infty]$ be a function. Then f is an $\mathcal{F}$-measurable function if and only if its "subgraph"*

$$\{\,(x, t) \in X \times \mathbb{R} : t < f(x)\,\}$$

belongs to $\mathcal{F} \otimes \mathcal{B}(\mathbb{R})$.

Proof. Write G for the subgraph. Suppose first that f is a measurable function. Then each set $U_t = \{\,x : f(x) > t\,\}$ belongs to $\mathcal{F}$. But

$$G = \{\,(x, t) \in X \times \mathbb{R} : t < f(x)\,\}$$
$$= \bigcup_{r \in \mathbb{Q}} U_r \times (-\infty, r),$$

so $G \in \mathcal{F} \otimes \mathcal{B}(\mathbb{R})$.

Conversely, suppose G is a measurable set. Then for each $t \in \mathbb{R}$, the set $\{\,x : f(x) > t\,\}$ is a cross-section of G as in Proposition 2.1.2.   ☺

(2.1.6) Exercise. Prove or disprove: a function $f\colon X \to \mathbb{R}$ is measurable if and only if its graph is measurable, $\{\,(x, t) : t = f(x)\,\} \in \mathcal{F} \otimes \mathcal{B}(\mathbb{R})$.

(2.1.7) Exercise. Let $(X, \mathcal{F})$ be a measurable space, and let S be a metric space. A function $f\colon X \to S$ is **measurable** iff $f^{-1}[E] \in \mathcal{F}$ for all Borel sets $E \subseteq S$. Is this equivalent to $\{\,(x, y) \in X \times S : y = f(x)\,\} \in \mathcal{F} \otimes \mathcal{B}(S)$?

An important property of a real-valued measurable function is that of "approximation by simple functions." Let $(X, \mathcal{F})$ be a Borel space. A function $f\colon X \to \mathbb{R}$ is **simple** iff it has the form

$$f(x) = \sum_{i=1}^{N} a_i \, \mathbb{1}_{A_i}(x),$$

where $a_i \in \mathbb{R}$ and $A_i \subseteq X$ for each i. We may assume that the sets A_i are disjoint. If f is measurable, we may assume that $A_i \in \mathcal{F}$.

(2.1.8) Theorem. *Let $(X, \mathcal{F})$ be a Borel space, and let $f\colon X \to [0, \infty]$. The following are equivalent:*

(a) *f is measurable;*
(b) *there exist measurable simple functions $f_n\colon X \to [0, \infty)$ with $f_n(x) \leq f_{n+1}(x)$ for all $x \in X$ and $n \in \mathbb{N}$, and $\lim_{n \to \infty} f_n(x) = f(x)$ for all $x \in X$.*

Proof. Assume that f is measurable. For $n \in \mathbb{N}$, define $f_n(x) = k\,2^{-n}$ on the set $\{\,x : k\,2^{-n} \leq f(x) < (k+1)\,2^{-n}\,\}$ for $0 \leq k < n\,2^n$, and $f_n(x) = n$ on the set $\{x : f(x) \geq n\}$. Then it can be verified that f_n is a measurable function, $f_n(x) \leq f_{n+1}(x)$, and $\lim_{n\to\infty} f_n(x) = f(x)$.

Conversely, assume that such measurable simple functions f_n exist. Then the subgraph $G_f = \{(x,t) : 0 < t < f(x)\}$ is the union of all of the sets G_{f_n}. The sets G_{f_n} are $\mathcal{M} \otimes \mathcal{L}$-measurable, so G_f is measurable, and therefore by Proposition 2.1.5, the function f is measurable.   ☺

Can we generalize to metric spaces other than $\mathbb{R}$?

(2.1.9) Exercise. Let S be a metric space. Let $f: X \to S$ be a function. Prove or disprove: f is a Borel function if and only if f is the pointwise limit of Borel functions f_n with countably many values.

Image Measure.

(2.1.10) Definition. Let $(X_1, \mathcal{F}_1)$ and $(X_2, \mathcal{F}_2)$ be Borel spaces, and let $f: X_1 \to X_2$ be a Borel function. Corresponding to each measure $\mathcal{M}$ on $\mathcal{F}_1$, there is a measure on $\mathcal{F}_2$, called the **image** of $\mathcal{M}$ under f. The image measure is denoted by $f_*(\mathcal{M})$ and defined by

$$f_*(\mathcal{M})(B) = \mathcal{M}\big(f^{-1}[B]\big) \qquad \text{for } B \in \mathcal{F}_2.$$

Because of the definition, it may sometimes be written $f_*(\mathcal{M}) = \mathcal{M}f^{-1}$.

On the other hand, if we are given a measure $\mathcal{N}$ on $(X_2, \mathcal{F}_2)$, there may or may not exist a measure $\mathcal{M}$ on $(X_1, \mathcal{F}_1)$ such that $f_*(\mathcal{M}) = \mathcal{N}$. If such an $\mathcal{M}$ exists, we may say that $\mathcal{N}$ may be **lifted** to X_1.

(2.1.11) Theorem. *Let S and T be metric spaces, S complete and separable. Let f be a Borel function from S onto T. If $\mathcal{N}$ is any finite Borel measure on T, then there is a Borel measure $\mathcal{M}$ on S such that $f_*(\mathcal{M}) = \mathcal{N}$.*

This result (and other more general results) may be found in texts on descriptive set theory, for example [38, §2, No. 4, Prop. 8], [174], or [266].

Product Measures. Let X_1 and X_2 be sets. Suppose a σ-algebra $\mathcal{F}_i$ on each X_i and a measure $\mathcal{M}_i$ on each $\mathcal{F}_i$ are given. We are interested in defining a measure on the Cartesian product $X_1 \times X_2$.

The product measure will be defined by Method I. On the collection of **measurable rectangles**

$$\mathcal{A} = \{A \times B : A \in \mathcal{F}_1, B \in \mathcal{F}_2\}$$

define the set-function

$$C(A \times B) = \mathcal{M}_1(A)\,\mathcal{M}_2(B).$$

(Recall the convention $0 \cdot \infty = 0$.) The product outer measure $\overline{\mathcal{M}_1 \otimes \mathcal{M}_2}$ is the corresponding method I outer measure—that is, the largest outer measure $\overline{\mathcal{M}}$

on $X_1 \times X_2$ satisfying $\overline{\mathcal{M}}(A \times B) \leq C(A \times B)$ for all $A \times B \in \mathcal{A}$ [MTFG, Theorem (5.2.2)]. As usual, we will write $\mathcal{M}_1 \otimes \mathcal{M}_2$ for the measure obtained by restricting this outer measure to its Carathéodory measurable sets.

Lebesgue measure is a good example. If $\mathcal{M}_1 = \mathcal{M}_2 = \mathcal{L}$ is Lebesgue measure on the line $\mathbb{R}$, then $\mathcal{M}_1 \otimes \mathcal{M}_2$ is two-dimensional Lebesgue measure $\mathcal{L}^2$ in the plane $\mathbb{R}^2 = \mathbb{R} \times \mathbb{R}$. [MTFG, §5.3]

By the general theory, $\overline{\mathcal{M}}$ is an outer measure; but we will show that it satisfies $\overline{\mathcal{M}}(A \times B) = \mathcal{M}_1(A)\mathcal{M}_2(B)$. The combinatorial fact we need is the "countable subadditivity" of the set-function C. This is proved by reducing to the case of intervals in the line and applying the known countable subadditivity of two-dimensional Lebesgue measure.

(2.1.12) Lemma. *Let* $(X_1, \mathcal{F}_1, \mathcal{M}_1)$ *and* $(X_2, \mathcal{F}_2, \mathcal{M}_2)$ *be as above. Let* $A \in \mathcal{F}_1$ *and* $B \in \mathcal{F}_2$. *If*

$$A \times B \subseteq \bigcup_{i=1}^{\infty} A_i \times B_i$$

is a countable cover of $A \times B$ *by measurable rectangles, then*

$$\mathcal{M}_1(A)\,\mathcal{M}_2(B) \leq \sum_{i=1}^{\infty} \mathcal{M}_1(A_i)\,\mathcal{M}_2(B_i).$$

Thus, $\mathcal{M}_1(A)\,\mathcal{M}_2(B) \leq \overline{\mathcal{M}_1 \otimes \mathcal{M}_2}(A \times B).$

Proof. Let $a < \mathcal{M}_1(A)$ and $b < \mathcal{M}_2(B)$ be given. It will suffice to show that

$$\sum_{i=1}^{\infty} \mathcal{M}_1(A_i)\,\mathcal{M}_2(B_i) \geq ab.$$

For each $x \in A$, we have

$$B \subseteq \bigcup_{x \in A_i} B_i,$$

so that

$$\sum_{x \in A_i} \mathcal{M}_2(B_i) > b.$$

So there is $n \in \mathbb{N}$ with

$(*)$ $$\sum_{\substack{i \leq n \\ x \in A_i}} \mathcal{M}_2(B_i) > b.$$

Write $n(x)$ for the least such n. Then (Exercise 2.1.13) for each $n \in \mathbb{N}$ the set

$$R_n = \{x \in A : n(x) \leq n\}$$

belongs to $\mathcal{F}_1$. The sets R_n increase to A, so there is n large enough that $\mathcal{M}_1(R_n) > a$. It will suffice to show that

$$\sum_{i=1}^{n} \mathcal{M}_1(A_i)\,\mathcal{M}_2(B_i) \geq ab.$$

Next, I reduce to the case where the sets A_i are disjoint. Let $\mathcal{D}$ be the set of atoms of the finite σ-algebra generated by $\{A_1, A_2, \cdots, A_n\}$. (See Exercise 1.1.5.) Then, for each $i \leq n$, the set A_i is the disjoint union of the atoms $D \in \mathcal{D}$ contained in A_i; and the measure of A_i is the sum of the measures of those atoms. So if the set $A_i \times B_i$ is replaced by the finite list of sets

$$D \times B_i \qquad (D \in \mathcal{D}, D \subseteq A_i),$$

we have

$$\sum_{\substack{D \in \mathcal{D} \\ D \subseteq A_i}} C(D \times B_i) = C(A_i \times B_i).$$

Making this substitution for each i, we get a new finite collection of measurable rectangles (which we will still call $A_i \times B_i$) such that for each pair i, j we have either $A_i \cap A_j = \varnothing$ or $A_i = A_j$, and such that the new set of rectangles has the same value for the sum

$$\sum_i \mathcal{M}_1(A_i)\,\mathcal{M}_2(B_i).$$

Also, $(*)$ tells us that for each i we have

$$\sum_{A_j = A_i} \mathcal{M}_2(B_j) > b.$$

Now I want to reduce to the case where the sets A_i and B_i are intervals in the line. Let $\mathcal{D}$ be the set of different A_i sets that appear in the sum. For each $D \in \mathcal{D}$, choose an interval $\widehat{D} = [c, d)$ with $\mathcal{L}(\widehat{D}) = d - c = \mathcal{M}_1(D)$; to be specific, take the sets of $\mathcal{D}$ in some definite order; let the first left endpoint be 0 and let each successive left endpoint be the previous right endpoint. Thus $\bigcup_{D \in \mathcal{D}} \widehat{D} = [0, d)$, where $d > a$. For each $D \in \mathcal{D}$, consider the sets B_j with $A_j = D$. Again choose intervals $\widehat{B_j}$ with length $\mathcal{M}_2(B_j)$ such that $\bigcup_{A_j = D} \widehat{B_j}$ is an interval $[0, d)$ with $d > b$. Then

$$\sum_i \mathcal{M}_1(A_i)\,\mathcal{M}_2(B_i) = \sum_i \mathcal{L}(\widehat{A_i})\,\mathcal{L}(\widehat{B_i}).$$

But

$$\bigcup \widehat{A_i} \times \widehat{B_i} \supseteq [0, a) \times [0, b),$$

so by the additivity of areas in the plane (that is, additivity of $\mathcal{L}^2$), we may conclude

$$\sum_i \mathcal{L}(\widehat{A_i})\,\mathcal{L}(\widehat{B_i}) \geq ab$$

as required. ☺

(2.1.13) Exercise. Show that $R_n \in \mathcal{F}_1$ in the preceding proof.

Now we prove the most important property of the product measure.

(2.1.14) Theorem. *Let $(X_1, \mathcal{F}_1, \mathcal{M}_1)$ and $(X_2, \mathcal{F}_2, \mathcal{M}_2)$ be measure spaces. If $A \in \mathcal{F}_1$ and $B \in \mathcal{F}_2$, then $A \times B$ is $\overline{\mathcal{M}_1 \otimes \mathcal{M}_2}$-measurable and*

$$\mathcal{M}_1 \otimes \mathcal{M}_2 (A \times B) = \mathcal{M}_1(A) \, \mathcal{M}_2(B).$$

Thus, every set in $\mathcal{F}_1 \otimes \mathcal{F}_2$ is $\overline{\mathcal{M}_1 \otimes \mathcal{M}_2}$-measurable, and $\overline{\mathcal{M}_1 \otimes \mathcal{M}_2}$ is a regular measure.

Proof. By definition, $\overline{\mathcal{M}_1 \otimes \mathcal{M}_2}$-measurability is specified by the condition of Carathéodory. Let $E \subseteq X_1 \times X_2$ be an arbitrary set. I must show that

$$\overline{\mathcal{M}_1 \otimes \mathcal{M}_2}(E) = \overline{\mathcal{M}_1 \otimes \mathcal{M}_2}(E \cap (A \times B)) + \overline{\mathcal{M}_1 \otimes \mathcal{M}_2}(E \setminus (A \times B)).$$

Since the inequality $\leq$ is true in general, it remains to prove the inequality $\geq$. Suppose

$$E \subseteq \bigcup_i A_i \times B_i$$

is a cover of E by measurable rectangles. Each rectangle $A_i \times B_i$ is a union of three disjoint rectangles:

$$A_i \times B_i = \big((A_i \cap A) \times (B_i \cap B)\big) \cup \big(A_i \times (B_i \setminus B)\big) \cup \big((A_i \setminus A) \times (B_i \cap B)\big),$$

and of course C adds:

$$
\begin{aligned}
C(A_i \times B_i) = \; & C\big((A_i \cap A) \times (B_i \cap B)\big) \\
& + C\big(A_i \times (B_i \setminus B)\big) + C\big((A_i \setminus A) \times (B_i \cap B)\big).
\end{aligned}
$$

So we have

$$
\begin{aligned}
\sum C(A_i \times B_i) = \; & \sum C\big((A_i \cap A) \times (B_i \cap B)\big) \\
& + \sum C\big(A_i \times (B_i \setminus B)\big) + \sum C\big((A_i \setminus A) \times (B_i \cap B)\big).
\end{aligned}
$$

Now, $E \cap (A \times B)$ is covered by the rectangles $A_i \times B_i$, and $E \setminus (A \times B)$ is covered by the rectangles $A_i \times (B_i \setminus B)$ and $(A_i \setminus A) \times (B_i \cap B)$. So

$$\sum C\big((A_i \cap A) \times (B_i \cap B)\big) \geq \overline{\mathcal{M}_1 \otimes \mathcal{M}_2}\big(E \cap (A \times B)\big),$$

$$
\begin{aligned}
\sum C\big(A_i \times (B_i \setminus B)\big) + \sum C\big((A_i \setminus A) \times (B_i \cap B)\big) & \\
\geq \overline{\mathcal{M}_1 \otimes \mathcal{M}_2}\big(E \setminus (A \times B)\big). &
\end{aligned}
$$

We may conclude that

$$\sum C(A_i \times B_i) \geq \overline{\mathcal{M}_1 \otimes \mathcal{M}_2}\big(E \cap (A \times B)\big) + \overline{\mathcal{M}_1 \otimes \mathcal{M}_2}\big(E \setminus (A \times B)\big).$$

This is true for any cover of E, so we have

$$\overline{\mathcal{M}_1 \otimes \mathcal{M}_2}(E) \geq \overline{\mathcal{M}_1 \otimes \mathcal{M}_2}(E \cap (A \times B)) + \overline{\mathcal{M}_1 \otimes \mathcal{M}_2}(E \setminus (A \times B))$$

as required. This completes the proof that $A \times B$ is measurable.

Next I will show that $\mathcal{M}_1 \otimes \mathcal{M}_2(A \times B) = \mathcal{M}_1(A)\,\mathcal{M}_2(B)$. One cover of $A \times B$ is $A \times B$ itself. So $\mathcal{M}_1 \otimes \mathcal{M}_2(A \times B) \leq C(A \times B)$. The opposite inequality $\geq$ is the content of Lemma 2.1.12.

Because every element of the Vitali cover used in the Method I definition is measurable, the measure is regular.   ☺

2.2 Integrals

We define an integral of a nonnegative function as "the area under the graph." If the function is not measurable, this will be the "upper integral." Then the definition of the integral will be extended by linearity.

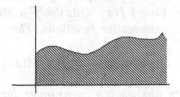

Figure 2.2.1. Area under the graph.

Let $(X, \mathcal{F}, \mathcal{M})$ be a measure space. Let $f \colon X \to [0, \infty]$ be a function. The **upper integral** of f with respect to $\mathcal{M}$ is

$$\overline{\int} f(x)\,\mathcal{M}(dx) = \overline{\mathcal{M} \otimes \mathcal{L}}(G_f),$$

where

$$G_f = \{\, (x, t) \in X \times \mathbb{R} : 0 < t < f(x) \,\}.$$

If f is a measurable function, then G_f is in $\mathcal{F} \otimes \mathcal{B}(\mathbb{R})$ and is therefore $\overline{\mathcal{M} \otimes \mathcal{L}}$-measurable; then the upper integral is simply called the **integral** of f with respect to $\mathcal{M}$ and written

$$\int f(x)\,\mathcal{M}(dx) = \mathcal{M} \otimes \mathcal{L}(G_f).$$

If $f \colon X \to [-\infty, \infty]$ is a measurable function, then its positive and negative parts

$$f^+(x) = f(x) \vee 0,$$
$$f^-(x) = (-f(x)) \vee 0$$

are also measurable, and they satisfy $f = f^+ - f^-$. The **integral** of f is defined as

$$\int f(x)\,\mathcal{M}(dx) = \int f^+(x)\,\mathcal{M}(dx) - \int f^-(x)\,\mathcal{M}(dx),$$

provided that this is not of the form $\infty - \infty$.

If $\int f(x)\,\mathcal{M}(dx)$ exists as a real number, then we say that f is **integrable**. Another way of saying this is $\int |f(x)|\,\mathcal{M}(dx) < \infty$.

Let $E \in \mathcal{F}$. The integral on the set E is defined as

$$\int_E f(x)\,\mathcal{M}(dx) = \int f(x)\,\mathbb{1}_E(x)\,\mathcal{M}(dx).$$

(2.2.2) Exercise. If $f \leq g$, then $\int f(x)\,\mathcal{M}(dx) \leq \int g(x)\,\mathcal{M}(dx)$.

(2.2.3) Proposition. *Let $(X, \mathcal{F}, \mathcal{M})$ be a measure space. Let $f\colon X \to [-\infty, \infty]$ and $g\colon X \to [-\infty, \infty]$ be functions.*

(a) *Suppose f has the form $\sum_{i=0}^{\infty} a_i\,\mathbb{1}_{A_i}$, where $a_i \geq 0$ and the A_i are disjoint sets. Then $\overline{\int} f(x)\,\mathcal{M}(dx) \leq \sum a_i \overline{\mathcal{M}}(A_i)$. If the sets A_i are measurable, so that f is measurable, then $\int f(x)\,\mathcal{M}(dx) = \sum a_i \mathcal{M}(A_i)$.*

(b) *Suppose f and g are nonnegative functions. Then*

$$\overline{\int} (f(x) + g(x))\,\mathcal{M}(dx) \leq \overline{\int} f(x)\,\mathcal{M}(dx) + \overline{\int} g(x)\,\mathcal{M}(dx).$$

Suppose f and g are nonnegative measurable functions. Then $f + g$ is measurable, and

$$\int (f(x) + g(x))\,\mathcal{M}(dx) = \int f(x)\,\mathcal{M}(dx) + \int g(x)\,\mathcal{M}(dx).$$

Suppose f and g are integrable functions. Then $f + g$ is integrable, and

$$\int (f(x) + g(x))\,\mathcal{M}(dx) = \int f(x)\,\mathcal{M}(dx) + \int g(x)\,\mathcal{M}(dx).$$

(c) *Suppose $f \geq 0$ and $a \geq 0$. Then $\overline{\int} af(x)\,\mathcal{M}(dx) = a \overline{\int} f(x)\,\mathcal{M}(dx)$. Suppose f is an integrable function and $a \in \mathbb{R}$. Then af is also integrable, and $\int af(x)\,\mathcal{M}(dx) = a \int f(x)\,\mathcal{M}(dx)$.*

Proof. (a) The outer measure $\overline{\mathcal{M}}$ used here is defined by

$$\overline{\mathcal{M}}(A) = \inf \{ \mathcal{M}(B) : B \in \mathcal{F}, B \supseteq A \}.$$

First, suppose $f = \mathbb{1}_A$. For this function f, we have $G_f = A \times (0,1)$. If $B \supseteq A$ is a measurable set, then

$$\overline{\mathcal{M} \otimes \mathcal{L}}(G_f) \leq C(B \times (0,1)) = \mathcal{M}(B)\mathcal{L}((0,1)) = \mathcal{M}(B).$$

Taking the infimum over all B, we get $\overline{\mathcal{M} \otimes \mathcal{L}}(G_f) \leq \overline{\mathcal{M}}(A)$. On the other hand, if A is measurable and if $G_f \subseteq \bigcup_i A_i \times B_i$ is any countable cover of G_f by measurable rectangles then $\bigcup_i A_i \times B_i$ contains the set $A \times (0,1)$. So $\sum_i C(A_i \times B_i) \geq \overline{\mathcal{M}}(A)$. Thus $\overline{\mathcal{M} \otimes \mathcal{L}}(G_f) \geq \overline{\mathcal{M}}(A)$, as required.

Now suppose

$$f = \sum_{i=1}^{\infty} a_i\,\mathbb{1}_{A_i}.$$

Then the subgraph $G_f = \{(x,t) : 0 < t < f(x)\}$ is the disjoint union of the rectangles

$$A_i \times (0, a_i).$$

These rectangles have outer measure $a_i \overline{\mathcal{M}}(A_i)$, as we have just seen, so by the subadditivity of the outer measure $\overline{\mathcal{M} \otimes \mathcal{L}}$, we may conclude that $\int f(x) \mathcal{M}(dx) \leq \sum a_i \overline{\mathcal{M}}(A_i)$. If the sets A_i are measurable, then the rectangles $A_i \times (0, a_i)$ are measurable rectangles, so we may use the countable additivity of the measure $\mathcal{M} \otimes \mathcal{L}$ to conclude that $\int f(x) \mathcal{M}(dx) = \sum a_i \mathcal{M}(A_i)$.

(b) We must show that $\overline{\mathcal{M} \otimes \mathcal{L}}(G_{f+g}) \leq \overline{\mathcal{M} \otimes \mathcal{L}}(G_f) + \overline{\mathcal{M} \otimes \mathcal{L}}(G_g)$, where

$$G_f = \{(x,t) \in X \times \mathbb{R} : 0 < t < f(x)\},$$

and similarly for other functions. Suppose $G_f \subseteq \bigcup_i A_i \times B_i$ is a countable cover of G_f by measurable rectangles, and $G_g \subseteq \bigcup_i C_i \times D_i$ is a countable cover of G_g by measurable rectangles. I will show that $\overline{\mathcal{M} \otimes \mathcal{L}}(G_{f+g}) \leq \sum C(A_i \times B_i) + \sum C(C_i \times D_i)$. This will establish $\overline{\mathcal{M} \otimes \mathcal{L}}(G_{f+g}) \leq \overline{\mathcal{M} \otimes \mathcal{L}}(G_f) + \overline{\mathcal{M} \otimes \mathcal{L}}(G_g)$, as required.

For a positive integer n, consider the finite union $\bigcup_{i=1}^{n} A_i \times B_i$. For each $x \in X$, let the measure of the cross-section

$$f_n(x) = \mathcal{L}\left\{t : (x,t) \in \bigcup_{i=1}^{n} A_i \times B_i\right\}$$

define $f_n(x)$. Then $f_n : X \to [0, \infty]$ is a function with finitely many values. Each value is taken on an $\mathcal{F}$-measurable set. The subgraph G_{f_n} consists of a finite union of measurable rectangles, and $\mathcal{M} \otimes \mathcal{L}(G_{f_n}) \leq \sum_{i=1}^{n} C(A_i \times B_i)$. Now, as n increases, the finite unions increase to $\bigcup_{i=1}^{\infty} A_i \times B_i$, which covers G_f, so $\lim_{n \to \infty} f_n(x) \geq f(x)$. In the same way, define $g_n : X \to [0, \infty]$ by

$$g_n(x) = \mathcal{L}\left\{t : (x,t) \in \bigcup_{i=1}^{n} C_i \times D_i\right\}.$$

The sum $h_n = f_n + g_n$ also has finitely many values, its subgraph $E_n = G_{f_n+g_n}$ is a finite union of rectangles, and

$$\mathcal{M} \otimes \mathcal{L}(E_n) = \mathcal{M} \otimes \mathcal{L}(G_{f_n}) + \mathcal{M} \otimes \mathcal{L}(G_{g_n}).$$

Now, $\bigcup_n E_n$ is a countable union of measurable rectangles that contains G_{f+g}. We have

$$\overline{\mathcal{M} \otimes \mathcal{L}}(G_{f+g}) \leq \mathcal{M} \otimes \mathcal{L}\left(\bigcup_n E_n\right)$$

$$= \lim_{n \to \infty} \mathcal{M} \otimes \mathcal{L}(E_n)$$

$$= \lim_{n \to \infty} \mathcal{M} \otimes \mathcal{L}(G_{f_n}) + \mathcal{M} \otimes \mathcal{L}(G_{g_n})$$

$$\leq \sum_{i=1}^{\infty} C(A_i \times B_i) + \sum_{i=1}^{\infty} C(C_i \times D_i),$$

as required.

Now suppose f and g are measurable. Then there exist nonnegative measurable simple functions f_n and g_n with $f_n \leq f_{n+1}$, $g_n \leq g_{n+1}$, $f_n \to f$, and $g_n \to g$ pointwise. Now, $f_n + g_n$ is again a measurable simple function, and $\int (f_n(x) + g_n(x)) \, \mathcal{M}(dx) = \int f_n(x) \, \mathcal{M}(dx) + \int g_n(x) \, \mathcal{M}(dx)$. But $f_n + g_n$ increases to $f + g$ pointwise, so the subgraph $G_{f_n+g_n}$ increases to the subgraph G_{f+g}. Therefore, $\int (f(x) + g(x)) \, \mathcal{M}(dx) = \lim_n \int (f_n(x) + g_n(x)) \, \mathcal{M}(dx) = \lim_n \int f(x) \, \mathcal{M}(dx) + \int g_n(x) \, \mathcal{M}(dx) = \int f(x) \, \mathcal{M}(dx) + \int g(x) \, \mathcal{M}(dx)$.

Next, suppose f and g are integrable. Let $h = f + g$. Now, $f = f^+ - f^-$, $g = g^+ - g^-$, and $h = h^+ - h^-$. So

$$h^+ - h^- = f^+ - f^- + g^+ - g^-,$$

and thus

$$h^+ + f^- + g^- = h^- + f^+ + g^+.$$

The additivity of the integral for nonnegative measurable functions (just proved) shows that

$$\int h^+(x) \, \mathcal{M}(dx) + \int f^-(x) \, \mathcal{M}(dx) + \int g^-(x) \, \mathcal{M}(dx)$$
$$= \int h^-(x) \, \mathcal{M}(dx) + \int f^+(x) \, \mathcal{M}(dx) + \int g^+(x) \, \mathcal{M}(dx).$$

Now, all these terms are finite, so we have

$$\int h^+(x) \, \mathcal{M}(dx) - \int h^-(x) \, \mathcal{M}(dx)$$
$$= \int f^+(x) \, \mathcal{M}(dx) - \int f^-(x) \, \mathcal{M}(dx)$$
$$+ \int g^+(x) \, \mathcal{M}(dx) - \int g^-(x) \, \mathcal{M}(dx),$$

or $\int h(x) \, \mathcal{M}(dx) = \int f(x) \, \mathcal{M}(dx) + \int g(x) \, \mathcal{M}(dx)$.

(c) If $A_i \times B_i$ is a measurable rectangle in $X \times \mathbb{R}$, and $a > 0$, then $A_i \times (aB_i)$ is also a measurable rectangle, and $C(A_i \times (aB_i)) = aC(A_i \times B_i)$. If $G_f \subseteq \bigcup A_i \times B_i$, then $G_{af} \subseteq \bigcup A_i \times (aB_i)$. These observations, together with the definitions, will prove the results.   ☺

It makes sense to integrate a function with value ∞. But note that if $f \colon X \to [0,\infty]$ is measurable and $\int f(x) \, \mathcal{M}(dx) < \infty$, then $f(x) < \infty$ for almost all x.

If two functions f, g agree almost everywhere, then they have the same integral. (If $f(x) = g(x)$ for all $x \notin N$, with $\mathcal{M}(N) = 0$, then the same is true for f^+, g^+, and the subgraphs G_f, G_g differ only by a subset of $N \times \mathbb{R}$ that has measure zero.) So in fact, it makes sense to integrate a function $f \colon X \to \mathbb{R}$ that is **defined almost everywhere**, that is, a function $f \colon X_0 \to \mathbb{R}$ where $\mathcal{M}(X \setminus X_0) = 0$. Extending the definition to all of X in an arbitrary way (say 0 outside X_0), we may then integrate it.

(2.2.4) Exercise. Prove the **change of variables formula**: Suppose $(X_1, \mathcal{F}_1, \mathcal{M})$ is a measure space, $(X_2, \mathcal{F}_2)$ is a measurable space, and $h\colon X_1 \to X_2$, $f\colon X_2 \to \mathbb{R}$ are Borel functions. Then

$$\int f(y)\,(h_*(\mathcal{M}))(dy) = \int f(h(x))\,\mathcal{M}(dx),$$

in the sense that if one side is defined, so is the other, and the values are equal. [Hint. Approximate by simple functions.]

Limit Theorems. Now we will consider how limits may be integrated. Countable additivity of the measure is used here, of course. We begin with nonnegative functions and then apply the results to integrable functions.

(2.2.5) The Monotone Convergence Theorem. *Let $(X, \mathcal{F}, \mathcal{M})$ be a measure space. Let $f_n\colon X \to [0, \infty]$ be functions $(n \in \mathbb{N})$ that increase: $f_1(x) \leq f_2(x) \leq \cdots$ for all x. Then $\overline{\int} \lim_n f_n(x)\,\mathcal{M}(dx) = \lim_n \overline{\int} f_n(x)\,\mathcal{M}(dx)$.*

Proof. The sets

$$G_n = \{\,(x, t) : 0 < t < f_n(x)\,\}$$

increase to the set

$$G = \{\,(x, t) : 0 < t < \lim f_n(x)\,\}.$$

Now, $\overline{\mathcal{M} \otimes \mathcal{L}}$ is a regular measure (Theorem 2.1.14). Even if the functions f_n are not measurable, so that the sets G_n are not measurable, we still have $\overline{\mathcal{M} \otimes \mathcal{L}}(G_n) \to \overline{\mathcal{M} \otimes \mathcal{L}}(G)$ by Theorem 1.1.3.   ☺

(2.2.6) Exercise. Let $(X, \mathcal{F}, \mathcal{M})$ be a measure space. Let $f_n\colon X \to \mathbb{R}$ be integrable functions $(n \in \mathbb{N})$ that increase: $f_1(x) \leq f_2(x) \leq \cdots$ for all x. Then $\int \lim_n f_n(x)\,\mathcal{M}(dx) = \lim_n \int f_n(x)\,\mathcal{M}(dx)$.

(2.2.7) Beppo Levi's Theorem. *Let $(X, \mathcal{F}, \mathcal{M})$ be a measure space, and let $\sum_{n=1}^{\infty} f_n(x)$ be an infinite series whose terms are functions from X to $[0, \infty]$. Then*

$$\overline{\int} \sum_{n=1}^{\infty} f_n(x)\,\mathcal{M}(dx) \leq \sum_{n=1}^{\infty} \overline{\int} f_n(x)\,\mathcal{M}(dx).$$

If the functions are measurable, then

$$\int \sum_{n=1}^{\infty} f_n(x)\,\mathcal{M}(dx) = \sum_{n=1}^{\infty} \int f_n(x)\,\mathcal{M}(dx).$$

Proof. Apply the Monotone Convergence Theorem to the sequence of partial sums $\sum_{k=1}^{n} f_k(x)$. The finite subadditivity in Proposition 2.2.3(b) is equality when the functions are measurable.   ☺

(2.2.8) Fatou's Lemma.   *Let $(X, \mathcal{F}, \mathcal{M})$ be a measure space, and let $f_n \colon X \to [0, \infty]$ be a sequence of functions. Then*

$$\overline{\int} \liminf_{n \to \infty} f_n(x)\, \mathcal{M}(dx) \leq \liminf_{n \to \infty} \overline{\int} f_n(x)\, \mathcal{M}(dx).$$

Proof. For each n, define
$$g_n(x) = \inf_{k \geq n} f_k(x).$$

Then g_n increases to $\liminf_k f_k(x)$, so by the Monotone Convergence Theorem 2.2.5,
$$\lim_n \overline{\int} g_n(x)\, \mathcal{M}(dx) = \overline{\int} \liminf_k f_k(x)\, \mathcal{M}(dx).$$

Now, $g_n(x) \leq f_n(x)$ for all x and all n, so

$$\liminf_n \overline{\int} f_n(x)\, \mathcal{M}(dx) \geq \liminf_n \overline{\int} g_n(x)\, \mathcal{M}(dx). \qquad \smiley$$

(2.2.9) Exercise. Give an example involving measurable functions in a finite measure space where the conclusion in Fatou's Lemma is a strict inequality.

(2.2.10)   Lebesgue's Dominated Convergence Theorem. *Suppose $(X, \mathcal{F}, \mathcal{M})$ is a measure space, $g \colon X \to [0, \infty]$ is an integrable function, and $f_n \colon X \to [-\infty, \infty]$ are measurable functions. Suppose $f_n(x)$ converges as $n \to \infty$ for almost all x, and $|f_n(x)| \leq g(x)$ for all n and almost all x. Then*

$$\int \lim_n f_n(x)\, \mathcal{M}(dx) = \lim_n \int f_n(x)\, \mathcal{M}(dx).$$

Proof. Since g is integrable and $|f_n| \leq g$, we know that all f_n are also integrable. Since $\int g(x)\, \mathcal{M}(dx) < \infty$, it follows that $g(x) < \infty$ for almost all x. By altering the functions on a set of measure zero (which does not alter their integrals), we may assume that the following hold everywhere, rather than merely almost everywhere:

(a) $f_n(x)$ converges (call the limit $f(x)$);
(b) $|f_n(x)| \leq g(x)$;
(c) $g(x) < \infty$.

Now, $g(x) + f_n(x)$ is a sequence of nonnegative, integrable functions, which converges to $g(x) + f(x)$. By Fatou's Lemma,

$$\int \big(g(x) + f(x)\big)\, \mathcal{M}(dx) \leq \liminf_n \int \big(g(x) + f_n(x)\big)\, \mathcal{M}(dx),$$

and therefore $\int f(x)\, \mathcal{M}(dx) \leq \liminf \int f_n(x)\, \mathcal{M}(dx)$. A similar argument applied to the sequence $g(x) - f_n(x)$ shows that

$$\int \big(g(x) - f(x)\big)\, \mathcal{M}(dx) \leq \liminf_n \int \big(g(x) - f_n(x)\big)\, \mathcal{M}(dx),$$

and therefore $\int f(x) \, \mathcal{M}(dx) \geq \limsup \int f_n(x) \, \mathcal{M}(dx)$. Thus,

$$\int f(x) \, \mathcal{M}(dx) \leq \liminf \int f_n(x) \, \mathcal{M}(dx)$$

$$\leq \limsup \int f_n(x) \, \mathcal{M}(dx) \leq \int f(x) \, \mathcal{M}(dx).$$

So we may conclude that $\int f(x) \, \mathcal{M}(dx) = \lim \int f_n(x) \, \mathcal{M}(dx)$.   ☺

(2.2.11) Exercise. Show that the dominating function g is needed. That is, give an example showing that

$$\int \lim_n f_n(x) \, \mathcal{M}(dx) = \lim_n \int f_n(x) \, \mathcal{M}(dx)$$

may be false, even when both sides are defined and finite.

Suppose $(X, \mathcal{F}, \mathcal{M})$ is a finite measure space. A sequence f_n of measurable functions is **mean-square bounded** iff there is a constant K such that $\int |f_n(x)|^2 \, \mathcal{M}(dx) \leq K$ for all n.[1] This may replace the assumption of the dominating function.

(2.2.12) Proposition. *Let $(X, \mathcal{F}, \mathcal{M})$ be a finite measure space, and let f_n be a mean-square bounded sequence of measurable functions. If $f_n(x) \to f(x)$ for almost all x, then $\int f_n(x) \, \mathcal{M}(dx) \to \int f(x) \, \mathcal{M}(dx)$.*

Proof. Let K be a constant such that $\int |f_n(x)|^2 \, \mathcal{M}(dx) \leq K$ for all n. Note that since $\mathcal{M}(X) < \infty$, the functions themselves are also integrable:

$$\int |f_n(x)| \, \mathcal{M}(dx) = \int_{\{|f_n| \leq 1\}} |f_n(x)| \, \mathcal{M}(dx) + \int_{\{|f_n| > 1\}} |f_n(x)| \, \mathcal{M}(dx)$$

$$\leq \mathcal{M}(X) + \int |f_n(x)|^2 \, \mathcal{M}(dx) < \infty.$$

Now by Fatou's Lemma, $\int |f(x)|^2 \, \mathcal{M}(dx) \leq \liminf \int |f_n(x)|^2 \, \mathcal{M}(dx) \leq K$. Let $\varepsilon > 0$ be given. Choose λ so large that $K/\lambda < \varepsilon/4$. Then for any n we have

$$\int_{\{|f_n| > \lambda\}} |f_n(x)| \, \mathcal{M}(dx) \leq \int \frac{|f_n(x)|^2}{\lambda} \, \mathcal{M}(dx) \leq \frac{K}{\lambda} < \frac{\varepsilon}{4},$$

and similarly, $\int_{\{|f| > \lambda\}} |f(x)| \, \mathcal{M}(dx) < \varepsilon/4$. So also, $\lambda \mathcal{M}\{|f| > \lambda\} < \varepsilon/4$. Now define $f_n'(x) = -\lambda \vee f_n(x) \wedge \lambda$. Then $f_n' \to f$ on the set $\{|f| \leq \lambda\}$. Also, $|f_n'|$ is dominated by the constant λ, which is integrable. So by the Lebesgue Dominated Convergence Theorem 2.2.10, $\int_{\{|f| \leq \lambda\}} |f_n'(x) - f(x)| \, \mathcal{M}(dx) \to 0$.

---

[1] The "mean square" of a measurable function f is $\int |f(x)|^2 \, \mathcal{M}(dx)/\mathcal{M}(X)$. A mean-square bounded sequence is an example of a **uniformly integrable** sequence.

Choose N so large that for all $n \geq N$ we have $\int_{\{|f| \leq \lambda\}} |f_n'(x) - f(x)| \, \mathcal{M}(dx) < \varepsilon/4$. Then we have for all such n

$$\int |f_n(x) - f(x)| \, \mathcal{M}(dx) \leq \int_{\{|f| \leq \lambda\}} |f_n'(x) - f(x)| \, \mathcal{M}(dx)$$
$$+ \int_{\{|f| > \lambda\}} |f(x)| \, \mathcal{M}(dx)$$
$$+ \int_{\{|f| > \lambda\}} \lambda \, \mathcal{M}(dx)$$
$$+ \int |f_n(x) - f_n'(x)| \, \mathcal{M}(dx)$$
$$< \frac{\varepsilon}{4} + \frac{\varepsilon}{4} + \frac{\varepsilon}{4} + \int_{\{|f_n| > \lambda\}} |f_n(x)| \, \mathcal{M}(dx)$$
$$< \frac{\varepsilon}{4} + \frac{\varepsilon}{4} + \frac{\varepsilon}{4} + \frac{\varepsilon}{4} = \varepsilon.$$

This completes the proof that $\int f_n(x) \, \mathcal{M}(dx) \to \int f(x) \, \mathcal{M}(dx)$.   ☺

Iterated Integrals. A product measure may be computed as an integral.

(2.2.13) Theorem. *Let $(X_1, \mathcal{F}_1, \mathcal{M}_1)$ and $(X_2, \mathcal{F}_2, \mathcal{M}_2)$ be finite measure spaces. For $E \in \mathcal{F}_1 \otimes \mathcal{F}_2$, write*

$$E_{[x]} = \{y \in X_2 : (x, y) \in E\} \qquad \text{for } x \in X_1,$$
$$E^{[y]} = \{x \in X_1 : (x, y) \in E\} \qquad \text{for } y \in X_2$$

for cross-sections. Then:

(a) *$E_{[x]} \in \mathcal{F}_2$ and $E^{[y]} \in \mathcal{F}_1$.*
(b) *The function $x \mapsto \mathcal{M}_2(E_{[x]})$ is $\mathcal{F}_1$-measurable, and the function $y \mapsto \mathcal{M}_1(E^{[y]})$ is $\mathcal{F}_2$-measurable.*
(c) *$\int \mathcal{M}_2(E_{[x]}) \, \mathcal{M}_1(dx) = \int \mathcal{M}_1(E^{[y]}) \, \mathcal{M}_2(dy) = \mathcal{M}_1 \otimes \mathcal{M}_2(E)$.*

Proof. (a) is Proposition 2.1.2.

(b) Let $\mathcal{A}$ be the collection of all sets $E \in \mathcal{F}_1 \otimes \mathcal{F}_2$ such that the function $h_E(x) = \mathcal{M}_2(E_{[x]})$ is $\mathcal{F}_2$-measurable. Now, for a measurable rectangle $A \times B$ we have

$$h_{A \times B}(x) = \mathcal{M}_2(B) \, \mathbb{1}_A(x),$$

so $A \times B \in \mathcal{A}$. Also, $\mathcal{A}$ is a lambda system: The whole space $X_1 \times X_2$ is a measurable rectangle, so it belongs to $\mathcal{A}$. If $E \subseteq F$ and both belong to $\mathcal{A}$, then

$$h_{F \setminus E}(x) = h_F(x) - h_E(x)$$

is the difference of two measurable functions, so it is measurable. If E_n increases to E, and each h_{E_n} is measurable, then

$$h_E(x) = \lim_{n \to \infty} h_{E_n}(x)$$

is the pointwise limit of a sequence of measurable functions, so it is measurable.

Now, by the Pi–Lambda Theorem, $\mathcal{A}$ includes the σ-algebra generated by the measurable rectangles, which is $\mathcal{F}_1 \otimes \mathcal{F}_2$.

The proof for $y \mapsto \mathcal{M}_1(E^{[y]})$ is the same.

(c) Define a set-function $\mathcal{N}$ by

$$\mathcal{N}(E) = \int \mathcal{M}_2(E_{[x]}) \, \mathcal{M}_1(dx)$$

for $E \in \mathcal{F}_1 \otimes \mathcal{F}_2$. I claim that $\mathcal{N}$ is a measure. Indeed, $\mathcal{N}(\varnothing) = \int 0 \, \mathcal{M}_1(dx) = 0$; and if E_n are disjoint sets and $E = \bigcup_n E_n$, then

$$E_{[x]} = \bigcup_n (E_n)_{[x]}$$

is also a disjoint union, so we have

$$\mathcal{N}(E) = \int \mathcal{M}_2(E_{[x]}) \, \mathcal{M}_1(dx)$$

$$= \int \mathcal{M}_2 \left(\bigcup (E_n)_{[x]} \right) \mathcal{M}_1(dx) = \int \sum \mathcal{M}_2 \left((E_n)_{[x]} \right) \mathcal{M}_1(dx)$$

$$= \sum \int \mathcal{M}_2 \left((E_n)_{[x]} \right) \mathcal{M}_1(dx) = \sum \mathcal{N}(E_n).$$

Now, $\mathcal{M}_1 \otimes \mathcal{M}_2$ and $\mathcal{N}$ are two measures on $\mathcal{F}_1 \otimes F_2$, and they agree on the pi system of measurable rectangles, so (by Theorem 1.1.7) they agree on all of $\mathcal{F}_1 \otimes \mathcal{F}_2$.

The proof for $\int \mathcal{M}_1(E^{[y]}) \, \mathcal{M}_2(dy)$ is the same.   ☺

(2.2.14) Exercise. Theorem 2.2.13 remains correct for σ-finite measures $(X_1, \mathcal{F}_1, \mathcal{M}_1)$ and $(X_2, \mathcal{F}_2, \mathcal{M}_2)$.

(2.2.15) Exercise. Does Theorem 2.2.13 remain correct for non-σ-finite measures?

Integration of a function with respect to a product measure is the next topic. Such a "double integral" may be computed as an "iterated integral" in either of the two orders. The hypotheses of measurability and integrability are important.

(2.2.16) Theorem. *Let $(X_1, \mathcal{F}_1, \mathcal{M}_1)$ and $(X_2, \mathcal{F}_2, \mathcal{M}_2)$ be σ-finite measure spaces.*

(a) *(**Tonelli's Theorem**) Let $f \colon X_1 \times X_2 \to [0, \infty]$ be $\mathcal{F}_1 \otimes \mathcal{F}_2$-measurable. Then for every $x \in X_1$, the function $y \mapsto f(x, y)$ is $\mathcal{F}_2$-measurable; for every $y \in X_2$, the function $x \mapsto f(x, y)$ is $\mathcal{F}_1$-measurable the function $x \mapsto \int f(x, y) \, \mathcal{M}_2(dy)$ is $\mathcal{F}_1$-measurable; the function $y \mapsto \int f(x, y) \, \mathcal{M}_1(dx)$ is $\mathcal{F}_2$-measurable; and*

$$\int f(u)\, \mathcal{M}_1 \otimes \mathcal{M}_2(du) = \int \left(\int f(x,y)\, \mathcal{M}_2(dy) \right) \mathcal{M}_1(dx)$$

$$= \int \left(\int f(x,y)\, \mathcal{M}_1(dx) \right) \mathcal{M}_2(dy).$$

(b) **(Fubini's Theorem)** *Let* $f\colon X_1 \times X_2 \to \mathbb{R}$ *be* $\mathcal{F}_1 \otimes \mathcal{F}_2$-*measurable and* $\mathcal{M}_1 \otimes \mathcal{M}_2$-*integrable. Then for almost every* $x \in X_1$, *the function* $y \mapsto f(x,y)$ *is* $\mathcal{M}_2$-*integrable; for almost every* $y \in X_2$, *the function* $x \mapsto f(x,y)$ *is* $\mathcal{M}_1$-*integrable; the function* $x \mapsto \int f(x,y)\, \mathcal{M}_2(dy)$ *is defined almost everywhere and is* $\mathcal{M}_1$-*integrable; the function* $y \mapsto \int f(x,y)\, \mathcal{M}_1(dx)$ *is defined almost everywhere and is* $\mathcal{M}_2$-*integrable; and*

$$\int f(u)\, \mathcal{M}_1 \otimes \mathcal{M}_2(du) = \int \left(\int f(x,y)\, \mathcal{M}_2(dy) \right) \mathcal{M}_1(dx)$$

$$= \int \left(\int f(x,y)\, \mathcal{M}_1(dx) \right) \mathcal{M}_2(dy).$$

Proof. (a) If f is an indicator function $f = \mathbb{1}_E$, then the results are Theorem 2.2.13 (and Exercise 2.2.14). For nonnegative simple functions $f = \sum a_i \mathbb{1}_{A_i}$ the result follows by linearity of the integral and the Monotone Convergence Theorem. For general nonnegative measurable f, approximate by simple functions (Theorem 2.1.8) and apply the Monotone Convergence Theorem.

(b) If $f\colon X_1 \times X_2 \to \mathbb{R}$ is $\mathcal{F}_1 \otimes \mathcal{F}_2$-measurable and $\mathcal{M}_1 \otimes \mathcal{M}_2$-integrable, then so are the positive and negative parts f^+, f^-. Apply part (a) to these nonnegative functions. Since f is integrable,

$$(*) \quad \int \left(\int f^+(x,y)\, \mathcal{M}_2(dy) \right) \mathcal{M}_1(dx) = \int f^+(u)\, \mathcal{M}_1 \otimes \mathcal{M}_2(du) < \infty,$$

and thus $\int f^+(x,y)\, \mathcal{M}_2(dy) < \infty$ for almost all x. Similarly, we have

$$\int f^-(x,y)\, \mathcal{M}_2(dy) < \infty$$

for almost all x. For almost all x we may subtract, so that

$$\int f(x,y)\, \mathcal{M}_2(dy) = \int f^+(x,y)\, \mathcal{M}_2(dy) - \int f^-(x,y)\, \mathcal{M}_2(dy).$$

But again by $(*)$, the result is an integrable function of x, so we may subtract to obtain

$$\int \left(\int f(x,y)\, \mathcal{M}_2(dy) \right) \mathcal{M}_1(dx) = \int f(u)\, \mathcal{M}_1 \otimes \mathcal{M}_2(du).$$

The other half is proved in the same way.   ☺

Full and Fine Variations. Suppose a measure $\mathcal{M}$ is constructed as a variation V^C on a metric space. Then the integral of a nonnegative function may also be computed as a variation.

Let S be a metric space. Let $C: S \times (0, \infty) \to [0, \infty)$ be a constituent function, and let $h: S \to [0, \infty)$ be a point function. Then we may define a new constituent function hC by

$$hC(x, r) = h(x)C(x, r).$$

Note that we have not allowed h to take the value ∞ (because ∞ is not allowed as a value for a constituent function).

(2.2.17) Proposition. *Let* $h: S \to [0, \infty)$ *be a Borel function. Let the measure* $\mathcal{M}$ *be the restriction of a full variation* V^C *to its measurable sets. Then*

$$\int_E h(x) \, \mathcal{M}(dx) = V^{hC}(E)$$

for all Borel sets E.

Proof. Since $1C = C$, the equation is true for the constant function 1—that is, $V^{1C}(E) = V^C(E) = \mathcal{M}(E)$ for all Borel sets E. It is also clear from the definitions that $V^{ahC}(E) = aV^{hC}(E)$ for a nonnegative constant a, so we have $V^{aC}(E) = a\mathcal{M}(E)$.

Now, V^{hC} is a metric outer measure, so all Borel sets are measurable. If $f = \sum a_n \mathbb{1}_{A_n}$, where the A_n are disjoint Borel sets and the a_n are nonnegative constants, then

$$V^{fC}(E) = \sum V^{fC}(E \cap A_n) = \sum V^{a_n C}(E \cap A_n)$$

$$= \sum a_n \mathcal{M}(E \cap A_n) = \int_E f(x) \, \mathcal{M}(dx).$$

So the conclusion is true for Borel simple functions.

Finally, let h be a nonnegative Borel function. There is a sequence f_n of nonnegative Borel simple functions that increases to h. If $c < 1$, then the sets

$$E_n = \{ x \in E : f_n(x) \geq ch(x) \}$$

increase to E. But $V^{f_n C}(E) \geq cV^{hC}(E_n)$, so $\lim_n V^{f_n C}(E) \geq cV^{hC}(E)$. Let $c \to 1$ to conclude that $V^{f_n C}(E) \to V^{hC}(E)$. Therefore, we have

$$V^{hC}(E) = \lim_n V^{f_n C}(E) = \lim_n \int_E f_n(x) \, \mathcal{M}(dx) = \int_E h(x) \, \mathcal{M}(dx),$$

as required.   ☺

A similar result is true for fine variation integrals. The proof is left to the reader.

(2.2.18) Exercise. Let $h\colon S \to [0, \infty)$ be a Borel function. Let the measure $\mathcal{M}$ be the restriction of a fine variation v^C to its measurable sets. Then

$$\int_E h(x)\, \mathcal{M}(dx) = v^{hC}(E)$$

for all Borel sets E.

An interesting example is the one-dimensional Lebesgue measure $\overline{\mathcal{L}}$. For the set-function $C(x, r) = 2r$, we get $\overline{\mathcal{L}} = V^C = v^C$. Therefore, for any nonnegative Borel function h we have

$$\int_E h(x)\, dx = v^{hC}(E) = V^{hC}(E).$$

The definition (in terms of gauges) can be used to define the so-called Henstock, or Kurtzweil, integral, which is more general than the Lebesgue integral [113, 18].

These results apply to some of the fractal measures of particular interest here. Let h be a nonnegative Borel function on a metric space S, and let $s > 0$. Then the "packing measure integral"

$$\int_E h(x)\, \mathcal{P}^s(dx)$$

may be computed as the full variation of the constituent function

$$C(x, r) = h(x)\, (2r)^s$$

on the set E. And the "covering measure integral"

$$\int_E h(x)\, \mathcal{C}^s(dx)$$

may be computed as the fine variation of the same constituent function C on the set E.

(2.2.19) Proposition. *Let S be a metric space, and let $\mathcal{N}$ be a finite Borel measure on S with the strong Vitali property; let h be a nonnegative Borel function on S, and let a constituent function C be defined by $C(x, r) = h(x)\mathcal{N}(\overline{B}_r(x))$. Then $v^C(F) = V^C(F) = \int_F h(x)\, \mathcal{N}(dx)$.*

Proof. Combine Proposition 2.2.17 with Proposition 1.3.18.   ☺

2.3 Radon–Nikodým Theorem

Let $(X, \mathcal{F}, \mathcal{N})$ be a measure space. Let $f\colon X \to [0, \infty]$ be a measurable function. Then $\mathcal{M}(E) = \int_E f(x)\, \mathcal{N}(dx)$ defines a new measure on the σ-algebra

$\mathcal{F}$ called the **indefinite integral** of f. Can we recognize when a measure $\mathcal{M}$ has this form? One important property of an indefinite integral $\mathcal{M}$ is that it is **absolutely continuous** with respect to $\mathcal{N}$—that is, if $E \in \mathcal{F}$ and $\mathcal{N}(E) = 0$, then also $\mathcal{M}(E) = 0$. We write $\mathcal{M} \ll \mathcal{N}$ for this condition.

(2.3.1) Exercise. Suppose $(X, \mathcal{F}, \mathcal{N})$ is a finite measure space, and let $\mathcal{M}$ be another finite measure on $\mathcal{F}$. Consider these conditions:

(a) $\mathcal{M} \ll \mathcal{N}$—that is, if $\mathcal{N}(E) = 0$, then $\mathcal{M}(E) = 0$.
(b) For every $\varepsilon > 0$, there is $\delta > 0$ such that for any $E \in \mathcal{F}$, if $\mathcal{N}(E) < \delta$, then $\mathcal{M}(E) < \varepsilon$.
(c) If E_n is a sequence of measurable sets, and $\mathcal{N}(E_n) \to 0$, then $\mathcal{M}(E_n) \to 0$.
(d) If E_n is a decreasing sequence of measurable sets, and $\mathcal{N}(E_n) \to 0$, then $\mathcal{M}(E_n) \to 0$.

Determine the implications among the conditions.

(2.3.2) Exercise. Repeat the previous exercise without the assumption that the measures are finite.

(2.3.3) Exercise. Let $\mathcal{M}(E) = \int_E f(x)\,\mathcal{N}(dx)$ for all $E \in \mathcal{F}$. For what functions g may we conclude that

$$\int g(x)\,\mathcal{M}(dx) = \int g(x)f(x)\,\mathcal{N}(dx)?$$

The next result asserts that (under the right conditions) absolute continuity is both necessary and sufficient for a measure $\mathcal{M}$ to be obtained as an indefinite integral.

(2.3.4) Radon–Nikodým Theorem. *Let $(X, \mathcal{F}, \mathcal{N})$ be a finite measure space. Let $\mathcal{M}$ be a finite measure on $\mathcal{F}$. Then the following are equivalent:*

(a) *$\mathcal{M} \ll \mathcal{N}$—that is, for any $E \in \mathcal{F}$, if $\mathcal{N}(E) = 0$, then also $\mathcal{M}(E) = 0$.*
(b) *$\mathcal{M}$ is an indefinite integral—that is, there is a nonnegative integrable function $f \colon X \to \mathbb{R}$ such that $\mathcal{M}(E) = \int_E f(x)\,\mathcal{N}(dx)$ for all $E \in \mathcal{F}$.*

Just as $\mathcal{M}$ is called the indefinite integral of f, we will say that f is the **Radon–Nikodým derivative** of $\mathcal{M}$ with respect to $\mathcal{N}$. When $\mathcal{N}$ is d-dimensional Lebesgue measure in $\mathbb{R}^d$, we will simply say that f is the Radon–Nikodým derivative of $\mathcal{M}$, or sometimes the **density** for $\mathcal{M}$.

(2.3.5) Proposition. *The Radon–Nikodým derivative is unique almost everywhere in the following sense. Let $(X, \mathcal{F}, \mathcal{N})$ and $\mathcal{M}$ be as in Theorem 2.3.4. Suppose f and g are both measurable functions, and $\mathcal{M}(E) = \int_E f(x)\,\mathcal{N}(dx) = \int_E g(x)\,\mathcal{N}(dx)$ for all $E \in \mathcal{F}$. Then $f(x) = g(x)$ for $\mathcal{N}$-almost all x.*

Proof. Let $E = \{x \in X : f(x) < g(x)\}$. Then we have $\int_E f(x)\,\mathcal{N}(dx) < \int_E g(x)\,\mathcal{N}(dx)$ unless $\mathcal{N}(E) = 0$. But both of the integrals are equal to $\mathcal{M}(E)$, and therefore $\mathcal{N}(E) = 0$. This shows that $f(x) \geq g(x)$ almost everywhere. Similarly, $g(x) \geq f(x)$ almost everywhere.   ☺

(2.3.6) Exercise. Use Theorem 2.3.4 to show that the Radon–Nikodým theorem remains true for σ-finite measures:

Let $(X, \mathcal{F}, \mathcal{N})$ be a σ-finite measure space. Let $\mathcal{M}$ be a σ-finite measure on $\mathcal{F}$. Then the following are equivalent:

(a) For any $E \in \mathcal{F}$, if $\mathcal{N}(E) = 0$, then also $\mathcal{M}(E) = 0$.
(b) There is a nonnegative measurable function $f\colon X \to \mathbb{R}$ such that $\mathcal{M}(E) = \int_E f(x)\,\mathcal{N}(dx)$ for all $E \in \mathcal{F}$.

(2.3.7) Exercise. Investigate two Hausdorff measures $\mathcal{H}^s$, $\mathcal{H}^t$ in the line $\mathbb{R}$. Show that if $s < t$, then $\mathcal{H}^t \ll \mathcal{H}^s$. Use this to show that the Radon–Nikodým theorem may fail for non-σ-finite measures.

Metric Densities. We do not include here a proof of the abstract form of the Radon–Nikodým theorem. It may be found in many texts on measure and integration, such as [48, 102, 111, 124, 135, 172, 200, 229, 230].[2] However, for measures on a metric space, the theorem may be proved in a geometric way in some cases. This will be explained next.

Let S be a metric space, and let $\mathcal{M}$, $\mathcal{N}$ be finite Borel measures on S. Consider the upper and lower densities of $\mathcal{M}$ with respect to $\mathcal{N}$:

$$\overline{D}(x) = \limsup_{r \to 0} \frac{\mathcal{M}(\overline{B}_r(x))}{\mathcal{N}(\overline{B}_r(x))}, \qquad \underline{D}(x) = \liminf_{r \to 0} \frac{\mathcal{M}(\overline{B}_r(x))}{\mathcal{N}(\overline{B}_r(x))}.$$

(These definitions make sense for x in the support of $\mathcal{N}$, so that the denominators are never zero. Write $\overline{D}(x) = \underline{D}(x) = 0$ by convention outside the support.) At a point x where $\overline{D}(x) = \underline{D}(x)$, we write $D(x) = \overline{D}(x)$ and call it the **density** of $\mathcal{M}$ with respect to $\mathcal{N}$.

(2.3.8) Exercise. Let $\mathcal{M}$ and $\mathcal{N}$ be finite Borel measures. Show that the upper density $\overline{D}(x)$ and the lower density $\underline{D}(x)$ are Borel functions of x.

Under the right conditions, the density $D(x)$ is a version of the Radon–Nikodým derivative of $\mathcal{M}$ with respect to $\mathcal{N}$. Recall that the strong Vitali property holds for many Borel measures, including all measures in Euclidean space.

(2.3.9) Proposition. *Let S be a metric space, let $\mathcal{M}$ and $\mathcal{N}$ be finite Borel measures on S. Let $E \subseteq S$ be a Borel set. Then,*

---

[2] It may even be proved using martingale theory [13, Theorem 1.3.2].

(a) $v^{hN}(E) \leq M(E)$ *for any nonnegative finite Borel function* $h \leq \overline{D}$.
(b) $V^{hN}(E) \leq M(E)$ *for any nonnegative finite Borel function* $h \leq \underline{D}$.

Assume also that N *has the strong Vitali property. Then*

(c) $\int_E \overline{D}(x)\,N(dx) \leq M(E)$.
(d) $\int_E \underline{D}(x)\,N(dx) \leq M(E)$.

Assume also that $M \ll N$. *Then*

(e) $\overline{D} = \underline{D}$ *a.e. and* $\int_E \overline{D}(x)\,N(dx) = M(E)$.

Proof. (a) Let $\varepsilon > 0$ be given. Let $k(x) = (h(x) - \varepsilon) \vee 0$, so that k is a non-negative Borel function, $k(x) \leq h(x)$ everywhere, and $k(x) < \overline{D}(x)$ wherever $\overline{D}(x) > 0$. Then

$$\beta = \left\{ (x,r) : x \in E, r > 0, \frac{M(\overline{B}_r(x))}{N(\overline{B}_r(x))} \geq k(x) \right\}$$

is a fine cover of E. Let $U \supseteq E$ be an open set. Now,

$$\beta' = \left\{ (x,r) \in \beta : \overline{B}_r(x) \subseteq U \right\}$$

is again a fine cover of E. But if $\pi \subseteq \beta'$ is any packing, we have

$$\sum_{(x,r)\in\pi} h(x)N(\overline{B}_r(x)) \leq \varepsilon \sum_{(x,r)\in\pi} N(\overline{B}_r(x)) + \sum_{(x,r)\in\pi} k(x)N(\overline{B}_r(x))$$

$$\leq \varepsilon N(U) + \sum M(\overline{B}_r(x)) \leq \varepsilon N(U) + M(U).$$

Take the supremum over π to conclude that $v^{hN}_{\beta'} \leq \varepsilon N(U) + M(U)$. Therefore, $v^{hN}(E) \leq \varepsilon N(U) + M(U)$. Take the infimum over ε and U to conclude that $v^{hN}(E) \leq M(E)$.

(b) Let $c < 1$, and let $U \supseteq E$ be an open set. Define a gauge Δ on E such that $\overline{B}_{\Delta(x)}(x) \subseteq U$ for all $x \in E$ and

$$\frac{M(\overline{B}_r(x))}{N(\overline{B}_r(x))} \geq ch(x)$$

for all $x \in E$ and all $r < \Delta(x)$. Now, if π is a Δ-fine packing of E, then

$$\sum_{\pi} h(x)\,N(\overline{B}_r(x)) \leq \frac{1}{c} \sum_{\pi} M(\overline{B}_r(x)) \leq \frac{1}{c}M(U).$$

Take the supremum over π to conclude that $V^{hN}_{\Delta}(E) \leq (1/c)M(U)$. Therefore, $V^{hN}(E) \leq (1/c)M(U)$. Take the infimum over c and U to conclude that $V^{hN}(E) \leq M(E)$.

(c) Because of the strong Vitali property, $N(E) = v^N(E)$ for all Borel sets E (1.3.18). Note also that any measure of the form $\int_E f(x)\,N(dx)$ also has the strong Vitali property, where f satisfies $0 \leq f < \infty$; so also $\int_E f(x) \, N(dx) = v^{fN}(E)$.

I claim first that $N(E_\infty) = 0$, where $E_\infty = \{ x \in E : \overline{D}(x) = \infty \}$. Suppose not. Let $R = (M(E_\infty) + 2)/N(E_\infty)$. There is an open set $U \supseteq E_\infty$ such that $M(U) \leq M(E_\infty) + 1$. Now,

$$\beta = \left\{ (x, r) : x \in E_\infty, r > 0, \frac{M(\overline{B}_r(x))}{N(\overline{B}_r(x))} > R, \overline{B}_r(x) \subseteq U \right\}$$

is a fine cover of E_∞. For any packing $\pi \subseteq \beta$,

$$M(U) \geq \sum_\pi M(\overline{B}_r(x)) \geq R \sum_\pi N(\overline{B}_r(x)).$$

Take the supremum over π to conclude that $M(U) \geq R v_\beta^N$. So we have

$$M(E_\infty) + 1 \geq M(U) \geq R v_\beta^N \geq R v^N(E_\infty) = M(E_\infty) + 2,$$

a contradiction. Therefore, $\overline{D}(x) < \infty$ for N-almost all $x \in E$.

Now we may apply part (a) using the set $E_1 = E \setminus E_\infty$ as follows: For any finite-valued Borel function $h \leq \overline{D}$, we have

$$\int_E h(x) \, N(dx) = \int_{E_1} h(x) \, N(dx) = v^{hN}(E_1) \leq M(E_1) \leq M(E).$$

Take the supremum over all such h to conclude that $\int_E \overline{D}(x) \, N(dx) \leq M(E)$.

(d) This is a consequence of (c). Or it may be deduced from (b) in the same way as (c) was deduced from (a).

(e) Again we have $N(E_\infty) = 0$, where $E_\infty = \{ x \in E : \overline{D}(x) = \infty \}$. But now, since $M \ll N$, this also means that $M(E_\infty) = 0$. Write $E_1 = E \setminus E_\infty$. Now, $N(E_1) < \infty$, so $V^N(E_1) < \infty$. Let Δ be a gauge such that $V_\Delta^N(E_1) < \infty$. Let $\varepsilon > 0$. Then

$$\beta = \left\{ (x, r) : x \in E_1, 0 < r < \Delta(x), \frac{M(\overline{B}_r(x))}{N(\overline{B}_r(x))} \leq \underline{D}(x) + \varepsilon \right\}$$

is a fine cover of E_1. So by the strong Vitali property, there is a packing $\pi \subseteq \beta$ with

$$N\left(E_1 \setminus \bigcup_\pi \overline{B}_r(x)\right) = 0,$$

so also

$$M\left(E_1 \setminus \bigcup_\pi \overline{B}_r(x)\right) = 0.$$

Then

$$M(E_1) \leq \sum_\pi M(\overline{B}_r(x)) \leq \sum (\underline{D}(x) + \varepsilon) \, N(\overline{B}_r(x))$$

$$= \sum \underline{D}(x) N(\overline{B}_r(x)) + \varepsilon \sum N(\overline{B}_r(x)) \leq V_\Delta^{DN}(E_1) + \varepsilon V_\Delta^N(E_1).$$

Take the infimum over ε to obtain $\mathcal{M}(E_1) \leq V_{\Delta}^{DN}(E_1)$. Then take the infimum over Δ to obtain $\mathcal{M}(E_1) \leq V^{DN}(E_1) = \int_{E_1} \underline{D}(x)\,\mathcal{N}(dx)$. Then add the null set E_∞ to obtain $\mathcal{M}(E) = \mathcal{M}(E_1) \leq \int_{E_1} \underline{D}(x)\,\mathcal{N}(dx) = \int_E \underline{D}(x)\,\mathcal{N}(dx)$.

Now we combine the inequalities:

$$\mathcal{M}(E) \leq \int_E \underline{D}(x)\,\mathcal{N}(dx) \leq \int_E \overline{D}(x)\,\mathcal{N}(dx) \leq \mathcal{M}(E).$$

So there is equality everywhere. In particular, $\underline{D}(x) = \overline{D}(x)$ for $\mathcal{N}$-almost all x, and $\int_E \overline{D}(x)\,\mathcal{N}(dx) = \mathcal{M}(E)$ as claimed.   ☺

The ideas used above will help you prove the following refinements of Theorems 1.5.11 and 1.5.13. Use the densities

$$\overline{D}_{\mathcal{M}}^s(x) = \limsup_{r\to 0} \frac{\mathcal{M}(\overline{B}_r(x))}{(2r)^s}, \qquad \underline{D}_{\mathcal{M}}^s(x) = \liminf_{r\to 0} \frac{\mathcal{M}(\overline{B}_r(x))}{(2r)^s}.$$

The upper density corresponds to the covering measure, and the lower density to the packing measure.

(2.3.10) Exercise. Let $\mathcal{M}$ be a finite Borel measure on the separable metric space S and let $E \subseteq S$ be a Borel set. Then

$$\mathcal{M}(E) \geq \int_E \overline{D}_{\mathcal{M}}^s(x)\,\mathcal{C}^s(dx).$$

If, in addition, $\mathcal{C}^s(E) < \infty$, and $\overline{D}_{\mathcal{M}}^s(x) < \infty$ on E, then

$$\mathcal{M}(E) = \int_E \overline{D}_{\mathcal{M}}^s(x)\,\mathcal{C}^s(dx).$$

(2.3.11) Exercise. Let $\mathcal{M}$ be a finite Borel measure on the metric space S, and let $E \subseteq S$ be a Borel set. Then

$$\mathcal{M}(E) \geq \int_E \underline{D}_{\mathcal{M}}^s(x)\,\mathcal{P}^s(dx).$$

If, in addition, $\mathcal{M}$ has the strong Vitali property, $\mathcal{P}^s(E) < \infty$, and $\underline{D}_{\mathcal{M}}^s(x) < \infty$ on E, then

$$\mathcal{M}(E) = \int_E \underline{D}_{\mathcal{M}}^s(x)\,\mathcal{P}^s(dx).$$

2.4 Measures as Linear Functionals

Integration with respect to a fixed measure is "linear" in a sense to be explored next. Here "linear" means that

$$\int \big(f(x) + g(x)\big)\, \mathcal{M}(dx) = \int f(x)\, \mathcal{M}(dx) + \int g(x)\, \mathcal{M}(dx),$$

$$\int t f(x)\, \mathcal{M}(dx) = t \int f(x)\, \mathcal{M}(dx).$$

Let's state this more precisely.

Linearity Conditions. Given an arbitrary measure and an arbitrary measurable function, the integral may or may not exist, and if it exists it may be finite or infinite.

(2.4.1) Definition. Let X be a set. The collection $\mathfrak{F}(X, \mathbb{R})$ of all real-valued functions defined on X is a "linear space," or "vector space," under the natural (pointwise) operations:

(a) The **zero function** is the function $z \colon X \to \mathbb{R}$ defined by $z(x) = 0$ for all $x \in X$; normally we write 0 for this function.

(b) If $f \colon X \to \mathbb{R}$ and $g \colon X \to \mathbb{R}$ are two functions, then their **sum** is the function $k \colon X \to \mathbb{R}$ defined by $k(x) = f(x) + g(x)$ for all $x \in X$; normally we write $f + g$ for this function.

(c) If $f \colon X \to \mathbb{R}$ is a function and $t \in \mathbb{R}$ is a scalar, then the **scalar multiple** of f by t is the function $h \colon X \to \mathbb{R}$ defined by $h(x) = t f(x)$ for all $x \in X$; normally we write $t f$ for this function.

(2.4.2) Definition. A subset $\mathfrak{U} \subseteq \mathfrak{F}(X, \mathbb{R})$ of this set is called a **linear space** iff it is closed under the operations described above; that is,

(a) $0 \in \mathfrak{U}$.

(b) if $f, g \in \mathfrak{U}$, then $f + g \in \mathfrak{U}$.

(c) if $f \in \mathfrak{U}$ and $t \in \mathbb{R}$, then $t f \in \mathfrak{U}$.

(2.4.3) Definition. A **linear functional** is a function $\Phi \colon \mathfrak{U} \to \mathbb{R}$ where the domain $\mathfrak{U}$ is a linear space and

(a) $\Phi(0) = 0$.

(b) $\Phi(f + g) = \Phi(f) + \Phi(g)$ for all $f, g \in \mathfrak{U}$.

(c) $\Phi(t f) = t \Phi(f)$ for all $f \in \mathfrak{U}$ and all $t \in \mathbb{R}$.

(2.4.4) Exercise. Can (a) be omitted in any of these definitions? In (2.4.3), can all three formulas be replaced by the single one $\Phi(f + tg) = \Phi(f) + t\Phi(g)$?

(2.4.5) Exercise. Let $(X, \mathfrak{F}, \mathcal{M})$ be a measure space. A function $f \colon X \to \mathbb{R}$ is called $\mathcal{M}$-**integrable** iff f is measurable and $\int |f(x)|\, \mathcal{M}(dx) < \infty$. Show that the set of all $\mathcal{M}$-integrable functions is a linear space, and

$$\Phi(f) = \int f(x)\, \mathcal{M}(dx)$$

defines a linear functional Φ on it. The functional Φ defined in this way will be called "integration with respect to $\mathcal{M}$."

(2.4.6) Exercise. Let S be a compact metric space, and let $\mathcal{M}$ be a finite Borel measure on S. Show that the collection $\mathcal{C}(S, \mathbb{R})$ of all continuous functions $S \to \mathbb{R}$ is a linear space, and integration with respect to $\mathcal{M}$ defines a linear functional on it.

(2.4.7) Exercise. Let S be a metric space, and let $\mathcal{M}$ be a finite Borel measure on S. Show that the collection $\mathcal{C}_b(S, \mathbb{R})$ of all bounded continuous functions $S \to \mathbb{R}$ is a linear space, and integration with respect to $\mathcal{M}$ defines a linear functional on it.

Riesz Representation Theorem. Next we want to study linear functionals on certain function spaces and try to recognize when a functional is integration with respect to a measure. This is a large topic, and we will only consider a few simple results here.

A function $f \colon X \to \mathbb{R}$ is called **nonnegative** iff $f(x) \geq 0$ for all $x \in X$. A linear functional Φ defined on a linear space $\mathfrak{U}$ is called **positive** iff $\Phi(f) \geq 0$ for all nonnegative functions $f \in \mathfrak{U}$.

Certainly, integration with respect to a (positive) measure defines a positive linear functional.

A sequence f_n of real-valued functions on $\mathbb{R}$ is said to **decrease to zero** iff $f_1(x) \geq f_2(x) \geq \cdots$ for all $x \in X$ and $\lim_{n \to \infty} f_n(x) = 0$ for all $x \in X$. A linear functional Φ defined on a linear space $\mathfrak{U}$ is called σ-**smooth** iff $\Phi(f_n) \to 0$ for every sequence $f_n \in \mathfrak{U}$ that decreases to zero.

Certainly, integration with respect to a measure defines a σ-smooth linear functional.

(2.4.8) Exercise. Give an example of a linear functional that is not σ-smooth.

The next proof uses the Stone method of extension.

(2.4.9) Riesz Representation Theorem. *Let S be a metric space, and let Φ be a positive σ-smooth linear functional on $\mathcal{C}_b(S, \mathbb{R})$. Then there is a unique finite Borel measure $\mathcal{M}$ on S such that*

$$\Phi(f) = \int f(x)\, \mathcal{M}(dx) \qquad \text{for all } f \in \mathcal{C}_b(S, \mathbb{R}).$$

Proof. (a) We begin by defining a value for open sets. If $U \subseteq S$ is open, define

$$\overline{\mathcal{M}}(U) = \sup \left\{ \Phi(f) : f \in \mathcal{C}_b(S, \mathbb{R}), 0 \leq f \leq \mathbb{1}_U \right\}.$$

Note that since Φ is positive, if $f \leq g$, then $\Phi(f) \leq \Phi(f) + \Phi(g - f) = \Phi(g)$. Thus we can see that the set-function $\overline{\mathcal{M}}$ is **monotone** in the sense that if

$U \subseteq V$ are open sets, then $\overline{\mathcal{M}}(U) \leq \overline{\mathcal{M}}(V)$. Also note that for all U, $0 \leq \overline{\mathcal{M}}(U) \leq \Phi(\mathbb{1}_U)$, so all values are finite.

(b) Now we claim that $\overline{\mathcal{M}}$ is subadditive on open sets; that is, if U and V are open, then $\overline{\mathcal{M}}(U \cup V) \leq \overline{\mathcal{M}}(U) + \overline{\mathcal{M}}(V)$. Indeed, given a continuous function f with $0 \leq f \leq \mathbb{1}_{U \cup V}$, the functions g, h defined by

$$g(x) = \begin{cases} \dfrac{\text{dist}(x, S \setminus U)}{\text{dist}(x, S \setminus U) + \text{dist}(x, S \setminus V)} \, f(x), & \text{if } x \in U, \\[2mm] 0, & \text{otherwise}, \end{cases}$$

$$h(x) = \begin{cases} \dfrac{\text{dist}(x, S \setminus V)}{\text{dist}(x, S \setminus U) + \text{dist}(x, S \setminus V)} \, f(x), & \text{if } x \in V, \\[2mm] 0, & \text{otherwise} \end{cases}$$

are continuous, $f = g + h$, $0 \leq g \leq \mathbb{1}_U$, and $0 \leq h \leq \mathbb{1}_V$. So $\Phi(f) = \Phi(g) + \Phi(h) \leq \overline{\mathcal{M}}(U) + \overline{\mathcal{M}}(V)$. This is true for all such f, so $\overline{\mathcal{M}}(U \cup V) \leq \overline{\mathcal{M}}(U) + \overline{\mathcal{M}}(V)$.

Note that if U and V are disjoint open sets, then $\overline{\mathcal{M}}(U \cup V) = \overline{\mathcal{M}}(U) + \overline{\mathcal{M}}(V)$, since for any continuous functions g, h with $0 \leq g \leq \mathbb{1}_U$, $0 \leq h \leq \mathbb{1}_V$, the sum f satisfies $0 \leq f \leq \mathbb{1}_{U \cup V}$.

(c) Next, we show that $\overline{\mathcal{M}}$ is countably subadditive on open sets; that is, if $U_1, U_2, \cdots$ are open, then

$$\overline{\mathcal{M}}\left(\bigcup_{k=1}^{\infty} U_k \right) \leq \sum_{k=1}^{\infty} \overline{\mathcal{M}}(U_k).$$

This will follow from (b) and the following "increasing limit" property: If $U_1 \subseteq U_2 \subseteq \cdots$ are open, then

$$\overline{\mathcal{M}}\left(\bigcup_{k=1}^{\infty} U_k \right) = \lim_{k \to \infty} \overline{\mathcal{M}}(U_k).$$

(The inequality $\geq$ and existence of the limit are clear from monotonicity (a).) Write $U = \bigcup U_k$. Let f be continuous and $0 \leq f \leq \mathbb{1}_U$. Then the functions f_k defined by

$$f_k(x) = f(x) \cdot \left(1 \wedge k \, \text{dist}(x, S \setminus U_k) \right)$$

are continuous, $0 \leq f_k \leq \mathbb{1}_{U_k}$, and $f_k(x) \to f(x)$ for all x. Now, $f - f_k$ decreases to zero, so $\Phi(f_k) = \Phi(f) - \Phi(f - f_k) \to \Phi(f)$. Thus $\Phi(f) \leq \lim_k \overline{\mathcal{M}}(U_k)$. This is true for all such f, so $\overline{\mathcal{M}}(U) \leq \lim \overline{\mathcal{M}}(U_k)$.

(d) Next we extend the definition to arbitrary subsets. Let $E \subseteq S$. Define

$$\overline{\mathcal{M}}(E) = \inf \left\{ \overline{\mathcal{M}}(U) : U \text{ is open and } U \supseteq E \right\}.$$

(Note that if E itself is open, then this definition yields the previously defined value of $\overline{\mathcal{M}}(E)$; we have kept the same notation.) Then $\overline{\mathcal{M}}$ is an outer measure on S: monotonicity follows from the monotonicity for open sets in (a), and countable subadditivity follows from (c).

In fact, $\overline{\mathcal{M}}$ is a metric outer measure: Suppose sets E, F have positive separation, that is, $\text{dist}(E, F) > 0$. Let W be any open set containing $E \cup F$. Define

$$U = \{\, x \in W : \text{dist}(x, E) < \text{dist}(x, F) \,\},$$
$$V = \{\, x \in W : \text{dist}(x, E) > \text{dist}(x, F) \,\}.$$

Then U and V are disjoint open sets satisfying $U \supseteq E$, $V \supseteq F$, and $U \cup V \subseteq W$. Thus $\overline{\mathcal{M}}(W) \geq \overline{\mathcal{M}}(U \cup V) = \overline{\mathcal{M}}(U) + \overline{\mathcal{M}}(V) \geq \overline{\mathcal{M}}(E) + \overline{\mathcal{M}}(F)$. This is true for all such W, so $\overline{\mathcal{M}}(E \cup F) \geq \overline{\mathcal{M}}(E) + \overline{\mathcal{M}}(F)$. Together with subadditivity, we get $\overline{\mathcal{M}}(E \cup F) = \overline{\mathcal{M}}(E) + \overline{\mathcal{M}}(F)$.

(e) Let $\mathcal{M}$ be the finite Borel measure obtained by restricting $\overline{\mathcal{M}}$ to the Borel sets. If f is continuous and $0 \leq f \leq 1$, then $\{f > 0\}$ is open and $0 \leq f \leq \mathbb{1}_{\{f > 0\}}$, so clearly $\Phi(f) \leq \mathcal{M}\{f > 0\}$. Applying this to the continuous function $1 - f$, we get $\Phi(f) \geq \mathcal{M}\{f = 1\}$.

(f) Now we must prove the integral property: $\int f(x)\mathcal{M}(dx) = \Phi(f)$. By the linearity of both sides, it is enough to prove this for continuous functions f with values between 0 and 1.

So let $f: S \to [0, 1]$ be continuous. Fix a positive integer n. Then $f = (1/n)\sum_{i=0}^{n-1} f_i$, where

$$f_i(x) = \begin{cases} 1, & \text{if } f(x) > \dfrac{i+1}{n}, \\[2mm] n\left(f(x) - \dfrac{i}{n}\right), & \text{if } \dfrac{i}{n} \leq f(x) \leq \dfrac{i+1}{n}, \\[2mm] 0, & \text{if } f(x) < \dfrac{i}{n}. \end{cases}$$

Each function f_i has values between 0 and 1, so we may apply the inequalities in (e). Thus

$$\Phi(f) = \frac{1}{n}\sum_{i=0}^{n-1} \Phi(f_i) \geq \frac{1}{n}\sum_{i=0}^{n-1} \mathcal{M}\left\{f \geq \frac{i+1}{n}\right\} = \sum_{k=0}^{n} \frac{k}{n}\mathcal{M}\left\{\frac{k}{n} \leq f < \frac{k+1}{n}\right\},$$

and similarly,

$$\Phi(f) \leq \sum_{k=0}^{n-1} \frac{k+1}{n}\mathcal{M}\left\{\frac{k}{n} < f \leq \frac{k+1}{n}\right\}.$$

As $n \to \infty$, both the upper and lower estimates converge to $\int f(x)\,\mathcal{M}(dx)$. So $\int f(x)\mathcal{M}(dx) = \Phi(f)$.

(g) Lastly, consider the uniqueness of $\mathcal{M}$. Suppose $\mathcal{N}$ is a Borel measure with $\mathcal{M}(E) \neq \mathcal{N}(E)$ for some Borel set E. Possibly replacing E by its complement, we may assume that $\mathcal{M}(E) > \mathcal{N}(E)$. Now (by Theorem 1.1.12), there is an open set $U \supseteq E$ with $\mathcal{N}(U)$ as close as we like to $\mathcal{N}(E)$; so in particular, there is such an open set with $\mathcal{N}(U) < \mathcal{M}(E)$. By our definition of $\mathcal{M}$, there is a continuous function f with $0 \leq f \leq \mathbb{1}_U$ and $\Phi(f) > \mathcal{N}(U)$. But then $\int f(x)\,\mathcal{N}(dx) \leq \mathcal{N}(U) < \Phi(f)$, so $\int f(x)\mathcal{N}(dx) \neq \Phi(f)$. ☺

When S is compact, the σ-smoothness assumption may be omitted. For this, we will use

(2.4.10) Dini's Theorem. *Let S be a compact space. Let $f_n \colon S \to \mathbb{R}$ be a sequence of continuous functions that decreases to zero. Then f_n converges to 0 uniformly on S.*

Proof. Let $\varepsilon > 0$ be given. For $n \in \mathbb{N}$, define

$$G_n = \{\, x \in S : f_n(x) < \varepsilon \,\}.$$

These sets are open, since the f_n are continuous; and $S = \bigcup_{n=1}^{\infty} G_n$, since f_n decreases to zero. Thus, by the compactness of S, there is a finite subcover:

$$S = \bigcup_{n=1}^{m} G_n.$$

Now, the sets G_n increase as n increases, so this means that $S = G_m$. So $f_m(x) < \varepsilon$ for all $x \in S$. And thus, for any $n \geq m$, also $f_n(x) < \varepsilon$ for all $x \in S$. This shows that f_n converges uniformly to 0.   ☺

(2.4.11) Theorem. *Let S be a compact metric space, and let Φ be a positive linear functional on $\mathfrak{C}(S, \mathbb{R})$. Then there is a unique finite Borel measure $\mathcal{M}$ on S such that*

$$\Phi(f) = \int f(x)\,\mathcal{M}(dx) \qquad \text{for all } f \in \mathfrak{C}(S, \mathbb{R}).$$

Proof. Suppose f_n decreases to zero. Then by Dini's Theorem, f_n converges uniformly to zero. Now, if $f_n(x) \leq \varepsilon$ for all $x \in S$, then $f_n \leq \varepsilon \mathbb{1}$, so

$$\Phi(f_n) \leq \varepsilon \Phi(\mathbb{1}).$$

Thus $\Phi(f_n) \to 0$. This shows that Φ is σ-smooth, so Theorem 2.4.9 may be applied.   ☺

Bounded Lipschitz Functions. Let S and T be metric spaces.[3] A function $f \colon S \to T$ is called a **Lipschitz** function[4] iff there is a constant M such that

$$\rho\big(f(x), f(y)\big) \leq M\,\rho(x, y)$$

for all $x, y \in S$. The least such constant M is the **Lipschitz constant** for f. An important linear space of functions for us will be the space $\mathrm{Lip}_b(S, \mathbb{R})$ of real-valued bounded Lipschitz functions on S.

There is a representation theorem for linear functionals on this space that is exactly like the one for bounded continuous functions.

---

[3] We write ρ for both of the metrics.
[4] "Function of bounded increase" in [MTFG].

(2.4.12) Theorem. *Let S be a metric space, and let Φ be a positive σ-smooth linear functional on $\mathrm{Lip}_b(S, \mathbb{R})$. Then there is a unique finite Borel measure $\mathcal{M}$ on S such that*

$$\Phi(f) = \int f(x)\, \mathcal{M}(dx) \qquad \text{for all } f \in \mathrm{Lip}_b(S, \mathbb{R}).$$

The proof follows the proof of Theorem 2.4.9 above, with a few appropriate changes. Replace $\mathfrak{C}_b(S, \mathbb{R})$ with $\mathrm{Lip}_b(S, \mathbb{R})$ everywhere. The next few exercises contain the information for these changes.

(2.4.13) Exercise. The product of two bounded Lipschitz functions is a bounded Lipschitz function. More explicitly, let $f\colon S \to \mathbb{R}$ and $g\colon S \to \mathbb{R}$. Suppose there are constants K, L, M, N such that for all $x, y \in S$, we have

$$|f(x) - f(y)| \le M\rho(x, y), \qquad |f(x)| \le K,$$
$$|g(x) - g(y)| \le N\rho(x, y), \qquad |g(x)| \le L.$$

Then

$$|f(x)g(x) - f(y)g(y)| \le (KN + LM)\rho(x, y), \qquad |f(x)g(x)| \le KL.$$

This fact, used in part (c) of the proof, can now easily be established: If f is a Lipschitz function with $0 \le f \le 1$, and U_k is an open set, then the function defined by

$$f_k(x) = f(x) \cdot \left(1 \wedge k\operatorname{dist}(x, S \setminus U_k)\right)$$

is a Lipschitz function.

For part (a), define

$$\overline{\mathcal{M}}(U) = \sup\left\{\, \Phi(f) : f \in \mathrm{Lip}_b(S, \mathbb{R}), 0 \le f \le \mathbb{1}_U \,\right\}.$$

(2.4.14) Exercise. A counterexample for part (b) of the proof: Find a metric space S, two open sets U, V, and a bounded Lipschitz function f with $0 \le f \le \mathbb{1}_{U \cup V}$ such that the function

$$g(x) = \begin{cases} \dfrac{\operatorname{dist}(x, S \setminus U)}{\operatorname{dist}(x, S \setminus U) + \operatorname{dist}(x, S \setminus V)}\, f(x), & \text{if } x \in U, \\[2mm] 0, & \text{otherwise} \end{cases}$$

is not a Lipschitz function.

We will need the following to carry out part (b).

(2.4.15) Exercise. Let S be a metric space, and let $f\colon S \to [0, \infty)$ be continuous. Then there is a sequence of Lipschitz functions that increases pointwise to f.

Now for part (b). If we have a bounded Lipschitz function f with $0 \leq f \leq \mathbb{1}_{U \cup V}$, define $g(x)$ and $h(x)$ as in part (b) of the proof. (They need not be Lipschitz.) Then construct sequences of Lipschitz functions $g_n \nearrow g$, $h_n \nearrow h$. Now, $g_n + h_n \in \mathrm{Lip}_b(S, \mathbb{R})$, and $g_n + h_n \nearrow g + h = f$, so by σ-smoothness, we have $\Phi(g_n + h_n) \to \Phi(f)$. But $\overline{\mathcal{M}}(U) + \overline{\mathcal{M}}(V) \geq \Phi(g_n) + \Phi(h_n)$ for all n, so $\overline{\mathcal{M}}(U) + \overline{\mathcal{M}}(V) \geq \Phi(f)$. Take the supremum over all f to conclude that $\overline{\mathcal{M}}(U) + \overline{\mathcal{M}}(V) \geq \overline{\mathcal{M}}(U \cup V)$.

(2.4.16) Exercise. Complete the proof of Theorem 2.4.12.

2.5 Spaces of Measures

Sometimes, mathematicians find it useful to consider an entire set of functions as a "space" in its own right. It might be used as a vector space, a metric space, or a measurable space, for example. The space $\mathfrak{C}(S, T)$ of all continuous functions from one metric space to another is an example. See [*MTFG*, §2.3]. Similarly, it may be useful to consider a set of sets as a space. The set $\mathfrak{K}(S)$ of all nonempty compact subsets of a metric space S is a good example. (Such a space of sets is called a "hyperspace.") This space, with the Hausdorff metric, is a complete metric space provided that S is, and constructions of fractal sets can often be viewed as limits in this space.

In this section we will consider a third sort of "space": this time a space $\mathfrak{P}(S)$ of measures on a metric space S.

Tight Measures. Let S be a metric space. A finite measure $\mathcal{M}$ defined on the Borel sets of S is automatically regular (Theorem 1.1.12): For any Borel set $A \subseteq S$, and any positive number ε, there exist a closed set $F \subseteq A$ and an open set $U \supseteq A$ with $\mathcal{M}(U \setminus F) < \varepsilon$. We will be interested in a stronger property.

(2.5.1) Definition. Let S be a metric space and let $\mathcal{M}$ be a finite Borel measure on S. We say that $\mathcal{M}$ is **tight** iff for every Borel set $E \subseteq S$,

$$\mathcal{M}(E) = \sup \{ \mathcal{M}(K) : K \text{ compact}, K \subseteq E \}.$$

(2.5.2) Exercise. Let $\mathcal{M}$ be a finite measure on a metric space S. Suppose

$$\mathcal{M}(S) = \sup \{ \mathcal{M}(K) : K \text{ compact}, K \subseteq S \}.$$

Then $\mathcal{M}$ is tight.

(Instead of "tight" the phrase "almost compactly supported" is sometimes seen. Also, the term "finite Radon measure" may be used for "finite tight Borel measure.")

In the metric spaces $\mathbb{R}^d$, every finite Borel measure is easily seen to be tight, since every closed set is an increasing union of a sequence of compact sets. But in fact, this property is true in many other metric spaces as well.

(2.5.3) Theorem. *Let S be a complete separable metric space, and let $\mathcal{M}$ be a finite Borel measure on S. Then $\mathcal{M}$ is tight.*

Proof. Combine (1.1.14) and Exercise 2.5.2.   ☺

A measure $\mathcal{M}$ is called a **probability measure** iff the measure of the whole space is 1.

(2.5.4) Definition. Let S be a metric space. Write $\mathfrak{P}(S)$ for the set of all tight Borel probability measures on S.

Topology on Measures. When we consider the set $\mathfrak{P}(S)$ as a "space," it needs more structure. We want to talk about convergence of a sequence of measures, for example.

We consider briefly a candidate for this job that is not useful for our purposes. Let $\mathcal{M}_n$ be a sequence in $\mathfrak{P}(S)$ and let $\mathcal{M} \in \mathfrak{P}(S)$. We say that $\mathcal{M}_n$ converges **setwise** to $\mathcal{M}$ iff

$$\lim_n \mathcal{M}_n(A) = \mathcal{M}(A)$$

for every Borel set A.

(2.5.5) Exercise. Let $\mathcal{M}_n$ be a sequence in $\mathfrak{P}(S)$. Suppose that for each Borel set A, the sequence $\mathcal{M}_n(A)$ is a Cauchy sequence of real numbers. Then is the limit, defined by
$$\mathcal{M}(A) = \lim_n \mathcal{M}_n(A),$$
an element of $\mathfrak{P}(S)$?

Here is another way to specify setwise convergence in $\mathfrak{P}(S)$. Note that the definition of $\mathcal{M}_n \to \mathcal{M}$ may be rephrased as $\int \mathbb{1}_A(x)\,\mathcal{M}_n(dx) \to \int \mathbb{1}_A(x)\,\mathcal{M}(dx)$ for all Borel sets $A \subseteq S$. But a larger class of functions may be used in place of the functions $\mathbb{1}_A$:

(2.5.6) Exercise. Suppose the sequence $\mathcal{M}_n$ converges setwise to $\mathcal{M}$. Then

$$\lim_{n \to \infty} \int h(x)\,\mathcal{M}_n(dx) = \int h(x)\,\mathcal{M}(dx)$$

for any bounded Borel function h.

It turns out that for our purposes, setwise convergence is too strong—setwise convergence occurs too rarely. Consider a simple example in the metric space $\mathbb{R}$. Let $\mathcal{M}_n$ be the Dirac measure $\mathcal{E}_{1/n}$ at the point $1/n$. It might seem natural that $\mathcal{M}_n$ converges to $\mathcal{M} = \mathcal{E}_0$. But that is false for setwise convergence, since for example if $E = (-\infty, 0]$, then $\mathcal{M}_n(E) = 0$ for all n, but $\mathcal{M}(E) = 1$.

Here is another example, also in the line $\mathbb{R}$. For each $n \in \mathbb{N}$, consider the "$1/n$-average measure"

$$\mathcal{M}_n = \frac{1}{n} \sum_{j=1}^{n} \mathcal{E}_{j/n}.$$

Equivalently, integrals are

$$\int h(x)\, \mathcal{M}_n(dx) = \frac{1}{n} \sum_{j=1}^{n} h\left(\frac{j}{n}\right).$$

Now it might seem natural that $\mathcal{M}_n$ converges to Lebesgue measure on $[0, 1]$. But if E is the set of all rational numbers, we have $\mathcal{M}_n(E) = 1$ for all n, while $\mathcal{L}(E) = 0$.

Narrow convergence. A weaker mode of convergence for measures (that is, one that occurs more often) may be defined based on Exercise 2.5.6. If a smaller class of functions h is used, then convergence is more likely.

(2.5.7) Definition. Let S be a metric space. We say that a sequence $\mathcal{M}_n \in \mathfrak{P}(S)$ **converges narrowly** to $\mathcal{M} \in \mathfrak{P}(S)$ iff

$$\lim_{n\to\infty} \int h(x)\, \mathcal{M}_n(dx) = \int h(x)\, \mathcal{M}(dx)$$

for all bounded continuous functions $h\colon S \to \mathbb{R}$.

Narrow (French "étroite") convergence is also sometimes known as "weak convergence"; but that term may also be used for other kinds of convergence, so I keep to the less ambiguous term "narrow."

By Exercise 2.5.6, if a sequence $\mathcal{M}_n$ converges setwise, then it converges narrowly. But in general, the converse is not true. Examine the examples above:

(2.5.8) Exercise. In the metric space $\mathbb{R}$, show that (a) $\mathcal{E}_{1/n}$ converges narrowly to $\mathcal{E}_0$, and (b) $(1/n) \sum_{j=1}^{n} \mathcal{E}_{j/n}$ converges narrowly to $\mathcal{L}{\upharpoonright}[0, 1]$.

But note the following facts on set convergence.

(2.5.9) Proposition. *Suppose $\mathcal{M}_n$ converges narrowly to $\mathcal{M}$. If F is a closed set, then*

$$\limsup_{n\to\infty} \mathcal{M}_n(F) \le \mathcal{M}(F).$$

Proof. Let $\varepsilon > 0$. The functions

$$h_k(x) = 0 \vee \left(1 - k\,\mathrm{dist}(x, F)\right)$$

decrease to 1_F as $k \to \infty$, so for large k we have $\int h_k(x)\,\mathcal{M}(dx) < \mathcal{M}(F) + \varepsilon$. Now, $\mathcal{M}_n(F) \leq \int h_k(x)\,\mathcal{M}_n(dx)$, and $\lim_n \int h_k(x)\,\mathcal{M}_n(dx) = \int h_k(x)\,\mathcal{M}(dx) < \mathcal{M}(F) + \varepsilon$. Thus for large enough n we have $\mathcal{M}_n(F) < \mathcal{M}(F) + \varepsilon$. This shows that $\limsup_n \mathcal{M}_n(F) \leq \mathcal{M}(F)$. ☺

(2.5.10) Exercise. Suppose $\mathcal{M}_n$ converges narrowly to $\mathcal{M}$. If U is an open set, then

$$\liminf_{n \to \infty} \mathcal{M}_n(U) \geq \mathcal{M}(U).$$

Narrow convergence may, in fact, be characterized in terms of sets. Because of the following characterization, sets E with $\mathcal{M}(\partial E) = 0$ are known as **sets of continuity** for $\mathcal{M}$.

(2.5.11) Theorem. *The sequence $\mathcal{M}_n$ converges narrowly to $\mathcal{M}$ if and only if*

$$\lim_{n \to \infty} \mathcal{M}_n(E) = \mathcal{M}(E)$$

for all sets $E \subseteq S$ satisfying $\mathcal{M}(\partial E) = 0$.

Proof. Suppose first that $\mathcal{M}_n$ converges narrowly to $\mathcal{M}$. Let E be a set with $\mathcal{M}(\partial E) = 0$. Now, the interior E° of E is open, and the closure $\overline{E}$ of E is closed. So

$$\limsup_n \mathcal{M}_n(E) \leq \limsup_n \mathcal{M}_n\left(\overline{E}\right) \leq \mathcal{M}\left(\overline{E}\right)$$
$$= \mathcal{M}\left(E^\circ\right) + \mathcal{M}\left(\partial E\right) = \mathcal{M}\left(E^\circ\right)$$
$$\leq \liminf_n \mathcal{M}_n\left(E^\circ\right) \leq \liminf_n \mathcal{M}_n(E) \leq \limsup_n \mathcal{M}_n(E).$$

Thus everything is equal, and $\lim_n \mathcal{M}_n(E) = \mathcal{M}(\overline{E}) = \mathcal{M}(E)$.

Conversely, suppose $\lim_n \mathcal{M}_n(E) = \mathcal{M}(E)$ for all sets E of continuity for $\mathcal{M}$. Let $h\colon S \to \mathbb{R}$ be a bounded continuous function. We must show that $\lim_n \int h(x)\,\mathcal{M}_n(dx) = \int h(x)\,\mathcal{M}(dx)$. Fix $\varepsilon > 0$. Now, h is bounded, say $a \leq h(x) \leq b$ for all x. For $\alpha \in \mathbb{R}$, write

$$L_\alpha = \left\{\, x \in S : h(x) = \alpha \,\right\}.$$

The sets L_α are disjoint closed (hence measurable) sets, so all except countably many of them satisfy $\mathcal{M}(L_\alpha) = 0$. In particular, the set of α such that $\mathcal{M}(L_\alpha) = 0$ is dense in the line $\mathbb{R}$. So we may choose numbers

$$\alpha_0 < \alpha_1 < \alpha_2 < \cdots < \alpha_p$$

where $\alpha_i - \alpha_{i-1} < \varepsilon$, $\mathcal{M}(L_{\alpha_i}) = 0$, $\alpha_0 < a$, and $\alpha_p > b$. Now write

$$A_i = \{ x \in S : \alpha_{i-1} < h(x) \le \alpha_i \} \qquad (1 \le i \le p).$$

But then $\mathcal{M}(\partial A_i) = 0$ because the sets L_{α_i} and $L_{\alpha_{i-1}}$ have measure zero. Now, all of S is the disjoint union of the sets A_i, and $h(x)$ differs from α_i by at most ε for all $x \in A_i$. So for any probability measure $\mathcal{N}$ we have

$$\left| \int h(x)\, \mathcal{N}(dx) - \sum_{i=1}^{p} \alpha_i \mathcal{N}(A_i) \right| \le \varepsilon$$

(in particular, for $\mathcal{N} = \mathcal{M}_n$ or $\mathcal{M}$). Since $\mathcal{M}_n(A_i) \to \mathcal{M}(A_i)$ for each i, if n is large enough we will have

$$\left| \sum_{i=1}^{p} \alpha_i \mathcal{M}_n(A_i) - \sum_{i=1}^{p} \alpha_i \mathcal{M}(A_i) \right| < \varepsilon,$$

and thus

$$\left| \int h(x)\, \mathcal{M}_n(dx) - \int h(x)\, \mathcal{M}(dx) \right| < 3\varepsilon.$$

So $\lim_n \int h(x)\, \mathcal{M}_n(dx) = \int h(x)\, \mathcal{M}(dx)$.   ☺

There are converses for (2.5.9) and (2.5.10).

(2.5.12) Proposition. *Let $\mathcal{M}_n$ be a sequence in $\mathfrak{P}(S)$ and let $\mathcal{M} \in \mathfrak{P}(S)$. If $\limsup \mathcal{M}_n(F) \le \mathcal{M}(F)$ for all closed sets F, then $\mathcal{M}_n$ converges narrowly to $\mathcal{M}$.*

Proof. Suppose $\limsup \mathcal{M}_n(F) \le \mathcal{M}(F)$ for all closed sets F. We will show that $\lim \mathcal{M}_n(E) = \mathcal{M}(E)$ for all sets of continuity E. Let E be a set of continuity, $\mathcal{M}(\partial E) = 0$. Now, the closure $\overline{E}$ of E is closed, so

$$\limsup_n \mathcal{M}_n(E) \le \limsup_n \mathcal{M}_n(\overline{E}) \le \mathcal{M}(\overline{E}) \le \mathcal{M}(E) + \mathcal{M}(\partial E) = \mathcal{M}(E).$$

Similarly, $\overline{S \setminus E}$ is closed, and

$$\liminf_n \mathcal{M}_n(E) = 1 - \limsup_n \mathcal{M}_n(S \setminus E) \ge 1 - \limsup_n \mathcal{M}_n(\overline{S \setminus E})$$

$$\ge 1 - \mathcal{M}(\overline{S \setminus E}) \ge 1 - \mathcal{M}(S \setminus E) - \mathcal{M}(\partial E) = \mathcal{M}(E).$$

Thus $\mathcal{M}(E) \ge \limsup \mathcal{M}_n(E) \ge \liminf \mathcal{M}_n(E) \ge \mathcal{M}(E)$, so $\lim \mathcal{M}_n(E) = \mathcal{M}(E)$. By Theorem 2.5.11, $\mathcal{M}_n \to \mathcal{M}$ narrowly.   ☺

(2.5.13) Corollary. *Let $\mathcal{M}_n$ be a sequence in $\mathfrak{P}(S)$ and let $\mathcal{M} \in \mathfrak{P}(S)$. If $\liminf \mathcal{M}_n(U) \ge \mathcal{M}(U)$ for all open sets U, then $\mathcal{M}_n$ converges narrowly to $\mathcal{M}$.*

Metrics for the Narrow Topology. It will be useful to have a metric on $\mathfrak{P}(S)$ that specifies the narrow topology. For that purpose we will use the bounded Lipschitz functions as test functions.

Let S be a metric space. Fix a positive number γ. Let us write V_γ for the set of all functions $h\colon S \to \mathbb{R}$ satisfying

$$|h(x) - h(y)| \leq \rho(x,y), \qquad |h(x)| \leq \gamma,$$

for all $x, y \in S$. If $\mathcal{M}, \mathcal{N} \in \mathfrak{P}(S)$, define

$$\rho_\gamma(\mathcal{M}, \mathcal{N}) = \sup\left\{ \left| \int h(x)\, \mathcal{M}(dx) - \int h(x)\, \mathcal{N}(dx) \right| : h \in V_\gamma \right\}.$$

(2.5.14) Proposition. *The function ρ_γ is a metric on the set $\mathfrak{P}(S)$.*

Proof. Clearly, $\rho_\gamma(\mathcal{M}, \mathcal{N}) \geq 0$.

Suppose $\mathcal{M} \neq \mathcal{N}$. We claim that $\rho_\gamma(\mathcal{M}, \mathcal{N}) > 0$. By Exercise 1.1.11 there is a closed set F with $\mathcal{M}(F) \neq \mathcal{N}(F)$. The functions

$$h_n(x) = 0 \vee \bigl(1 - n\operatorname{dist}(x, F)\bigr)$$

decrease to $\mathbb{1}_F$, so for some n,

$$\int h_n\, \mathcal{M}(dx) \neq \int h_n\, \mathcal{N}(dx).$$

If $c > 0$ is so small that $c \leq 1/n$ and $c \leq \gamma$, then $ch_n \in V_\gamma$. Thus $\rho_\gamma(\mathcal{M}, \mathcal{N}) > 0$.

Clearly, $\rho_\gamma(\mathcal{M}, \mathcal{N}) = \rho_\gamma(\mathcal{N}, \mathcal{M})$.

Finally, the triangle inequality: Suppose $\mathcal{M}_1, \mathcal{M}_2, \mathcal{M}_3 \in \mathfrak{P}(S)$. Let $h \in V_\gamma$. Now,

$$\left| \int h(x)\, \mathcal{M}_1(dx) - \int h(x)\, \mathcal{M}_3(dx) \right|$$

$$\leq \left| \int h(x)\, \mathcal{M}_1(dx) - \int h(x)\, \mathcal{M}_2(dx) \right|$$

$$+ \left| \int h(x)\, \mathcal{M}_2(dx) - \int h(x)\, \mathcal{M}_3(dx) \right|$$

$$\leq \rho_\gamma(\mathcal{M}_1, \mathcal{M}_2) + \rho_\gamma(\mathcal{M}_2, \mathcal{M}_3).$$

This is true for all $h \in V_\gamma$, so $\rho_\gamma(\mathcal{M}_1, \mathcal{M}_3) \leq \rho_\gamma(\mathcal{M}_1, \mathcal{M}_2) + \rho_\gamma(\mathcal{M}_2, \mathcal{M}_3)$.   ☺

(2.5.15) Exercise. Let $a, b \in S$. Compute $\rho_\gamma(\mathcal{E}_a, \mathcal{E}_b)$ in terms of $\rho(a, b)$ and γ.

(2.5.16) Exercise. In the metric space $[0, 1]$, compute $\rho_\gamma(\mathcal{M}_n, \mathcal{M})$, where

$$\mathcal{M}_n = \frac{1}{n} \sum_{j=1}^{n} \mathcal{E}_{j/n} \qquad \text{and} \qquad \mathcal{M} = \mathcal{L}{\restriction}[0, 1].$$

(2.5.17) Theorem. *Let S be a metric space, let $\gamma > 0$, let $\mathcal{M}_n$ be a sequence in $\mathfrak{P}(S)$ and let $\mathcal{M} \in \mathfrak{P}(S)$. The following are equivalent.*

(a) $\rho_\gamma(\mathcal{M}_n, \mathcal{M}) \to 0$.
(b) $\int h(x)\,\mathcal{M}_n(dx) \to \int h(x)\,\mathcal{M}(dx)$ *for all* $h \in \mathrm{Lip}_b(S, \mathbb{R})$.
(c) $\mathcal{M}_n$ *converges narrowly to* $\mathcal{M}$.

Proof. Suppose that (a) $\rho_\gamma(\mathcal{M}_n, \mathcal{M}) \to 0$. If h is any bounded Lipschitz function, there is a positive constant c such that $ch \in V_\gamma$. Then $|\int h(x)\,\mathcal{M}_n(dx) - \int h(x)\,\mathcal{M}(dx)| \le (1/c)\rho_\gamma(\mathcal{M}_n, \mathcal{M}) \to 0$.

Next, suppose that (b) holds. The function h_k in Proposition 2.5.9 is bounded Lipschitz, so from (b) we may conclude that $\limsup \mathcal{M}_n(F) \le \mathcal{M}(F)$ for all closed sets F. So by Proposition 2.5.12, we conclude that $\mathcal{M}_n$ converges narrowly to $\mathcal{M}$.

Finally, suppose (c), that $\mathcal{M}_n \to \mathcal{M}$ narrowly. We must show that $\rho_\gamma(\mathcal{M}_n, \mathcal{M}) \to 0$. Fix $\varepsilon > 0$. We must show that for large n we have $|\int h(x)\,\mathcal{M}_n(dx) - \int h(x)\,\mathcal{M}(dx)| \le \varepsilon$ for all $h \in V_\gamma$.

I claim that S may be written as a disjoint union,

$$S = \bigcup_{i=0}^p E_i,$$

where each E_i is a set of continuity for $\mathcal{M}$, $\mathcal{M}(E_0) < \varepsilon/(5\gamma)$, and diam $E_i \le \varepsilon/5$ for $i \ge 1$. Here is how to do it: Since $\mathcal{M}$ is tight, there is a compact set K with $\mathcal{M}(S \setminus K) < \varepsilon/(5\gamma)$. For each $x \in K$, there is an open ball $B_r(x)$ with $0 < r < \varepsilon/10$ and $\mathcal{M}(\partial B_r(x)) = 0$. These open sets cover K, so there is a finite subcover,

$$K \subseteq \bigcup_{i=1}^p B_{r_i}(x_i).$$

Now let $E_1 = B_{r_1}(x_1)$, $E_2 = B_{r_2}(x_2) \setminus E_1, \cdots, E_p = B_{r_p}(x_p) \setminus (E_1 \cup \cdots \cup E_{p-1})$, and $E_0 = S \setminus (E_1 \cup \cdots \cup E_p)$. Then each of the sets is a set of continuity, since the boundary of each E_i is a subset of $\bigcup_i \partial B_{r_i}(x_i)$. Any E_i with $i \ge 1$ is a subset of the corresponding $B_{r_i}(x_i)$, so it has diameter $\le \varepsilon/5$. And $E_0 \subseteq S \setminus K$, so $\mathcal{M}(E_0) < \varepsilon/(5\gamma)$.

Choose $\eta > 0$ such that $p\eta < 1$ and $\gamma\eta < \varepsilon/5$. Since each E_i is a set of continuity for $\mathcal{M}$, there is N such that if $n \ge N$, then $|\mathcal{M}_n(E_i) - \mathcal{M}(E_i)| < \eta$ for $0 \le i \le p$. Choose a point $x_i \in E_i$. (Empty sets E_i all contribute 0 to the sums below, so throw them out.)

Now let $h \in V_\gamma$. For $1 \le i \le p$, if $x \in E_i$, then $|h(x) - h(x_i)| < \varepsilon/5$; and for $x \in E_0$, we have $|h(x)| \le \gamma$. So

$$\int h(x)\,\mathcal{M}_n(dx) \le \gamma\mathcal{M}_n(E_0) + \sum_{i=1}^p \left(h(x_i) + \frac{\varepsilon}{5}\right)\mathcal{M}_n(E_i)$$

$$\le \gamma\big(\mathcal{M}(E_0) + \eta\big) + \sum_{i=1}^p \left(h(x_i) + \frac{\varepsilon}{5}\right)\big(\mathcal{M}(E_i) + \eta\big)$$

$$\leq \gamma \left(\frac{\varepsilon}{5\gamma} + \eta \right) + \sum_{i=1}^{p} \left(h(x_i) - \frac{\varepsilon}{5} \right) \mathcal{M}(E_i)$$

$$+ 2 \frac{\varepsilon}{5} \sum_i \mathcal{M}(E_i) + p \frac{\varepsilon}{5} \eta$$

$$\leq \int h(x) \, \mathcal{M}(dx) + \varepsilon.$$

A similar argument proves that $\int h(x) \, \mathcal{M}_n(dx) \geq \int h(x) \, \mathcal{M}(dx) - \varepsilon$. This shows that $\left| \int h(x) \, \mathcal{M}_n(dx) - \int h(x) \, \mathcal{M}(dx) \right| \leq \varepsilon$, as required.   ☺

Note that although different values γ yield different metrics ρ_γ, they all correspond to the same (narrow) topology. If the metric space S is bounded, for example compact, then we will normally let γ be larger than the diameter of S. Then we may use in place of V_γ the collection of all functions $h \colon S \to \mathbb{R}$ satisfying

$$\left| h(x) - h(y) \right| \leq \rho(x, y)$$

for all $x, y \in S$. This metric is often called the "Hutchinson" metric, although Hutchinson himself calls it the "Monge-Kantorovich" metric [143].

(2.5.18) Exercise. In $\mathfrak{P}(\mathbb{R})$, let $\mathcal{N}_{a,s}$ denote the normal distribution with mean a and variance s^2. That is, the measure has density

$$n_{a,s}(x) = \frac{1}{\sqrt{2\pi} s} \exp \left(- \frac{(x - a)^2}{2s^2} \right).$$

Compute

$$\rho_\gamma (\mathcal{N}_{a,s}, \mathcal{N}_{b,t}).$$

Show that $\mathcal{N}_{0,s}$ converges to $\mathcal{E}_0$ as $s \to 0$.

(2.5.19) Exercise. Let F be a finite metric space. For $n \in \mathbb{N}$, let $\mathfrak{P}_n$ be the set of all measures $\mathcal{M} \in \mathfrak{P}(F)$ such that

$$\mathcal{M}(A) \in \left\{ \frac{0}{n}, \frac{1}{n}, \frac{2}{n}, \cdots, \frac{n}{n} \right\}$$

for all $A \subseteq F$. (Note that $\mathfrak{P}_n$ is a finite set.) For any $\varepsilon > 0$, show that there is an $n \in \mathbb{N}$ such that $\mathfrak{P}_n$ is an ε-net in $\mathfrak{P}(F)$.

Completeness. Under what circumstances is the space $\mathfrak{P}(S)$ complete? The next exercise tells us that completeness does not depend on which metric we use.

(2.5.20) Exercise. Let $\gamma > 0$. A sequence in $\mathfrak{P}(S)$ is a Cauchy sequence with respect to ρ_γ if and only if it is a Cauchy sequence with respect to ρ_1.

Now, S is embedded into $\mathfrak{P}(S)$ by identifying the point $a \in S$ with the Dirac measure $\mathcal{E}_a$. This is an isometry (for distances less than γ, when we use the metric ρ_γ in $\mathfrak{P}(S)$), and its range is closed (see the following exercise). So if $\mathfrak{P}(S)$ is complete, then so is S.

(2.5.21) Exercise. Suppose $\mathcal{M} \in \mathfrak{P}(S)$ is not a Dirac measure. Show that there is a closed set A with $0 < \mathcal{M}(A) < 1$. Show that $\mathcal{M}$ is not the limit of a sequence of Dirac measures.

The converse (if S is complete, then $\mathfrak{P}(S)$ is complete) will use another definition.

(2.5.22) Definition. Let S be a metric space, and let $\mathfrak{U}$ be a subset of $\mathfrak{P}(S)$. We say that $\mathfrak{U}$ is **uniformly tight** iff for every $\varepsilon > 0$ there is a compact set $K \subseteq S$ such that $\mathcal{M}(K) \geq 1 - \varepsilon$ for all $\mathcal{M} \in \mathfrak{U}$.

Here is a simple example. The set $\{\, \mathcal{E}_a : a \in \mathbb{R} \,\}$ is not uniformly tight in $\mathfrak{P}(\mathbb{R})$, since for any compact set K, there is $a \notin K$ such that $\mathcal{E}_a(K) = 0$.

(2.5.23) Exercise. In $\mathfrak{P}(\mathbb{R})$, consider the family

$$\mathfrak{U} = \{\, \mathcal{N}_{0,s} : 0 < s \leq 1 \,\},$$

where $\mathcal{N}_{0,s}$ is the normal distribution with mean 0 and variance s^2. Is $\mathfrak{U}$ uniformly tight?

(2.5.24) Proposition. *Let S be a complete metric space. A Cauchy sequence in $\mathfrak{P}(S)$ is uniformly tight.*

Proof. Let $\mathcal{M}_n$ be a sequence, Cauchy in the metric ρ_1. For each $\mathcal{M}_n$, there is a σ-compact set of measure 1. Combining all such sets, we obtain a single σ-compact set S_0 that has measure 1 for all $\mathcal{M}_n$ in the entire sequence. In particular, S_0 is separable.

First I claim that for $\varepsilon > 0$, there is a closed set F that admits a finite ε-net, and $\mathcal{M}_n(F) \geq 1 - \varepsilon$ for all n. Here is the construction. Let $\{x_1, x_2, \cdots\}$ be a countable set dense in S_0. For $k \in \mathbb{N}$, let

$$U_k = \bigcup_{i=1}^{k} \overline{B}_\varepsilon(x_i),$$

and define

$$f_k(x) = 0 \vee \left(1 - \frac{\varepsilon}{2}\, \mathrm{dist}(x, U_k)\right).$$

Thus, $f_k = 0$ on $\bigcup_{i=1}^{k} B_{\varepsilon/2}(x_i)$, $f_k = 1$ outside U_k, and $(2/\varepsilon)f_k \in V_1$ (we may assume $\varepsilon < 1$). Note that f_k decreases to zero on S_0. Because the sequence $\mathcal{M}_n$ is Cauchy, there is m such that for all $n \geq m$, we have $\rho_1(\mathcal{M}_n, \mathcal{M}_m) < \varepsilon^2/4$. Next, since f_k decreases to 0 on S_0, there is k such that $\int f_k(x)\, \mathcal{M}_m(dx) < \varepsilon/2$.

Therefore, for $n \geq m$, we have $\int f_k \, \mathcal{M}_n(dx) < \varepsilon/2 + (2/\varepsilon)(\varepsilon^2/4) = \varepsilon$. For each $n = 1, 2, \cdots, m-1$ there is $k_n \in \mathbb{N}$ such that $\int f_{k_n}(x) \, \mathcal{M}_n(dx) < \varepsilon$. Now let $j = \max\{k, k_1, k_2, \cdots, k_{m-1}\}$, so that $\int f_j(x) \, \mathcal{M}_n(dx) < \varepsilon$ for all n. Thus $\mathcal{M}_n(S \setminus U_j) \leq \int f_j(x) \, \mathcal{M}_n(dx) < \varepsilon$ for all n. So $\mathcal{M}_n(U_j) \geq 1 - \varepsilon$ for all n. And of course, $\{x_1, x_2, \cdots, x_j\}$ is a finite ε-net in U_j.

Again let $\varepsilon > 0$. We claim there is a compact set K with $\mathcal{M}_n(K) \geq 1 - \varepsilon$ for all n. For each $k \in \mathbb{N}$, by the preceding argument there is a closed set F_k with a finite $1/k$-net such that $\mathcal{M}_n(F_k) > 1 - \varepsilon 2^{-k}$ for all n. The set $K = \bigcap_{k=1}^{\infty} F_k$ is closed and has a $1/k$-net for every k. Therefore (since S is complete), K is compact. And we have $\mathcal{M}_n(S \setminus K) \leq \sum_{k=1}^{\infty} \varepsilon 2^{-k} = \varepsilon$. ☺

(2.5.25) Theorem. *Let S be a complete metric space. Then $\mathfrak{P}(S)$ is also complete.*

Proof. Let $\mathcal{M}_n \in \mathfrak{P}(S)$ be a Cauchy sequence. By the preceding proposition, the set $\{\mathcal{M}_n\}$ is uniformly tight. For each $h \in \mathrm{Lip}_b(S, \mathbb{R})$, the sequence $\int h(x) \, \mathcal{M}_n(dx)$ is a Cauchy sequence of real numbers. So

$$\Phi(h) = \lim_n \int h(x) \, \mathcal{M}_n(dx)$$

defines a function $\Phi \colon \mathrm{Lip}_b(S, \mathbb{R}) \to \mathbb{R}$. Now, Φ is a positive linear functional. It remains to show that (a) Φ is σ-smooth, so that it is represented by a measure $\mathcal{M}$, and (b) $\mathcal{M}$ is tight.

Let h_k decrease to zero in $\mathrm{Lip}_b(S, \mathbb{R})$. Say $h_1 \leq 1$. Let $\varepsilon > 0$ be given. Since $\mathcal{M}_n$ is uniformly tight, there is a compact set K with $\mathcal{M}_n(K) > 1 - \varepsilon$ for all n. By Dini's Theorem 2.4.10, there is k such that $h_k \leq \varepsilon$ on K. Now, for any n, we have $\int h_k(x) \, \mathcal{M}_n(dx) \leq \varepsilon \mathcal{M}_n(K) + 1 \mathcal{M}_n(S \setminus K) \leq \varepsilon + \varepsilon = 2\varepsilon$. This is true for all n, so $\Phi(h_k) \leq 2\varepsilon$. Thus Φ is σ-smooth and is represented by a measure $\mathcal{M}$.

We have a compact set K with $\mathcal{M}_n(K) \geq 1 - \varepsilon$ for all n. Now, I claim that $\mathcal{M}(K) \geq 1 - \varepsilon$ as well. Suppose $\mathcal{M}(K) < 1 - \varepsilon$. Then there is $\delta > 0$ so small that $\mathcal{M}(K_\delta) < 1 - \varepsilon$, where $K_\delta = \{x \in S : \mathrm{dist}(x, K) < \delta\}$. Now, the function

$$h(x) = 0 \vee \big(1 - \mathrm{dist}(x, K)/\delta\big)$$

belongs to V_1. But

$$\mathcal{M}(K_\delta) \geq \int h(x) \, \mathcal{M}(dx) = \lim \int h(x) \, \mathcal{M}_n(dx)$$
$$\geq \liminf \mathcal{M}_n(K) \geq 1 - \varepsilon.$$

So in fact, $\mathcal{M}(K) \geq 1 - \varepsilon$. ☺

Compactness. Next we come to the assertion that $\mathfrak{P}(S)$ is compact when S is compact. Actually, we will discuss a stronger result that characterizes the compact subsets of $\mathfrak{P}(S)$ even when S itself is not compact. This characterization is left to the reader with the following hints.

(2.5.26) Exercise. Let S be a metric space and $\mathfrak{U} \subseteq \mathfrak{P}(S)$ a uniformly tight subset. Let $\varepsilon > 0$. Show that that there is a finite set $F \subseteq S$ such that for every $\mathcal{M} \in \mathfrak{U}$, there is $\mathcal{M}' \in \mathfrak{P}(F)$ with $\rho_1(\mathcal{M}, \mathcal{M}') < \varepsilon$. Conclude that $\mathfrak{U}$ has a finite ε-net (for the metric ρ_1).

(2.5.27) Exercise. Let S be a metric space and $\mathfrak{U} \subseteq \mathfrak{P}(S)$ a totally bounded subset. Show that $\mathfrak{U}$ is uniformly tight.

(2.5.28) Exercise. Let S be a complete metric space. Let $\mathfrak{U}$ be a subset of $\mathfrak{P}(S)$. Then $\mathfrak{U}$ is compact (in the narrow topology) if and only if it is closed and uniformly tight.

(2.5.29) Corollary. *Let S be a compact metric space. Then $\mathfrak{P}(S)$ is also compact (in the narrow topology).*

*2.6 Remarks

As I said in [MTFG, §5.6], Henri Lebesgue's theory of measure and integration is one of the cornerstones of twentieth-century mathematics. Further material on the topic can be found in many graduate-level texts; for example [48, 102, 111, 124, 135, 172, 200, 229, 230]. Other references of related or historical interest include [38, 42, 113, 233].

Product measures began with Lebesgue measure in the plane. Computing double integrals as iterated integrals goes back to the Riemann integral. Measurable functions are studied in the texts on measure and integration; see also [131, 38, 157].

The limit theorems are the most important parts of the Lebesgue theory of the integral. In the Riemann theory of the integral, countable operations are not always possible, or they have more difficult proofs.

The Radon–Nikodým Theorem is sometimes known as the Lebesgue–Radon–Nikodým Theorem. H. Lebesgue proved a result of this type for integrals in the line; J. Radon generalized to Euclidean space; O. Nikodým [201] further generalized to a more abstract setting. See [32, (32.14)] for the Radon–Nikodým derivative computed as a density (in the line).

More on the use of the metric densities for the Radon–Nikodým Theorem may be found in [186, pp. 35–42].

Use of the full and fine variation for integrals involving fractal measures is found in [76, 77].

For our proof of the Riesz Representation Theorem, we have used the approximation method of P. Daniell; see [124, 6]. For additional material on representing linear functionals as integrals, see [110, 148]. I learned some of this from lectures of A. M. Gleason [111].

---

* An optional section.

Material on spaces of tight measures may be found in [157, 110, 111]. For the "Hutchinson" metric, compare [66]. Convergence of measures is the topic of [31].

When everything is restricted to a single bounded set, it is convenient to use a single metric rather than the family ρ_γ. But I have not made that simplification here for two reasons. When a measure $\mathcal{J}$ with compact support is computed as the limit of a sequence $\mathcal{J}_n$, it may be useful to allow the approximating measures $\mathcal{J}_n$ to have unbounded support (see Exercise 3.3.17 or Figure 3.3.23 where we begin with a "normal" measure). In connection with a random measure (see §4.2), even when all of the instances involved have compact support, there may be no single bound for all of them at once.

On the Exercises. *(2.1.13)*:

$$R_n = \bigcup_{I \in \mathcal{J}} \bigcap_{i \in I} A_i,$$

where $\mathcal{J} = \left\{ I \subseteq \{1, 2, \cdots, n\} : \sum_{i \in I} \mathcal{M}_2(B_i) > b \right\}$ is a finite set of sets.

(2.2.11): On the measure space $[0, 1]$ with Lebesgue measure, let $f_n(x) = n \mathbb{1}_{(0,1/n)}$ and $f(x) = 0$. Then $f_n(x) \to f(x)$ for all x, but $\int f_n(x)\,dx = 1 \not\to 0 = \int f(x)\,dx$. Or, perhaps, let $f_n(x) = (-1)^n n \mathbb{1}_{(0,1/n)}$ so that $\int f_n(x)\,dx$ does not converge at all.

(2.3.10): [77].

(2.3.11): [76].

(2.4.15): Hint. (From [230, Chapter 2].) Let

$$f_n(x) = \inf \left\{ f(y) + n\rho(x, y) : y \in S \right\}.$$

(2.5.5): Yes: Nikodým Theorem, [67, III.7.4].

(2.5.15): $\rho(a, b) \wedge 2\gamma$.

3. Integrals and Fractals

Here we will discuss some of the applications of integration to questions relating to fractals.

We begin with the inequality between the topological dimension and the Hausdorff dimension. Then we consider Frostman's idea to generalize potential theory so that it applies to any dimension. In [MTFG] was some material on iterated function systems and self-similar sets; here some variants are discussed: self-similar measures that correspond to iterated function systems with weights, functions with self-affine graphs, and graph self-similar measures.

*3.1 Topological and Fractal Dimension

There is a relation between the topological dimension and the fractal dimension of a metric space. The small inductive dimension ind S and the Hausdorff dimension dim S satisfy
$$\text{ind } S \leq \text{dim } S.$$
This is proved for compact spaces S in [MTFG, (6.2.10)]. For another proof, relying only on the theory of topological dimension, see [217]. That proof shows that the "covering dimension" Cov S is $\leq$ the "metric order" of S (which is called the "lower entropy index" in [MTFG, §6.5] and also sometimes known as the "box dimension"). But the argument actually proves that Cov $S \leq$ dim S (as noted in the commentary in [74]).

Here we will present a proof that ind $S \leq$ dim S for general metric spaces S that relies on Lebesgue integration. This proof is due to E. Szpilrajn [252].

Recall the definition of the small inductive dimension ind S (see [MTFG, §3.1]). The empty set has dimension ind $\varnothing = -1$. Recursively, ind $S \leq n$ if and only if there is a base for the open sets of S consisting of sets V with ind $\partial V \leq n - 1$. Finally, if ind $S \leq n$ is false for all $n \in \mathbb{N}$, then ind $S = \infty$.

The next two lemmas establish the way in which ind S and dim S will be used in the proof. Let us write
$$S(x_0, r) = \{\, x \in S : \rho(x, x_0) = r \,\}$$
for the "sphere" with center x_0 and radius r.

---

* An optional section.

(3.1.1) Lemma. *Let S be a metric space. Suppose* ind $S > n$, *where $n \in \mathbb{N}$.* *Then there is $x_0 \in S$ and $r > 0$ such that* ind $S(x_0, \varepsilon) \geq n$ *for all ε with* $0 < \varepsilon < r$.

Proof. Since ind $S > n$, there is no base for the open sets of S made up of sets V with ind $\partial V \leq n - 1$. In particular, the collection of open balls B with ind $\partial B \leq n - 1$ is not a base for the open sets. So there is a point $x_0 \in S$ and a positive number r such that ind $\partial B_\varepsilon(x_0) \geq n$ for all ε with $0 < \varepsilon < r$. But

$$\partial B_\varepsilon(x_0) \subseteq S(x_0, \varepsilon),$$

so ind $S(x_0, \varepsilon) \geq n$ for all such ε.   ☺

(3.1.2) Lemma. *Let $s \geq 1$, let $E \subseteq S$ be a set with $\mathcal{H}^s(E) = 0$, and let $x_0 \in E$. Then*

$$\mathcal{H}^{s-1}(S(x_0, r) \cap E) = 0$$

for (Lebesgue) almost all $r > 0$. In particular, the set of such r is dense in $(0, \infty)$.

Proof. Let $A \subseteq S$ be any set. Write

$$r_1 = \inf_{x \in A} \rho(x_0, x), \qquad r_2 = \sup_{x \in A} \rho(x_0, x)$$

for the least and greatest distance from x_0 to the set A. Then, of course, $r_2 - r_1 \leq \operatorname{diam} A$. Also

$$\overline{\int}_{(0,\infty)} \operatorname{diam}\left(S(x_0, r) \cap A\right)^{s-1} dr = \overline{\int}_{(r_1, r_2)} \operatorname{diam}\left(S(x_0, r) \cap A\right)^{s-1} dr$$

$$\leq (\operatorname{diam} A)^{s-1} \overline{\int}_{(r_1, r_2)} dr \leq (\operatorname{diam} A)^s.$$

(We have used upper integrals, since we do not know that the integrand is a measurable function of r.) When $s = 1$, we must interpret $(\operatorname{diam} \varnothing)^0 = 0$ as usual.

Now we are given a set E with $\mathcal{H}^s(E) = 0$. For each $n \in \mathbb{N}$, there is a cover $\{A_i^n\}_{i \in \mathbb{N}}$ of E with diam $A_i^n \leq 2^{-n}$ and

$$\sum_i \operatorname{diam}(A_i^n)^s \leq 2^{-n}.$$

Applying the above inequality, we see that

$$\sum_i \overline{\int}_{(0,\infty)} \operatorname{diam}\left(S(x_0, r) \cap A_i^n\right)^{s-1} dr \leq 2^{-n}.$$

Thus, by Beppo Levy's Theorem 2.2.7,

$$\overline{\int}_{(0,\infty)} \sum_i \operatorname{diam}\left(S(x_0, r) \cap A_i^n\right)^{s-1} dr \leq 2^{-n}.$$

If we write
$$D_n = \left\{ r \in (0, \infty) : \sum_i \operatorname{diam} \left(S(x_0, r) \cap A_i^n \right)^{s-1} > \frac{1}{n} \right\},$$

then $(1/n)\overline{\mathcal{L}}(D_n) \leq 2^{-n}$. Next, if we write $F_N = \bigcup_{n=N}^{\infty} D_n$, then $\overline{\mathcal{L}}(F_N) \leq \sum_{n=N}^{\infty} n2^{-n}$. Now, $\sum n2^{-n}$ is a convergent series, so $\lim_{N \to \infty} \sum_{n=N}^{\infty} n2^{-n} = 0$. But for all $N \in \mathbb{N}$,

$$\left\{ r \in (0, \infty) : \sum_i \operatorname{diam} \left(S(x_0, r) \cap A_i^n \right)^{s-1} \not\to 0 \right\} \subseteq F_N,$$

so we may conclude that
$$\lim_n \sum_i \operatorname{diam} \left(S(x_0, r) \cap A_i^n \right)^{s-1} = 0$$

for almost all r. Thus, we have $\mathcal{H}^{s-1}\big(S(x_0, r) \cap E \big) = 0$ for almost all r. ☺

(3.1.3) Corollary. *Let $n \geq 0$ be an integer, and let E be a set. If $\mathcal{H}^n(E) = 0$, then* ind $E \leq n - 1$.

Proof. By induction on n. If $n = 0$, then from $\mathcal{H}^0(E) = 0$ we conclude that $E = \varnothing$, and thus ind $E = -1$, which satisfies the inequality. Suppose $n \geq 1$ and the result is known for smaller values. Assume $\mathcal{H}^n(E) = 0$. Then by Lemma 3.1.2,
$$\mathcal{H}^{n-1}\big(S(x_0, r) \cap E \big) = 0$$

for r dense in $(0, \infty)$. Then by the induction hypothesis, ind $\big(S(x_0, r) \cap E \big) \leq n - 2$ for all such r. By Lemma 3.1.1, we have ind $E \leq n - 1$. ☺

(3.1.4) Corollary. *If S is a metric space, then* ind $S \leq \dim S$.

Proof. If $\dim S = \infty$, there is nothing to prove. Suppose $\dim S = s < \infty$. Let n be the least integer greater than s, so that $n - 1 \leq s < n$. Then $\mathcal{H}^n(S) = 0$, so by the preceding corollary, ind $S \leq n - 1 \leq s = \dim S$. ☺

3.2 Potential Theory

Abstract potential theory can shed some light on questions of fractal dimension. The classical Newtonian potential (corresponding to the inverse-square law) originated in physics. Other potentials can be used mathematically.

Newtonian Potential.* We begin with a simplified look at gravitational attraction, or electrical attraction and repulsion.

---

* Optional material. Anyone frightened of physics may skip ahead to the heading "Abstract Potential" on page 118.

Suppose mass (or charge) is distributed in space somehow. Its effects may be felt, even in regions of space with no mass (or charge). To understand this effect, we imagine placing a test particle with unit mass (or charge) at a given point of space, and measuring the force on the test particle. The force has both magnitude and direction: it is a "vector." We imagine trying this experiment at each point of space. We obtain a vector corresponding to each point—a "vector field." Once our Cartesian coordinate system has been fixed (and our units chosen), this vector field can be considered to be a function $\mathbf{F} \colon \mathbb{R}^3 \to \mathbb{R}^3$. To each point $\mathbf{x} \in \mathbb{R}^3$ there corresponds a force vector $\mathbf{F}(\mathbf{x})$. For example, consider the electric field corresponding to a unit charge at the origin. The magnitude of $\mathbf{F}(\mathbf{x})$ is $|\mathbf{x}|^{-2}$, according to the "inverse-square law." The direction of $\mathbf{F}(\mathbf{x})$ is the same as the direction of $\mathbf{x}$ itself: a unit vector with that direction is $\mathbf{x}/|\mathbf{x}|$. Thus, for the unit charge at the origin, we have

$$\mathbf{F}_0(\mathbf{x}) = |\mathbf{x}|^{-2} \frac{\mathbf{x}}{|\mathbf{x}|} = \frac{\mathbf{x}}{|\mathbf{x}|^3}.$$

(In practice, there might also be a constant factor multiplying the formula— the magnitude determined by the units used and the sign determined by whether the force is attracting or repelling.)

Now, physicists tell us that this vector field may be written as the gradient of a certain scalar field, the "potential energy." We can imagine a scalar $U(\mathbf{x})$ associated to each point $\mathbf{x} \in \mathbb{R}^3$ and compute $\mathbf{F}$ by

$$\mathbf{F}(\mathbf{x}) = -\nabla U(\mathbf{x}).$$

The minus sign will allow us to work with positive energies. For example, the potential corresponding to a unit charge at the origin is $U_0(\mathbf{x}) = |\mathbf{x}|^{-1}$. (Verify that the gradient of this is $-\mathbf{F}_0$.) Similarly, the potential corresponding to a unit charge at the point $\mathbf{x}_0$ is $U(\mathbf{x}) = |\mathbf{x} - \mathbf{x}_0|^{-1}$.

Suppose now the charge (or mass) is not all concentrated at a single point. It is distributed throughout a portion of space. This kind of distribution can be described by a measure $\mathcal{M}$. For each region $G \subseteq \mathbb{R}^3$ of space, $\mathcal{M}(G)$ is the amount of charge (or mass) in G. We allow measures that have points of positive measure (point charges or point masses), but we also allow measures with no points of positive measure (continuous distributions). The potential for such a charge distribution is obtained by the superposition of the individual potentials for each component part of the whole distribution, weighted by the charge of the parts. More formally, the **Newtonian potential** corresponding to a finite measure $\mathcal{M}$ is the function $U \colon \mathbb{R}^3 \to [0, \infty]$ defined by

$$U(\mathbf{y}) = U_{\mathcal{M}}(\mathbf{y}) = \int \frac{\mathcal{M}(d\mathbf{x})}{|\mathbf{x} - \mathbf{y}|}.$$

What is the meaning of the scalar field $U(\mathbf{y})$? If a distribution of charge is described by a measure $\mathcal{M}$, then that charge distribution would cause a "potential energy" of $U(\mathbf{y})$ for a unit charge placed at the point $\mathbf{y}$. More generally, the potential energy arising from the repulsion between one charged body (with charge distribution given by a measure $\mathcal{M}$) and a second charged body (with distribution $\mathcal{N}$) is

$$\int U_{\mathcal{M}}(\mathbf{y})\,\mathcal{N}(d\mathbf{y}) = \int\int \frac{\mathcal{M}(d\mathbf{x})\,\mathcal{N}(d\mathbf{y})}{|\mathbf{x} - \mathbf{y}|}.$$

As a special case, taking $\mathcal{N} = \mathcal{M}$, we may compute the energy of a given charge distribution itself, arising since each part of the distribution repels the rest of the distribution. The energy of the charge distribution $\mathcal{M}$ is

$$I(\mathcal{M}) = \int U_{\mathcal{M}}(\mathbf{y})\,\mathcal{M}(d\mathbf{y}) = \int\int \frac{\mathcal{M}(d\mathbf{x})\,\mathcal{M}(d\mathbf{y})}{|\mathbf{x} - \mathbf{y}|}.$$

Note in particular that if $\mathcal{M}$ contains any point charges, then the energy is infinite.

Now suppose we fix a compact set $K \subseteq \mathbb{R}^3$. Suppose we give it a unit charge. If it is a conductor, the charge can (in our simplified model) distribute itself within K in any way, so the actual charge distribution should be one that minimizes the energy. The reciprocal of this minimum energy is called the "capacity" of the set K:

$$\left(\inf\{\,I(\mathcal{M}) : \mathcal{M} \text{ is a probability measure concentrated on } K\,\}\right)^{-1}.$$

If $I(\mathcal{M}) = \infty$ for all measures on K, then we say the set K has capacity zero.

We will see below that the question of whether a set K has capacity zero is closely related to the Hausdorff dimension of the set: If $\dim K > 1$, then K has positive capacity, while if $\dim K < 1$, then K has capacity zero. We will also see below how this connection between potential theory and Hausdorff dimension can be generalized to characterize Hausdorff dimensions other than 1.

(3.2.1) Exercise. Let K be a solid ball of radius 1:

$$K = \{\,\mathbf{x} \in \mathbb{R}^3 : |\mathbf{x}| \leq 1\,\}.$$

(a) Let $\mathcal{M}_1$ be the probability measure on K proportional to the volume (3-dimensional Lebesgue measure). Compute the energy $I(\mathcal{M}_1)$. (b) Let $\mathcal{M}_2$ be the probability measure proportional to surface area on the boundary of K. Compute the energy $I(\mathcal{M}_2)$. (c) What estimate for the capacity of K do we obtain from the preceding two computations?

This ends our motivational excursion into physics.[1] We now return to pure mathematics.

---

[1] Mathematical models are often used in physics. But remember Topsøe's Thesis 22: "Those who seek a phenomenon which exactly follows a mathematical model, seek in vain." [262]

Abstract Potential. Let a real number $s \geq 0$ be given. Let (S, ρ) be a metric space. Let $\mathcal{M}$ be a finite Borel measure on S. The s-**potential** of $\mathcal{M}$ is a function $U^s_{\mathcal{M}} : S \to [0, \infty]$ defined by

$$U^s_{\mathcal{M}}(x) = \int \frac{\mathcal{M}(dy)}{\rho(x, y)^s}.$$

The s-**energy** of $\mathcal{M}$ is

$$I^s(\mathcal{M}) = \int U^s_{\mathcal{M}}(x)\, \mathcal{M}(dx) = \int \int \frac{\mathcal{M}(dx)\, \mathcal{M}(dy)}{\rho(x, y)^s}.$$

(3.2.2) Exercise. Let $S = \mathbb{R}$, and let $\mathcal{M} = \mathcal{L}\lceil[0, 1]$ be Lebesgue measure restricted to the unit interval. Compute $U^s_{\mathcal{M}}(x)$ for $x \in \mathbb{R}$.

(3.2.3) Exercise. Let $S = \{0, 1\}^{(\omega)}$ be the space of infinite strings of **0**'s and **1**'s [$MTFG$, §1.3]. Let ρ_r be the metric, $0 < r < 1$ [$MTFG$, §2.5]. Let $\mathcal{M} = \mathcal{M}_{1/2}$ be the measure of [$MTFG$, §5.5]. Recall that the Hausdorff dimension of this metric space is $\log(1/2)/\log r$, and $\mathcal{M}$ is the corresponding Hausdorff measure. Show that for all $\sigma \in S$,

$$U^s_{\mathcal{M}}(\sigma) = \begin{cases} \dfrac{r^s}{2r^s - 1}, & \text{if } s < \dfrac{\log(1/2)}{\log r}, \\[2ex] \infty, & \text{if } s \geq \dfrac{\log(1/2)}{\log r}. \end{cases}$$

Potential in String Spaces. Recall the **string models**, or **path models**, for iterated function systems used in [$MTFG$, §4.2]. Let E be a finite alphabet, and let $(r_e)_{e \in E}$ be a contracting ratio list: $0 < r_e < 1$ for each $e \in E$. If $\alpha = e_1 e_2 \cdots e_k$ is a string of length k made up of letters from E (we write $\alpha \in E^{(k)}$), define the **ratio** to be $r(\alpha) = r_{e_1} r_{e_2} \cdots r_{e_k}$, the product of the individual ratios. The space $S = E^{(\omega)}$ of infinite strings is a metric space in a natural way. For $\sigma \in S$, write $\sigma\lceil k$ for the **prefix** string consisting of the first k letters of σ. If $\alpha \in E^{(k)}$ is a finite string, the **cylinder** (or **basic open set**)

$$[\alpha] = \left\{ \sigma \in E^{(\omega)} : \sigma\lceil k = \alpha \right\}$$

consists of all infinite strings σ that begin with the finite string α. The metric ρ on S is defined such that $\operatorname{diam}[\alpha] = r(\alpha)$ for all finite strings α [$MTFG$, §4.2]. In particular, $\operatorname{diam} S = \operatorname{diam}[\Lambda] = 1$, where Λ is the empty string. The space $S = E^{(\omega)}$ is a compact metric space. The measure $\mathcal{M}$ on S is defined such that $\mathcal{M}([\alpha]) = r(\alpha)^{s_0}$ for all α, where s_0 is the **similarity dimension**, the solution of the equation

$$\sum_{e \in E} r_e^{s_0} = 1$$

[$MTFG$, §6.3]. With this metric and measure, we have

$$\mathcal{M}(B) = (\operatorname{diam} B)^{s_0}$$

for all balls B (since every ball B is of the form $B = [\alpha]$ for some α). So for upper and lower densities we have

$$\frac{r_{\min}}{2^{s_0}} \leq \underline{D}_{\mathcal{M}}^{s_0}(\sigma) \leq \overline{D}_{\mathcal{M}}^{s_0}(\sigma) \leq \frac{1}{2^{s_0}}.$$

Thus, by (1.5.11) and (1.5.13), $\dim S = \operatorname{Dim} S = s_0$.

Let us estimate the energy $I^s(\mathcal{M})$ for this measure. Let $\sigma, \tau \in S$ be two strings. If their first letters disagree, $\sigma{\restriction}1 \neq \tau{\restriction}1$, then $\rho(\sigma, \tau) = 1$. If their first letters agree, we must look at the second letters. And so on. The measure of the set of strings τ such that $\sigma{\restriction}1 \neq \tau{\restriction}1$ is $\mathcal{M}\big(S \setminus [\sigma{\restriction}1]\big) = 1 - \mathcal{M}\big([\sigma{\restriction}1]\big) = 1 - r(\sigma{\restriction}1)^{s_0}$. The measure of the set of strings τ such that $\sigma{\restriction}k = \tau{\restriction}k$ but $\sigma{\restriction}(k+1) \neq \tau{\restriction}(k+1)$ is $\mathcal{M}\big([\sigma{\restriction}k]\big) - \mathcal{M}\big(\sigma{\restriction}(k+1)\big) = r(\sigma{\restriction}k)^{s_0} - r(\sigma{\restriction}(k+1))^{s_0} = r(\sigma{\restriction}k)^{s_0}\big(1 - r_{\sigma_{k+1}}^{s_0}\big)$, where σ_{k+1} denotes the $(k+1)$st letter of σ. Thus

$$\begin{aligned}
U_{\mathcal{M}}^s(\sigma) &= \int \frac{\mathcal{M}(d\tau)}{\rho(\sigma,\tau)^s} \\
&= \sum_{k=0}^{\infty} \frac{r(\sigma{\restriction}k)^{s_0}\big(1 - r_{\sigma_{k+1}}^{s_0}\big)}{r(\sigma{\restriction}k)^s} \\
&= \sum_{k=0}^{\infty} r(\sigma{\restriction}k)^{s_0-s}\big(1 - r_{\sigma_{k+1}}^{s_0}\big).
\end{aligned}$$

Now, if we write $r_{\min} = \min r_e$ and $r_{\max} = \max r_e$, then $0 < r_{\min} \leq r_{\max} < 1$. We may compare the series with two geometric series:

$$\sum_k r_{\min}^{(s_0-s)k}\big(1 - r_{\max}^{s_0}\big) \leq U_{\mathcal{M}}^s(\sigma) \leq \sum_k r_{\max}^{(s_0-s)k}\big(1 - r_{\min}^{s_0}\big).$$

Thus, if $s \geq s_0$, we have $U_{\mathcal{M}}^s(\sigma) = \infty$ for all σ; while if $s < s_0$, we have $U_{\mathcal{M}}^s(\sigma) \leq A_s < \infty$, where A_s is the sum of a certain geometric series. Thus

$$\begin{aligned}
I^s(\mathcal{M}) &= \infty, &&\text{if } s \geq s_0, \\
I^s(\mathcal{M}) &< \infty, &&\text{if } s < s_0.
\end{aligned}$$

Potential and Hausdorff Dimension. Let (S, ρ) be a metric space, and $s > 0$. The s-**capacity** of S is

$$\big(\inf \{ I^s(\mathcal{M}) : \mathcal{M} \text{ is a Borel probability measure on } S \}\big)^{-1/s}.$$

We say that S has s-**capacity zero** iff $I^s(\mathcal{M}) = \infty$ for all nonzero finite measures $\mathcal{M}$ on S. Note that S has **positive** s-**capacity** iff there is a finite nonzero Borel measure $\mathcal{M}$ on S with finite s-energy, $I^s(\mathcal{M}) < \infty$. We will be concerned with whether the s-capacity is zero or positive, and not with the actual value of the s-capacity.

In fact, for any Borel set E in a complete metric space S:

(1) E has s-capacity zero for all $s > \dim E$.
(2) E has positive s-capacity for all $s < \dim E$.

The upper density theorem (1.5.13) is used for this.

(3.2.4) Theorem. *Let* $\mathcal{M}$ *be a finite Borel measure on the metric space* (S, ρ), *and let* $s > 0$. *If* $I^s(\mathcal{M}) < \infty$, *then* $\overline{D}^s_{\mathcal{M}}(x) = 0$ *for almost all* $x \in S$.

Proof. I must show that $\mathcal{M}(E) = 0$, where $E = \left\{ x \in S : \overline{D}^s(x) > 0 \right\}$. But E is a countable union of sets $E_\varepsilon = \left\{ x \in S : \overline{D}^s(x) > \varepsilon \right\}$, so it is enough to prove $\mathcal{M}(E_\varepsilon) = 0$ for $\varepsilon > 0$. Given $x \in E_\varepsilon$, there exists a sequence (r_i), decreasing to 0, with

$$\mathcal{M}\left(\overline{B}_{r_i}(x)\right) > \varepsilon(2r_i)^s.$$

Now, $I^s(\mathcal{M}) < \infty$, so $\mathcal{M}(\{x\}) = 0$, and therefore $\lim_{q \to 0} \mathcal{M}(\overline{B}_q(x)) = 0$. So there is q_i with $0 < q_i < r_i$ such that

$$\mathcal{M}\left(\overline{B}_{r_i}(x) \setminus \overline{B}_{q_i}(x)\right) > \varepsilon(2r_i)^s.$$

By replacing r_i with a subsequence, we may assume that $r_{i+1} < q_i$ for all i, so that the differences $A_i = \overline{B}_{r_i}(x) \setminus \overline{B}_{q_i}(x)$ are disjoint. Now, for $y \in \overline{B}_{r_i}(x)$, we have $\rho(x, y) \leq r_i < 2r_i$, so

$$U^s_{\mathcal{M}}(x) = \int \frac{\mathcal{M}(dy)}{\rho(x, y)^s} \geq \sum_i \int_{A_i} \frac{\mathcal{M}(dy)}{(2r_i)^s}$$

$$= \sum_i \frac{\mathcal{M}(A_i)}{(2r_i)^s} > \sum_{i=1}^{\infty} \varepsilon = \infty.$$

But $\int U^s_{\mathcal{M}}(x)\, \mathcal{M}(dx) < \infty$, so the set of x with $U^s_{\mathcal{M}}(x) = \infty$ has measure zero. Therefore, $\mathcal{M}(E_\varepsilon) = 0$.   ☺

(3.2.5) Corollary. *S has s-capacity zero for all $s > \dim S$.*

Proof. Suppose S has positive s-capacity. Then there is a finite nonzero measure $\mathcal{M}$ with $I^s(\mathcal{M}) < \infty$. By the theorem, $\overline{D}^s_{\mathcal{M}}(x) = 0$ for almost all $x \in S$. So by Theorem 1.5.13, $\mathcal{H}^s(S) = \infty$, and therefore $\dim S \geq s$.   ☺

(3.2.6) Theorem. *Let* (S, ρ) *be a metric space, let* $s < t$, *and let* $\mathcal{M}$ *be a finite nonzero Borel measure on* S. *If* $\overline{D}^t_{\mathcal{M}}(x) = 0$ *almost everywhere, then there is a Borel set* $E \subseteq S$ *with* $\mathcal{M}(E) > 0$ *and* $I^s(\mathcal{M}{\restriction}E) < \infty$.

Proof. For $k \in \mathbb{N}$, write

$$E_k = \left\{ x \in S : \frac{\mathcal{M}(\overline{B}_r(x))}{(2r)^t} < 1 \text{ for all } r \leq 2^{-k} \right\}.$$

Now $\overline{D}^t_{\mathcal{M}}(x) = 0$ almost everywhere, so E_k increases to (almost all of) S, and thus $\mathcal{M}(E_k) \to \mathcal{M}(S)$. Since $\mathcal{M}(S) > 0$, there is k such that $\mathcal{M}(E_k) > 0$. Choose such a k, and let $E = E_k$. Now if we write $\mathcal{M}_1 = \mathcal{M} \!\restriction\! E$ and $B_i = \overline{B}_{2^{-i}}(x)$, we have

$$
U^s_{\mathcal{M}_1}(x) = \int \frac{\mathcal{M}_1(dy)}{\rho(x,y)^s} \leq \int_{S \setminus B_k} \frac{\mathcal{M}_1(dy)}{\rho(x,y)^s} + \sum_{i=k}^{\infty} \int_{B_i \setminus B_{i+1}} \frac{\mathcal{M}_1(dy)}{\rho(x,y)^s}
$$

$$
\leq \frac{\mathcal{M}(E_k)}{2^{-ks}} + \sum_{i=k}^{\infty} \frac{\mathcal{M}(B_i)}{(2^{-i-1})^s} = \frac{\mathcal{M}(E_k)}{2^{-ks}} + \sum_{i=k}^{\infty} \frac{\mathcal{M}(B_i)}{(2^{-i+1})^t} \frac{(2^{-i+1})^t}{(2^{-i-1})^s}
$$

$$
\leq \frac{\mathcal{M}(E_k)}{2^{-ks}} + 4^t \sum_{i=k}^{\infty} \frac{\mathcal{M}(B_i)}{(2 \cdot 2^{-i})^t} \, (2^{i+1})^{s-t}
$$

$$
\leq \frac{\mathcal{M}(E_k)}{2^{-ks}} + 4^t \sum_{i=k}^{\infty} (2^{i+1})^{s-t}.
$$

This geometric series converges, so $U^s_{\mathcal{M}_1}(x)$ has an upper bound independent of x, and therefore $I^s(\mathcal{M}_1) < \infty$.   ☺

(3.2.7) Corollary. *Let E be a Borel set in a complete metric space S. Then E has positive s-capacity for all $s < \dim E$.*

Proof. Suppose $s < \dim E$. By $(1.5.16(\gamma))$ there is a finite measure with $\mathcal{M}(E) > 0$ such that $\overline{D}^t_{\mathcal{M}}(x) = 0$ almost everywhere. So by the theorem, there is a Borel set $F \subseteq A$ such that $\mathcal{M}(F) > 0$ and $I^s(\mathcal{M} \!\restriction\! F) < \infty$. Therefore, E has positive s-capacity.   ☺

Self-Similar Fractals. Next we consider self-similar fractals, as in [MTFG, Chap. 4]. For a self-similar fractal K there corresponds a string space $E^{(\omega)}$ and a continuous "model map" $h \colon E^{(\omega)} \to S$ such that $h\big[E^{(\omega)}\big] = K$ and

$$
\rho\big(h(\sigma), h(\tau)\big) \leq \rho(\sigma, \tau).
$$

The abstract potential theory can be used to provide a proof of the connection between the Hausdorff dimension and the similarity dimension of a self-similar fractal. Although the upper bound computation can be done this way, it is more difficult than the direct approach.

(3.2.8) Lemma. *Let S and T be metric spaces. Let $h \colon S \to T$ satisfy*

$$
\rho\big(h(x), h(y)\big) \leq \rho(x, y)
$$

for all $x, y \in S$. If $\mathcal{M}$ is a finite Borel measure on S, then $\mathcal{M}_1 = h_*(\mathcal{M})$ is a finite Borel measure on T,

$$
U^s_{\mathcal{M}_1}\big(h(x)\big) \geq U^s_{\mathcal{M}}(x)
$$

for all $x \in S$, and

$$
I^s(\mathcal{M}_1) \geq I^s(\mathcal{M}).
$$

Proof. Write $u = h(x)$. Then by the change of variables formula 2.2.4,

$$U^s_{\mathcal{M}_1}(u) = \int \frac{h_*(\mathcal{M})(dv)}{\rho(u,v)^s} = \int \frac{\mathcal{M}(dy)}{\rho\big(h(x),h(y)\big)^s}$$
$$\geq \int \frac{\mathcal{M}(dy)}{\rho(x,y)^s} = U^s_{\mathcal{M}}(x).$$

For the s-energy, apply the change of variables formula again.

$$I^s(\mathcal{M}_1) = \int U^s_{\mathcal{M}_1}(u)\,\mathcal{M}_1(du) = \int U^s_{\mathcal{M}_1}\big(h(x)\big)\,\mathcal{M}(dx)$$
$$\geq \int U^s_{\mathcal{M}}(x)\,\mathcal{M}(dx) = I^s(\mathcal{M}). \qquad ☺$$

(3.2.9) Exercise. Let S, T, and h be as before. Suppose every finite Borel measure on T can be lifted to S. If S has s-capacity zero, then so does T.

(3.2.10) Exercise. Let K be a self-similar fractal with similarity dimension s_0. Use capacity to prove that $\dim K \leq s_0$.

For the lower bound in Euclidean space $\mathbb{R}^d$, the computation depends on a combinatorial lemma. (In [MTFG], it appears in the proof of (6.3.12), or preceding (6.4.8); in this book, see Lemma 5.4.2, below.) If the "open set condition" is satisfied, then the finite measure $\mathcal{M}$ on the string space $E^{(\omega)}$ and the model map h satisfy an inequality: There is a constant C such that for all Borel sets $A \subseteq K$,

$$\mathcal{M}\big(h^{-1}[A]\big) \leq C\,(\text{diam } A)^{s_0},$$

where s_0 is the similarity dimension. From this, we see that the measure $\mathcal{M}_1 = h_*(\mathcal{M})$ satisfies that for any $s < s_0$,

$$\frac{\mathcal{M}_1\big(\overline{B}_\varepsilon(x)\big)}{(2\varepsilon)^s} \leq C\,(2\varepsilon)^{s_0-s},$$

so that the upper density $\overline{D}^s_{\mathcal{M}_1}(x)$ is 0. Then by Theorem 3.2.6, we may conclude that K has positive capacity for all $s < s_0$, and therefore, $\dim K \geq s_0$.

Cartesian Product of Fractals. Potential theory will enable us to prove very simply an inequality for the Hausdorff dimension of a product of metric spaces.

(3.2.11) Theorem. *Let S_1 and S_2 be nonempty complete metric spaces. Then the Cartesian product $S_1 \times S_2$ with the maximum metric satisfies*

$$\dim (S_1 \times S_2) \geq \dim S_1 + \dim S_2.$$

Proof. If $S_1 \times S_2$ has infinite Hausdorff dimension, there is nothing to prove. So assume it has finite dimension. Therefore, $S_1 \times S_2$, S_1, S_2 are separable.

Let $s_1 < \dim S_1$ and $s_2 < \dim S_2$ be given. Then there exist Borel probability measures $\mathcal{M}_1$ on S_1 and $\mathcal{M}_2$ on S_2 with $I^{s_1}(\mathcal{M}_1) < \infty$, $I^{s_2}(\mathcal{M}_2) < \infty$. The product measure $\mathcal{M}_1 \otimes \mathcal{M}_2$ is a finite measure on the product space, and

$$U^{s_1+s_2}_{\mathcal{M}_1 \otimes \mathcal{M}_2}(x,y) = \int_{S_1 \times S_2} \frac{\mathcal{M}_1 \otimes \mathcal{M}_2(d(u,v))}{\rho\big((x,y),(u,v)\big)^{s_1} \, \rho\big((x,y),(u,v)\big)^{s_2}}$$

$$\leq \int_{S_1} \frac{\mathcal{M}_1(du)}{\rho(x,u)^{s_1}} \int_{S_2} \frac{\mathcal{M}_2(dv)}{\rho(y,v)^{s_2}}$$

$$= U^{s_1}_{\mathcal{M}_1}(x) \, U^{s_2}_{\mathcal{M}_2}(y).$$

Now integrate:

$$I^{s_1+s_2}(\mathcal{M}_1 \otimes \mathcal{M}_2) = \int U^{s_1+s_2}_{\mathcal{M}_1 \otimes \mathcal{M}_2}(x,y) \, \mathcal{M}_1 \otimes \mathcal{M}_2(d(x,y))$$

$$\leq \int U^{s_1}_{\mathcal{M}_1}(x) \, \mathcal{M}_1(dx) \int U^{s_2}_{\mathcal{M}_2}(y) \, \mathcal{M}_2(dy)$$

$$= I^{s_1}(\mathcal{M}_1) I^{s_2}(\mathcal{M}_2).$$

But $I^{s_1}(\mathcal{M}_1) < \infty$ and $I^{s_2}(\mathcal{M}_2) < \infty$, so $I^{s_1+s_2}(\mathcal{M}_1 \otimes \mathcal{M}_2) < \infty$. Therefore, $\dim(S_1 \times S_2) \geq s_1 + s_2$. This is true for any $s_1 < \dim S_1$ and $s_2 < \dim S_2$, so $\dim(S_1 \times S_2) \geq \dim S_1 + \dim S_2$, as required. ☺

(3.2.12) Exercise. Let E_1 and E_2 be Borel sets in complete metric spaces S_1 and S_2. Then $\dim(E_1 \times E_2) \geq \dim E_1 + \dim E_2$.

(3.2.13) Exercise. Suppose S_1 and S_2 are arbitrary metric spaces. Can we still conclude that $\dim(S_1 \times S_2) \geq \dim S_1 + \dim S_2$?

3.3 Fractal Measures

Hausdorff Dimension of a Measure. Let S be a metric space, and let $\mathcal{M} \in \mathfrak{P}(S)$ be a tight probability measure. We may define the **Hausdorff dimension** of the measure $\mathcal{M}$ in a way quite similar to the definition used for sets.

(3.3.1) Definition. Let $\varepsilon > 0$. An $\mathcal{M}$-almost ε-cover is a countable collection $\{E_1, E_2, \cdots\}$ of Borel sets with $\operatorname{diam} E_i < \varepsilon$ for all i and

$$\mathcal{M}\left(S \setminus \bigcup_{i=1}^{\infty} E_i\right) = 0.$$

Let $s > 0$. Define

$$\mathcal{H}^s_\varepsilon(\mathcal{M}) = \inf \sum_{i=1}^{\infty} (\operatorname{diam} E_i)^s,$$

where the infimum is over all $\mathcal{M}$-almost ε-covers $\{E_i\}$. Then define

$$\mathcal{H}^s(\mathcal{M}) = \lim_{\varepsilon \to 0} \mathcal{H}^s_\varepsilon(\mathcal{M}).$$

As in the previous version of the Hausdorff measure, we may require the sets E_i to be closed sets.

Some of the basic properties are easily proved. We say two measures $\mathcal{M}_1, \mathcal{M}_2 \in \mathfrak{P}(S)$ are **equivalent**, and write $\mathcal{M}_1 \sim \mathcal{M}_2$, iff they have the same null sets (that is, each is absolutely continuous with respect to the other).

(3.3.2) Exercise. If $F \subseteq S$ is the support[2] of $\mathcal{M} \in \mathfrak{P}(S)$, then $\mathcal{H}^s(\mathcal{M}) \le \mathcal{H}^s(F)$. Give an example with strict inequality.

(3.3.3) Exercise. Monotone: If $\mathcal{M}_1 \ll \mathcal{M}_2$, then $\mathcal{H}^s(\mathcal{M}_1) \le \mathcal{H}^s(\mathcal{M}_2)$. If $\mathcal{M}_1 \sim \mathcal{M}_2$, then $\mathcal{H}^s(\mathcal{M}_1) = \mathcal{H}^s(\mathcal{M}_2)$.

It may be convenient to define $\mathcal{H}^s(\mathcal{M})$ for finite measures $\mathcal{M}$ that are not probability measures. To do this, choose some probability measure $\mathcal{M}_1$ equivalent to $\mathcal{M}$ and define $\mathcal{H}^s(\mathcal{M}) = \mathcal{H}^s(\mathcal{M}_1)$. By the preceding exercise, this does not depend on the choice. In fact, using the same method, we may define $\mathcal{H}^s(\mathcal{M})$ for a σ-finite measure.

(3.3.4) Exercise. Subadditive: If $\mathcal{M} \ll \sum_n \mathcal{M}_n$, then $\mathcal{H}^s(\mathcal{M}) \le \sum_n \mathcal{H}^s(\mathcal{M}_n)$.

(3.3.5) Exercise. Let $\mathcal{M}$ be a finite Borel measure. Define a set-function $\mathcal{N}$ as follows: Given any Borel set E, let $\mathcal{M}{\restriction}E$ be the restriction of $\mathcal{M}$ to E, and define $\mathcal{N}(E) = \mathcal{H}^s(\mathcal{M}{\restriction}E)$. Then $\mathcal{N}$ is a Borel measure.

(3.3.6) Exercise. Additive: Let $\mathcal{M}_1$ and $\mathcal{M}_2$ be finite measures. Suppose $\mathcal{M}_1 \perp \mathcal{M}_2$. Then $\mathcal{H}^s(\mathcal{M}_1 + \mathcal{M}_2) = \mathcal{H}^s(\mathcal{M}_1) + \mathcal{H}^s(\mathcal{M}_2)$. Generalize to countable additivity.

As usual, we get a "critical value" for s:

(3.3.7) Proposition. *Let $s < t$. If $\mathcal{H}^s(\mathcal{M}) < \infty$, then $\mathcal{H}^t(\mathcal{M}) = 0$. If $\mathcal{H}^t(\mathcal{M}) > 0$, then $\mathcal{H}^s(\mathcal{M}) = \infty$.*

Proof. Fix $\varepsilon > 0$. For any $\mathcal{M}$-almost ε-cover $\{E_i\}$, we have

$$\sum (\operatorname{diam} E_i)^t \le \sum \varepsilon^{t-s} (\operatorname{diam} E_i)^s = \varepsilon^{t-s} \sum (\operatorname{diam} E_i)^s.$$

Thus, $\mathcal{H}^t_\varepsilon(\mathcal{M}) \le \varepsilon^{t-s} \mathcal{H}^s_\varepsilon(\mathcal{M})$. Now, if $\mathcal{H}^s(\mathcal{M}) < \infty$, when we let $\varepsilon \to 0$, the right-hand side converges to 0, so $\mathcal{H}^t(\mathcal{M}) = 0$. Or if $\mathcal{H}^t(\mathcal{M}) > 0$, then when

---

[2] Recall that the support of $\mathcal{M}$ is the smallest closed set F with $\mathcal{M}(S \setminus F) = 0$.

$\varepsilon \to 0$, the left-hand side has positive limit, and $\varepsilon^{t-s} \to 0$, so $\mathcal{H}_\varepsilon^s(\mathcal{M}) \to \infty$.

☺

From this proposition, it follows that there is a critical value s_0 such that

$$\mathcal{H}^s(\mathcal{M}) = \begin{cases} \infty & \text{if } s < s_0, \\ 0 & \text{if } s > s_0. \end{cases}$$

We will sometimes call s_0 the **upper Hausdorff dimension** of the measure $\mathcal{M}$ and write $s_0 = \dim^* \mathcal{M}$.

There is an alternate possibility for defining the "Hausdorff dimension" of a finite measure $\mathcal{M}$: the smallest dimension of a set that carries $\mathcal{M}$.

(3.3.8) Proposition. *Let S be a metric space, and let $\mathcal{M} \in \mathfrak{P}(S)$. Then*

$$\dim^* \mathcal{M} = \inf \{ \dim E : E \text{ is a Borel set}, \mathcal{M}(S \setminus E) = 0 \}.$$

Proof. Write

$$s_1 = \inf \{ \dim E : E \text{ is a Borel set}, \mathcal{M}(S \setminus E) = 0 \}.$$

Let $E \subseteq S$ with $\mathcal{M}(S \setminus E) = 0$. An ε-cover of E is an almost ε-cover of S, so for any s we have $\mathcal{H}_\varepsilon^s(\mathcal{M}) \leq \mathcal{H}_\varepsilon^s(E)$. Let $\varepsilon \to 0$ to obtain $\mathcal{H}^s(\mathcal{M}) \leq \mathcal{H}^s(E)$. Thus $\dim^* \mathcal{M} \leq \dim E$. This shows that $\dim^* \mathcal{M} \leq s_1$.

For the reverse inequality, let $s > \dim^* \mathcal{M}$ be given, so that $\mathcal{H}^s(\mathcal{M}) = 0$. For each $n \in \mathbb{N}$, there is an $\mathcal{M}$-almost 2^{-n}-cover $\{A_{ni}\}_{i=1}^\infty$ with $\sum_i (\operatorname{diam} A_{ni})^s < 2^{-n}$. Then the set

$$F = \bigcap_n \bigcup_i A_{ni}$$

satisfies $\mathcal{M}(S \setminus F) = 0$. But for each n, the family $\{A_{ni}\}_{i=1}^\infty$ covers F. So $\mathcal{H}_{2^{-n}}^s(F) \leq 2^{-n}$, so that $\mathcal{H}^s(F) = 0$. Therefore, $\dim F \leq s$, so $s_1 \leq s$. This is true for every $s > \dim^* \mathcal{M}$, so we have $s_1 \leq \dim^* \mathcal{M}$.   ☺

(3.3.9) Exercise. Let $\mathcal{M} \in \mathfrak{P}(S)$. There is a set E with $\mathcal{M}(S \setminus E) = 0$ and $\dim^* \mathcal{M} = \dim E$.

There is also a "lower" Hausdorff dimension for a measure—the smallest dimension of a set charged by $\mathcal{M}$.

(3.3.10) Definition. Let S be a metric space, and let $\mathcal{M}$ be a finite nonzero Borel measure on S. The **lower Hausdorff dimension** of $\mathcal{M}$ is

$$\dim_* \mathcal{M} = \inf \{ \dim E : E \text{ is a Borel set}, \mathcal{M}(E) > 0 \}.$$

Clearly, $\dim_* \mathcal{M} \leq \dim^* \mathcal{M}$ (since $\mathcal{M}$ is not the zero measure). The lower Hausdorff dimension is sometimes called simply the Hausdorff dimension, and written $\dim \mathcal{M}$.

Similar definitions may be made for the packing dimension of a measure.

(3.3.11) Definition. Let S be a metric space, and let $\mathcal{M}$ be a finite Borel measure on S. The **upper packing dimension** of $\mathcal{M}$ is

$$\mathrm{Dim}^* \mathcal{M} = \inf \{ \mathrm{Dim}\, E : E \text{ is a Borel set, } \mathcal{M}(S \setminus E) = 0 \},$$

and the **lower packing dimension** of $\mathcal{M}$ is

$$\mathrm{Dim}_* \mathcal{M} = \inf \{ \mathrm{Dim}\, E : E \text{ is a Borel set, } \mathcal{M}(E) > 0 \}.$$

Pointwise Dimension. When the Hausdorff dimension varies from one point to another within a set, it may be useful to have a different dimension associated to each point. (We use lim sup and lim inf, since the desired limit may not exist.)

(3.3.12) Definition. Let S be a metric space, and let $\mathcal{M}$ be a finite Borel measure on S. For $x \in S$, the **upper pointwise dimension** of $\mathcal{M}$ at x is

$$\overline{L}_{\mathcal{M}}(x) = \limsup_{r \to 0} \frac{\log \mathcal{M}\big(\overline{B}_r(x)\big)}{\log r},$$

and the **lower pointwise dimension** of $\mathcal{M}$ at x is

$$\underline{L}_{\mathcal{M}}(x) = \liminf_{r \to 0} \frac{\log \mathcal{M}\big(\overline{B}_r(x)\big)}{\log r}.$$

If the two are equal, their common value is the **pointwise dimension** of $\mathcal{M}$ at x.

Note some inequalities involving the pointwise dimensions and the densities (Definition 1.5.8).

(3.3.13) Proposition. *Let S be a metric space. Let $\mathcal{M} \in \mathcal{P}(S)$, let $x \in S$, and let $s > 0$.*

(a) *If $\underline{L}_{\mathcal{M}}(x) < s$, then $\overline{D}_{\mathcal{M}}^s(x) \geq 2^{-s}$.*
(b) *If $\underline{L}_{\mathcal{M}}(x) > s$, then $\overline{D}_{\mathcal{M}}^s(x) \leq 2^{-s}$.*
(c) *If $\overline{L}_{\mathcal{M}}(x) < s$, then $\underline{D}_{\mathcal{M}}^s(x) \geq 2^{-s}$.*
(d) *If $\overline{L}_{\mathcal{M}}(x) > s$, then $\underline{D}_{\mathcal{M}}^s(x) \leq 2^{-s}$.*

Proof. (a) Assume $\underline{L}_{\mathcal{M}}(x) < s$. Let $\varepsilon > 0$. There is r with $0 < r < \varepsilon \wedge 1$ such that

$$\frac{\log \mathcal{M}(\overline{B}_r(x))}{\log r} < s,$$

$$\log \mathcal{M}(\overline{B}_r(x)) > s \log r,$$

$$\mathcal{M}(\overline{B}_r(x)) > r^s,$$

$$\frac{\mathcal{M}(\overline{B}_r(x))}{(2r)^s} > \frac{r^s}{(2r)^s} = \frac{1}{2^s}.$$

Therefore, $\overline{D}_M^s(x) \geq 2^{-s}$.

The other parts are done in the same way. ☺

The upper and lower packing and Hausdorff dimensions are actually the upper and lower bounds of the appropriate pointwise dimensions. That is, they are the best constants such that $\dim_* \mathcal{M} \leq \underline{L}_\mathcal{M}(x) \leq \dim^* \mathcal{M}$ almost everywhere:

(3.3.14) Theorem. *Let S be a metric space, and let $\mathcal{M}$ be a Borel probability measure on S. Then*

(a) $\dim^* \mathcal{M} = \inf \{ \alpha : \underline{L}_\mathcal{M}(x) \leq \alpha$ *for $\mathcal{M}$-almost all x* $\}$.
(b) $\dim_* \mathcal{M} = \sup \{ \alpha : \underline{L}_\mathcal{M}(x) \geq \alpha$ *for $\mathcal{M}$-almost all x* $\}$.

Suppose in addition that $\mathcal{M}$ has the strong Vitali property. Then

(c) $\mathrm{Dim}^* \mathcal{M} = \inf \{ \alpha : \overline{L}_\mathcal{M}(x) \leq \alpha$ *for $\mathcal{M}$-almost all x* $\}$.
(d) $\mathrm{Dim}_* \mathcal{M} = \sup \{ \alpha : \overline{L}_\mathcal{M}(x) \geq \alpha$ *for $\mathcal{M}$-almost all x* $\}$.

Proof. (a) Let $\alpha > 0$ be such that $\underline{L}_\mathcal{M}(x) \leq \alpha$ for $\mathcal{M}$-almost all x. Let $E = \{ x : \underline{L}_\mathcal{M}(x) \leq \alpha \}$, so that $\mathcal{M}(S \setminus E) = 0$. Now, if $s > \alpha$, then by part (a) of the preceding proposition, $\overline{D}_M^s(x) \geq 2^{-s}$ everywhere on E. But then by Theorem 1.5.13, $\mathcal{C}^s(E) < \infty$, so $\dim E \leq s$. Therefore, $\dim^* \mathcal{M} \leq s$. This is true for all $s > \alpha$, so $\dim^* \mathcal{M} \leq \alpha$.

Therefore, we have $\dim^* \mathcal{M}$ is a lower bound for

$$\{ \alpha : \underline{L}_\mathcal{M}(x) \leq \alpha \text{ for } \mathcal{M}\text{-almost all } x \}.$$

To show that $\dim^* \mathcal{M}$ is the *greatest* lower bound, we may show that $\underline{L}_M(x) \leq \dim^* \mathcal{M}$ for $\mathcal{M}$-almost all x. If not, there is $\beta > \dim^* \mathcal{M}$ such that $\mathcal{M}(F) > 0$, where $F = \{ x : \underline{L}_\mathcal{M}(x) > \beta \}$. Now, if E is any set with $\mathcal{M}(S \setminus E) = 0$, then $\mathcal{M}(F \cap E) > 0$. Then for $s < \beta$ we have $\overline{D}_M^s(x) \leq 2^{-s}$ everywhere on $F \cap E$. So by Theorem 1.5.13, $\mathcal{C}^s(F \cap E) > 0$, and $\dim F \cap E \geq s$. This is true for all $s < \beta$, so $\dim F \cap E \geq \beta$. So $\dim E \geq \beta$. Take the infimum over E to conclude that $\dim^* \mathcal{M} \geq \beta$. This contradicts the choice of β.

(b) is done in the same way. For (c) and (d), use Theorem 1.5.11 (which requires the strong Vitali property). ☺

IFS with Weights. Let S be a complete metric space. An **iterated function system** on S (indexed by a finite alphabet E) is a finite list $(f_e)_{e \in E}$ of functions $f_e \colon S \to S$. An iterated function system is **contracting** (or **hyperbolic**) iff there is a Lipschitz constant $r < 1$ such that

$$\rho\big(f_e(x), f_e(y)\big) \leq r\,\rho(x, y)$$

for all $x, y \in S$ and all $e \in E$. The **attractor** (or invariant set) for a contracting iterated function system is the unique nonempty compact set K such that

$$K = \bigcup_{e \in E} f_e[K].$$

Now we will assign a "weight" or "probability" p_e to each letter e in such a way that $p_e > 0$ and $\sum p_e = 1$. The entire list

$$\big(f_e, p_e\big)_{e \in E}$$

is an **iterated function system with weights**. The **attractor** (or invariant measure) of such a list is a probability measure $\mathcal{T} \in \mathfrak{P}(S)$ such that

$$\mathcal{T} = \sum_{e \in E} p_e\, f_{e*}(\mathcal{T}).$$

That is (by the definition of the "image measure," Definition 2.1.10),

$$\mathcal{T}(A) = \sum_{e \in E} p_e\, \mathcal{T}\big(f_e^{-1}[A]\big)$$

for all Borel sets A.

Under appropriate conditions, we will prove existence and uniqueness of the invariant measure (attractor) $\mathcal{T}$ in a manner similar to the method used for invariant set attractors. We first consider the case when the space S is bounded.

(3.3.15) Proposition. *Let S be a nonempty complete metric space. Suppose there is a constant t such that $\rho(x, y) \leq t$ for all $x, y \in S$. Let (f_e, p_e) be a contracting iterated function system with weights. Then there is a unique tight probability measure $\mathcal{T}$ such that*

$$\mathcal{T} = \sum_{e \in E} p_e\, f_{e*}(\mathcal{T}).$$

Proof. Let γ be a constant larger than t. We will use the contraction mapping theorem on the space $\mathfrak{P}(S)$ with the metric ρ_γ.

Define a function $F \colon \mathfrak{P}(S) \to \mathfrak{P}(S)$ by

$$F(\mathcal{M}) = \sum_{e \in E} p_e\, f_{e*}(\mathcal{M}).$$

First note that $F(\mathcal{M})$ is a Borel measure since the f_e are continuous. We next claim that $F(\mathcal{M})$ is tight: Given $\varepsilon > 0$, choose a compact set K with

$\mathcal{M}(K) > 1 - \varepsilon$. Let $K_1 = \bigcup_{e \in E} f_e[K]$. Then K_1 is compact (since the f_e are continuous), and for each e we have $f_e^{-1}[K_1] \supseteq K$, so $\mathcal{M}(f_e^{-1}[K_1]) \geq 1 - \varepsilon$. Thus

$$F(\mathcal{M})(K_1) = \sum p_e \, \mathcal{M}(f_e^{-1}[K_1]) \geq \sum p_e(1 - \varepsilon) = 1 - \varepsilon.$$

This shows that $F(\mathcal{M})$ is tight, so in fact, the function F maps $\mathfrak{P}(S)$ into itself.

We have postulated a constant $r < 1$ with $\rho(f_e(x), f_e(y)) \leq r \, \rho(x, y)$. For any $\mathcal{M}, \mathcal{N} \in \mathfrak{P}(S)$, we claim that $\rho_\gamma(F(\mathcal{M}), F(\mathcal{N})) \leq r \, \rho_\gamma(\mathcal{M}, \mathcal{N})$. Let $h \in V_\gamma$. Then the composite function $h \circ f_e$ satisfies

$$\left| h(f_e(x)) - h(f_e(y)) \right| \leq \rho\big(f_e(x), f_e(y)\big) \leq r \, \rho(x, y).$$

Fix a point $x_0 \in S$. Then for all $x \in S$, we have $\left| h(f_e(x)) - h(f_e(x_0)) \right| \leq r\gamma$. So

$$h_e(x) = \frac{1}{r}\big(h(f_e(x)) - h(f_e(x_0))\big)$$

defines a function $h_e \in V_\gamma$. Now

$$\left| \int h(x) \, F(\mathcal{M})(dx) - \int h(x) \, F(\mathcal{N})(dx) \right|$$

$$= \left| \sum p_e \left(\int h(f_e(x)) \, \mathcal{M}(dx) - \int h(f_e(x)) \, \mathcal{N}(dx) \right) \right|$$

$$= \left| \sum p_e \left(\int h(f_e(x)) \, \mathcal{M}(dx) - h(f_e(x_0)) - \int h(f_e(x)) \, \mathcal{N}(dx) + h(f_e(x_0)) \right) \right|$$

$$\leq r \sum p_e \left| \int h_e(x) \, \mathcal{M}(dx) - \int h_e(x) \, \mathcal{N}(dx) \right|$$

$$\leq r \, \rho_\gamma(\mathcal{M}, \mathcal{N}).$$

Thus $\rho_\gamma(F(\mathcal{M}), F(\mathcal{N})) \leq r \, \rho_\gamma(\mathcal{M}, \mathcal{N})$.

Now we may apply the contraction mapping theorem. There is a unique $\mathcal{T} \in \mathfrak{P}(S)$ with $F(\mathcal{T}) = \mathcal{T}$; that is,

$$\mathcal{T} = \sum_{e \in E} p_e \, f_{e*}(\mathcal{T}),$$

as required.   ☺

The next result eliminates the assumption of boundedness of S.

(3.3.16) Theorem. *Let (f_e, p_e) be a contracting iterated function system with weights on a nonempty complete metric space S. Then there is a unique tight probability measure $\mathcal{T}$ such that*

$$\mathcal{T} = \sum_{e \in E} p_e \, f_{e*}(\mathcal{T}).$$

Proof. There is a constant $r < 1$ such that $\rho(f_e(x), f_e(y)) \leq r\,\rho(x, y)$ for all e. Define $F \colon \mathfrak{P}(S) \to \mathfrak{P}(S)$ as before. Let x_e be the fixed point of f_e. Write x_1 for any one of them. Let

$$U = \max_e \rho(x_e, x_1) \qquad \text{and} \qquad T = \frac{(1+r)U + 1}{1 - r}.$$

Note that if $q \geq T$, then $rq + (1+r)U \leq q - 1$. For $q \geq T$ we claim that $f_e\big[\overline{B}_q(x_1)\big] \subseteq \overline{B}_{q-1}(x_1)$. Indeed, if $\rho(y, x_1) \leq q$, then

$$\begin{aligned}
\rho\big(f_e(y), x_1\big) &\leq \rho\big(f_e(y), f_e(x_e)\big) + \rho(x_e, x_1)\\
&\leq r\rho(y, x_e) + \rho(x_e, x_1)\\
&\leq r\rho(y, x_1) + (r+1)\rho(x_e, x_1)\\
&\leq rq + (1+r)U \leq q - 1.
\end{aligned}$$

So, in particular, $f_e\big[\overline{B}_T(x_1)\big] \subseteq \overline{B}_T(x_1)$ for all $e \in E$. The bounded case now shows that there is $\mathfrak{I} \in \mathfrak{P}\big(\overline{B}_T(x_1)\big)$ with $F(\mathfrak{I}) = \mathfrak{I}$.

Next, uniqueness. Suppose $F(\mathfrak{I}) = \mathfrak{I}$. For $q \geq T$, we have $f_e^{-1}\big[\overline{B}_{q-1}(x_1)\big] \supseteq \overline{B}_q(x_1)$ for all e. So

$$\mathfrak{I}\big(\overline{B}_q(x_1)\big) \leq F(\mathfrak{I})\big(\overline{B}_{q-1}(x_1)\big) = \mathfrak{I}\big(\overline{B}_{q-1}(x_1)\big).$$

Thus $\mathfrak{I}\big(\overline{B}_T(x_1)\big) = 1$.

We claim that $\mathfrak{I} \in \mathfrak{P}\big(\overline{B}_T(x_1)\big)$; that is, $\mathfrak{I}\big(S \setminus \overline{B}_T(x_1)\big) = 0$. To prove this, write F^n for the nth iterate of F. For any $\varepsilon > 0$, there is q large enough that $\mathfrak{I}\big(\overline{B}_q(x_1)\big) > 1 - \varepsilon$. But if $n \in \mathbb{N}$ is the least integer $> q - T$, we have $q - n < T$, and

$$\mathfrak{I}\big(\overline{B}_T(x_1)\big) = F^n(\mathfrak{I})\big(\overline{B}_T(x_1)\big) \geq \mathfrak{I}\big(\overline{B}_q(x_1)\big) \geq 1 - \varepsilon.$$

This is true for all $\varepsilon > 0$, so $\mathfrak{I}\big(\overline{B}_T(x_1)\big) = 1$. Now the uniqueness from the contraction mapping theorem proves the uniqueness as stated here.   ☺

Of course, the construction in the proof of the contraction mapping theorem shows (at least in the bounded case) that if $\mathfrak{I}_0$ is any element of $\mathfrak{P}(S)$ and $\mathfrak{I}_{n+1} = F(\mathfrak{I}_n)$ for $n \geq 0$, then $\mathfrak{I}_n$ converges to the fixed point $\mathfrak{I}$ with $F(\mathfrak{I}) = \mathfrak{I}$.

(3.3.17) Exercise. Suppose S is an unbounded complete metric space. Let $(f_e, p_e)_{e \in E}$ be a contracting iterated function system with weights. Let $\mathfrak{I}_0 \in \mathfrak{P}(S)$, and $\mathfrak{I}_{n+1} = \sum_e p_e\, f_{e*}(\mathfrak{I}_n)$ for $n \geq 0$. Does it necessarily follow that $\mathfrak{I}_n$ converges (narrowly) to the fixed point $\mathfrak{I}$?

(3.3.18) Exercise. (The "Collage Theorem" [13, p. 360].) Let S be a bounded complete metric space. Suppose the iterated function system with weights (f_e, p_e) admits contractivity factor $r < 1$, so that $\rho(f_e(x), f_e(y)) \leq r\,\rho(x, y)$ for all e. Define F as before. If $\mathfrak{I}$ is the fixed measure and $\mathcal{N} \in \mathfrak{P}(S)$ is any measure, then

$$\rho_\gamma\left(\mathcal{N}, \mathcal{T}\right) \le \frac{\rho_\gamma\left(\mathcal{N}, F(\mathcal{N})\right)}{1 - r}$$

for any γ larger than the diameter of S. What can we say for unbounded spaces S?

(3.3.19) Exercise. Let K be the attractor (fixed set) for (f_e) and $\mathcal{T}$ be the attractor (fixed measure) for (f_e, p_e). Then K is the support of $\mathcal{T}$. That is, (a) $\mathcal{T}(K) = 1$, and (b) for any open set V with $V \cap K \ne \varnothing$, we must have $\mathcal{T}(V) > 0$.

The same iterated function system (f_e) may be paired with different systems (p_e) of weights. When the weights are changed, the attractor K of the iterated function system (f_e) is unchanged, but the attractor $\mathcal{T}$ of the iterated function system with weights (f_e, p_e) may change.

For a self-similar fractal (with open set condition) there is a privileged choice of weights (the choice that was used implicitly in [$MTFG$, §6.4], for example). Let the maps f_e be similarities with ratios r_e. The **similarity dimension** s is the solution of the equation

$$\sum_{e \in E} r_e^s = 1.$$

Then the weights $p_e = r_e^s$ have sum 1. The weights (r_e^s) are known as the **uniform** weights for (r_e), and the invariant measure $\mathcal{T}$ for (f_e, r_e^s) is known as the **uniform** measure on the self-similar set K.

Examples. Let us take as an example the Sierpiński Gasket [$MTFG$, p. 7] with its uniform measure. Let the alphabet be $E = \{\mathbf{L}, \mathbf{R}, \mathbf{U}\}$ (for "left," "right," and "upper"). Begin with three points $a_\mathbf{L}, a_\mathbf{R}, a_\mathbf{U}$ in the plane $\mathbb{R}^2$, corners of an equilateral triangle with side 1. Define three similarities f_e with fixed point a_e, ratio $1/2$ and no rotation. [So $f_e(x) = (a_e + x)/2$.] The attractor K of this iterated function system is the **Sierpiński gasket**.

Now, the similarity dimension is the solution s of $3(1/2)^s = 1$, or $s = \log 3 / \log 2$. So the weights for the uniform measure are $p_e = (1/2)^s = 1/3$. This means that when the Sierpiński gasket is constructed, each of the three parts receives equal measure $1/3$.

One reason for calling this measure "uniform" is the analogy with the uniform measure on an interval in the line or a region in $\mathbb{R}^d$. Another is the observation that the density is constant:

(3.3.20) Exercise. Let $\mathcal{T}$ be the uniform measure on the Sierpiński gasket K. Let s be its similarity dimension. Then

$$\limsup_{r \searrow 0} \frac{\mathcal{T}\left(\overline{B}_r(x)\right)}{(2r)^s} \quad \text{and} \quad \liminf_{r \searrow 0} \frac{\mathcal{T}\left(\overline{B}_r(x)\right)}{(2r)^s}$$

are constant for $\mathcal{T}$-almost all $x \in K$.

A second example is the unit interval $[0, 1]$ with its "dyadic" representation. (The set $[0, 1]$ is not a "fractal in the sense of Mandelbrot" because $\dim[0, 1] = \mathrm{ind}[0, 1] = 1$. It is a "fractal in the sense of Taylor" because $\dim[0, 1] = \mathrm{Dim}[0, 1] = 1$.) Let us use the alphabet $E = \{\mathbf{0}, \mathbf{1}\}$ and similarities

$$f_0(x) = \frac{x}{2}, \qquad f_1(x) = \frac{1+x}{2}.$$

This will mean that the address $\sigma \in E^{(\omega)}$ corresponding to a point $x \in [0, 1]$ is its expansion in base 2.

Of course, the uniform measure on $[0, 1]$ corresponds to weights $1/2, 1/2$. But other weights are possible. We may write $p = p_0$ and $1 - p = p_1$. The self-similar measure $\mathcal{T}$ corresponding to this system of weights was called $\mathcal{M}_p$ in [MTFG, p. 208].

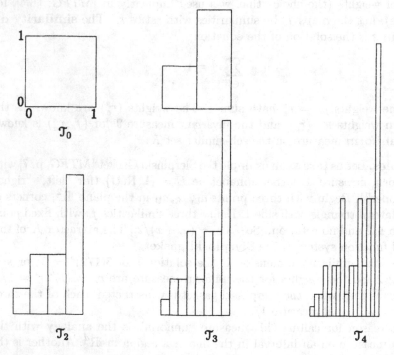

Figure 3.3.21. Densities for $\mathcal{T}_n$.

Let us examine the approximation process that we get from the Contraction Mapping Theorem. Let $\mathcal{T}_0$ be Lebesgue measure on $[0, 1]$. Recursively, let $\mathcal{T}_{n+1} = (1/2)f_{0*}(\mathcal{T}_n) + (1/2)f_{1*}(\mathcal{T}_n)$. Densities for a few of the $\mathcal{T}_n$ are shown in Figure 3.3.21. (I have used $p = 0.3$ for the figure.)

In the language of Chapter 4, the measure $\mathcal{M}_p$ is the distribution of a random variable

$$\sum_{i=1}^{\infty} \frac{X_i}{2^i},$$

where the X_i are i.i.d. Bernoulli random variables with $\mathbb{P}\{X_i = 0\} = p$ and $\mathbb{P}\{X_i = 1\} = 1 - p$. Figure 3.4.9 shows the c.d.f. for $\mathcal{M}_{0.3}$; see Exercise 3.4.10.

Now consider an example where the measures do not have compact support. Let us use the iterated function system above with

$$f_0(x) = \frac{x}{2}, \qquad f_1(x) = \frac{1+x}{2},$$

and let us take $p_0 = p_1 = 1/2$. The attractor $\mathcal{T}$ is Lebesgue measure on $[0, 1]$. But let us apply the Contraction Mapping Theorem starting with $\mathcal{T}_0$ the standard normal distribution; this measure is absolutely continuous with density

$$\frac{1}{\sqrt{2\pi}} \exp\left(\frac{-x^2}{2}\right).$$

(3.3.22) Exercise. If a measure $\mathcal{M}$ has density $g(x)$, then the measure $(1/2)f_{0*}(\mathcal{M}) + (1/2)f_{1*}(\mathcal{M})$ has density $g(2x) + g(2x - 1)$.

Figure 3.3.23 shows the densities for a few steps. The limit has density 1 on $[0, 1]$ and 0 elsewhere.

Similarity Dimension. Suppose an iterated function system $(f_e)_{e \in E}$ consists of similarities. Let r_e be the ratio for f_e; that is,

$$\rho\big(f_e(x), f_e(y)\big) = r_e \rho(x, y)$$

for $x, y \in S$. Also suppose $r_e < 1$ for all e. Then the attractor K may be called **self-similar**. The list (r_e) of ratios determines a **similarity dimension**, namely the solution s of the equation

$$\sum_{e \in E} r_e^s = 1.$$

Then $\dim K \leq \mathrm{Dim}\, K \leq s$, and if the iterated function system satisfies an **open set condition**, in fact $\dim K = \mathrm{Dim}\, K = s$. This material may be found in texts on fractal geometry, in particular [MTFG, §6.3].

Now suppose the iterated function system (f_e) of similarities with ratio list (r_e) is also given weights (p_e). As usual, we assume $p_e > 0$ and $\sum_e p_e = 1$. Let $\mathcal{T}$ be the attractor for this system. Because the maps f_e are similarities, we may call $\mathcal{T}$ a **self-similar measure.** We define the **similarity dimension** of the ratio list with weights (r_e, p_e) to be

$$\frac{\sum_{e \in E} p_e \log p_e}{\sum_{e \in E} p_e \log r_e}.$$

We may also (somewhat inaccurately) call this number the similarity dimension of the measure $\mathcal{T}$.

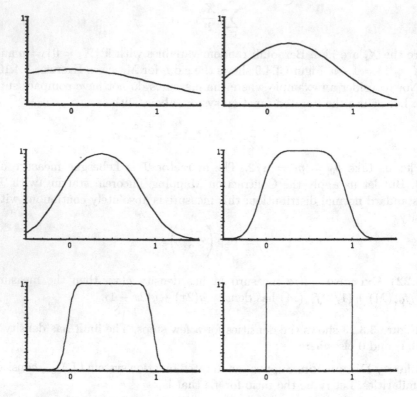

Figure 3.3.23. Densities for $\mathcal{T}_0$, $\mathcal{T}_1$, $\mathcal{T}_2$, $\mathcal{T}_3$, $\mathcal{T}_4$, and the limit.

(3.3.24) Proposition. *Suppose s is the similarity dimension for the ratio list (r_e), and let $p_e = r_e^s$ be the "uniform" weights. Then the similarity dimension for the ratio list with weights (r_e, p_e) is also s.*

Proof. Simply plug in the definition:

$$\frac{\sum p_e \log p_e}{\sum p_e \log r_e} = \frac{\sum p_e \log(r_e^s)}{\sum p_e \log r_e} = \frac{s \sum p_e \log r_e}{\sum p_e \log r_e} = s. \qquad ☺$$

(3.3.25) Exercise. Compute the similarity dimension for the following examples:

(a) Sierpiński gasket with uniform weights: $r_e = 1/2$ and $p_e = 1/3$ for $e \in \{\mathsf{L}, \mathsf{R}, \mathsf{U}\}$. Sometimes we write the ratio list with weights in order like this: $(1/2, 1/2, 1/2; 1/3, 1/3, 1/3)$ with the ratios first, then the weights.

(b) Dyadic unit interval with general weights: for $0 < p < 1$ this is the ratio list with weights $(1/2, 1/2; p, 1-p)$.

We will see later (Theorem 5.2.4) that when the self-similar measure $\mathcal{T}$ satisfies certain "nonoverlap" conditions, the (upper and lower) Hausdorff and packing dimensions coincide with this similarity dimension. (We postpone the proof until after our discussion of the Strong Law of Large Numbers.)

3.4 Self-Affine Graphs

Let $f: \mathbb{R} \to \mathbb{R}$ be a function. The **graph** of f is then the set of ordered pairs

$$G = \{ (x, f(x)) : x \in \mathbb{R} \}.$$

When $\mathbb{R}^2$ is considered as 2-dimensional Euclidean space in the usual way, the set G becomes an object that may have geometric properties. In this book, it would be natural to ask whether G is self-similar.

The graph of a function has this property: Each vertical line meets the graph in at most one point. But a self-similar set in the plane almost never has this property.

(3.4.1) Exercise. Let f be a continuous function defined on the interval $[0, 1]$. Suppose that the graph $G = \{ (x, f(x)) : 0 \leq x \leq 1 \}$ is the invariant set of an iterated function system consisting of similarities in the plane. Show that the graph is a line segment, so that f has the form $f(x) = ax + b$.

Now, an **affine** transformation of the plane may map vertical lines to vertical lines but still be nontrivial. The general affine transformation of the plane has the form

$$F(x, y) = (ax + by + e, cx + dy + f).$$

In matrix form, this is written

$$F \begin{bmatrix} x \\ y \end{bmatrix} = \begin{bmatrix} a & b \\ c & d \end{bmatrix} \begin{bmatrix} x \\ y \end{bmatrix} + \begin{bmatrix} e \\ f \end{bmatrix}.$$

To save space, we often just list the coefficients (a, b, c, d, e, f) in that order (and call them "Barnsley coefficients").

Now suppose we want an affine transformation F that maps vertical lines to vertical lines. That is, if (x_1, y_1) and (x_2, y_2) have the same x-coordinate, then so do the images $F(x_1, y_1)$, $F(x_2, y_2)$. Suppose F is the affine transformation with Barnsley coefficients (a, b, c, d, e, f). Writing x for the common value of the x-coordinate, this means

$$ax + by_1 + e = ax + by_2 + e.$$

The condition is $b = 0$. So the affine transformation of the plane that preserves vertical lines has Barnsley coefficients of the form $(a, 0, c, d, e, f)$.

The coefficients of the transformation have individual interpretations, illustrated in Figure 3.4.2. The parameter a governs the horizontal scaling. (If

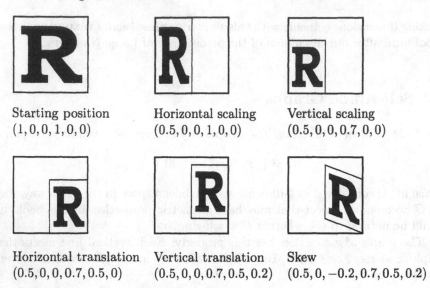

Figure 3.4.2. Affine transformation.

a is negative, the horizontal order is reversed.) The parameter d similarly governs the vertical scaling. The parameter e governs the horizontal translation. The parameter f governs the vertical translation. Finally, the parameter c is a "skewing" term; when $c = 0$ the transformation maps horizontal lines to horizontal lines, but when $c \neq 0$ horizontal lines map to sloping lines.

We will say that a function $f \colon \mathbb{R} \to \mathbb{R}$ has **self-affine graph** iff the graph of f is the invariant set of an iterated function system consisting of affine transformations.

Kiesswetter's Function. An example of such a function is Kiesswetter's function. The iterated function system consists of four affine transformations:

$$F_1 \text{ with Barnsley coefficients } (1/4, 0, 0, -1/2, 0, 0),$$
$$F_2 \text{ with Barnsley coefficients } (1/4, 0, 0, 1/2, 1/4, -1/2),$$
$$F_3 \text{ with Barnsley coefficients } (1/4, 0, 0, 1/2, 1/2, 0),$$
$$F_4 \text{ with Barnsley coefficients } (1/4, 0, 0, 1/2, 3/4, 1/2).$$

These affine transformations are illustrated in Figure 3.4.3. The large rectangle ranges horizontally from $x = 0$ to $x = 1$ and vertically from $y = -1$ to $y = 1$. The four transformations map that rectangle onto four smaller rectangles as shown.

How do we know that the invariant set really is the graph of a function f? It may be realized as a uniform limit of a sequence of continuous functions f_n. The graphs f_n are piecewise linear: f_0 is linear, with $f_0(0) = 0$ and $f_0(1) = 1$; the graph of each f_{n+1} is the image of the graph of f_n under the iterated function system. See Figure 3.4.4.

The reason that the endpoints of the four parts of f_{n+1} match up is seen in the first step: $F_1(0,0) = (0,0)$, $F_1(0,1) = F_2(0,0)$, $F_2(0,1) = F_3(0,0)$,

Figure 3.4.3. Affine transformations for Kiesswetter's function.

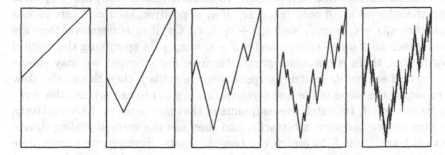

Figure 3.4.4. Graphs for f_n with $n = 0, 1, 2, 3$ and the limit f.

$F_3(0, 1) = F_4(0, 0)$, $F_4(0, 1) = (0, 1)$. So we may see by induction that $f_n(0) = 0$ for all n; $f_n(1) = 1$ for all n; and $F_i(1, f_n(1)) = F_{i+1}(0, f_n(0))$ for all n and $i = 1, 2, 3$.

(3.4.5) Exercise. Prove that the sequence f_n converges uniformly by verifying the Cauchy criterion for uniform convergence. For example, your estimate might show (by induction) that the uniform distance between f_n and f_m is at most 2^{-n+1} if $n < m$.

Kiesswetter's function f is continuous but nowhere differentiable. To see that f is nowhere differentiable, we may proceed as follows. For any integers k, j with $k \geq 0$ and $0 \leq j < 4^k$,

$$\left| f\left(\frac{j}{4^k}\right) - f\left(\frac{j+1}{4^k}\right) \right| = \frac{1}{2^k}.$$

(3.4.6) Exercise. If a function f is differentiable at a point x, and $x_k \leq x \leq y_k$ with $x_k \neq y_k$ and $\lim x_k = x = \lim y_k$, then

$$\lim_{k \to \infty} \frac{f(y_k) - f(x_k)}{y_k - x_k} = f'(x) \quad \text{exists.}$$

Now, in the case of the Kiesswetter function, if x is any point in $(0, 1)$, we may choose the pairs x_k, y_k of the form $j/4^k, (j + 1)/4^k$, and in that case

$$\left| \frac{f(y_k) - f(x_k)}{y_k - x_k} \right| = \frac{1/2^k}{1/4^k} \to \infty.$$

So in fact, f is not differentiable at the point x.

General Self-Affine Graphs. In the general case we will have an iterated function system determined as follows. First, an interval $[\alpha, \beta]$ will be the domain of our function. Next, a partition of it, $\alpha = x_0 < x_1 < \cdots < x_n = \beta$, and corresponding vertical coordinates $y_0, y_1, \cdots, y_n$. There will be n affine transformations, say F_i with Barnsley coefficients $(a_i, 0, c_i, d_i, e_i, f_i)$. We number the transformations from left to right to make it easier. Now, $a_i x + e_i$ must map the interval $[\alpha, \beta]$ onto $[x_{i-1}, x_i]$. If a_i is positive, then a_i, e_i are chosen such that $a_i \alpha + e_i = x_{i-1}$ and $a_i \beta + e_i = x_i$. Or, if a_i is negative, they are chosen such that $a_i \alpha + e_i = x_i$ and $a_i \beta + e_i = x_{i-1}$. In specifying the vertical coordinates, there is one more parameter free. For example, we may choose the vertical scaling d_i arbitrarily (positive or negative), then choose the skew c_i to match the slope of the line segment (x_{i-1}, y_{i-1}) to (x_i, y_i) and the vertical translation f_i to match the endpoints of that line segment. (Alternatively, we may choose the skew arbitrarily, and then use the vertical scaling d_i and vertical translation f_i to match the two endpoints. However, we should note that this method cannot change zero slope into nonzero slope, or vice versa.) The coordinates $x_0, x_1, \cdots, x_n$ are sometimes known as **knots**.

When we choose our iterated function system in this way, we will get a sequence of approximating functions f_k. We start with a linear function f_0 with $f_0(\alpha) = y_0$ and $f_0(\beta) = y_n$. Then the graph of each f_{k+1} is obtained by applying the affine transformations F_i to the graph of f_k. Now in general, this sequence f_k need not converge. When will it converge uniformly?

(3.4.7) Exercise. Suppose an affine iterated function system and corresponding sequence f_k of functions are defined as above. Prove or disprove: the sequence f_k converges uniformly if and only if $|d_i| < 1$ for all i.

(3.4.8) Exercise. Investigate the affine iterated function system with these two transformations:

$$F_1 \text{ with Barnsley coefficients } (1/2, 0, 10, 1/2, 0, 0),$$
$$F_2 \text{ with Barnsley coefficients } (-1/2, 0, 10, 1/2, 1, 0).$$

Are these transformations F_i contractions of $\mathbb{R}^2$? Does the sequence of approximating functions f_k converge uniformly to a limit f? Is that limit the *unique* nonempty compact invariant set for this iterated function system?

Bold Play. The next example goes back to Cesàro, 1906. In the literature (for example Patrick Billinglsey's textbook [32, §7]), it is described in terms of a gambling system known as "bold play." The gambler wants to increase his holdings to a certain amount by repeatedly playing a game at even money, but under unfavorable odds. He attempts to do this by always placing the maximum sensible bet. The probability of eventual success is a function $Q(x)$ of the fraction x of the goal that the gambler currently holds. Let p be the probability of winning on any given play; we are told the odds are unfavorable, that is, $0 < p < 1/2$. To analyze the function Q, consider two cases. If $x \geq 1/2$, then the bet to be placed should be the fraction $1 - x$ of the goal; if he wins he has reached the goal, and if he loses, he continues with stakes reduced to the fraction $x - (1 - x) = 2x - 1$ of the goal. Thus

$$Q(x) = p + (1 - p)Q(2x - 1), \qquad \text{if } x \geq 1/2.$$

On the other hand, if $x < 1/2$, then the bet to be placed should be the fraction x of the goal; if he wins, he increases his stake to fraction $2x$ of the goal and continues; if he loses, he is broke and that is that. Thus

$$Q(x) = pQ(2x), \qquad \text{if } x < 1/2.$$

The two equations for Q show that its graph is self-affine, with two transformations:

F_0 with Barnsley coefficients $(1/2, 0, 0, p, 0, 0)$,

F_1 with Barnsley coefficients $(1/2, 0, 0, 1 - p, 1/2, p)$.

The graph is shown in Figure 3.4.9.

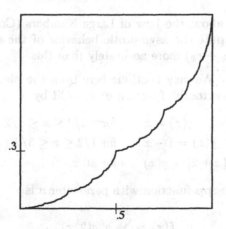

Figure 3.4.9. Cesàro's graph, with $p = 0.3$.

(3.4.10) Exercise. There is a Borel probability measure $\mathcal{M}$ such that $Q(x) = \mathcal{M}((-\infty, x])$ for all $x \in \mathbb{R}$. Describe this self-similar measure $\mathcal{M}$.

Cesàro's function Q is strictly increasing.

(3.4.11) Exercise. Let f be a strictly increasing function $f\colon [\alpha, \beta] \to \mathbb{R}$. The graph of f has Hausdorff dimension 1.

It is known that an increasing function is differentiable almost everywhere [250, p. 207]. But in this example of Cesàro, in fact $Q'(x) = 0$ almost everywhere.[3] To prove this, we will consider dyadic intervals. The portion of the graph on the small interval $[j/2^k, (j+1)/2^k]$ is the image of the entire graph under the composite function

$$F_{e_1} \circ F_{e_2} \circ \cdots \circ F_{e_k},$$

where j has expansion $\sum_{i=0}^{k-1} e_{k-i} 2^i$ in base 2. So

$$(*) \qquad Q\left(\frac{j+1}{2^k}\right) - Q\left(\frac{j}{2^k}\right) = p_{e_1} p_{e_2} \cdots p_{e_k},$$

where we have used the conventions $p_0 = p$, $p_1 = 1 - p$. Now, if $x \in (0,1)$ is a point where $Q'(x)$ exists and is not zero, then we may choose intervals $[x_k, y_k]$ of the form $[j/2^k, (j+1)/2^k]$ with $x \in [x_k, y_k]$, such that

$$\lim_k \frac{Q(y_k) - Q(x_k)}{y_k - x_k} = Q'(x),$$

and thus

$$\lim_k \frac{Q(y_{k+1}) - Q(x_{k+1})}{Q(y_k) - Q(x_k)} = \frac{1}{2}.$$

But by $(*)$, the ratio on the left is always either p or $1 - p$, so it certainly does not converge to $1/2$.

After you learn about the Law of Large Numbers (Corollary 4.4.7), you will be able to compute the asymptotic behavior of the difference quotient $(Q(y_k) - Q(x_k))/(y_k - x_k)$ more accurately than this.

Knopp Functions. We may continue here from the discussion in [MTFG, p. 202].[4] Define a "sawtooth" function $g\colon \mathbb{R} \to \mathbb{R}$ by

$$\begin{aligned}
g(x) &= x && \text{for } -1/2 \le x \le 1/2, \\
g(x) &= 1 - x && \text{for } 1/2 \le x \le 3/2, \\
g(x + 2) &= g(x) && \text{for all } x.
\end{aligned}$$

If $0 < a < 1$, the **Knopp** function with parameter a is

$$f(x) = \sum_{k=0}^{\infty} a^k g(2^k x).$$

---

[3] Such a function is said to be **singular**.

[4] The functions were called Besicovitch–Ursell functions in [MTFG], since they are a special case of functions defined in a 1937 paper by Besicovitch and Ursell [29]. But I now call them Knopp functions since they can be found in a 1918 paper of Knopp [153].

Pictures for various values of a may be found in [$MTFG$, p. 204].

The graph of a Knopp function is not self-similar. But it is self-affine.

(3.4.12) Exercise. Find an affine iterated function system for the portion of the graph of the Knopp function where $0 \leq x \leq 1$.

Let $p \in (0, 1]$. A function $f \colon [\alpha, \beta] \to \mathbb{R}$ satisfies a **Hölder condition** of order p iff there is a constant M such that

$$|f(x) - f(y)| \leq M|x - y|^p$$

for all $x, y \in [\alpha, \beta]$.

(3.4.13) Exercise. A function $f \colon [\alpha, \beta] \to \mathbb{R}$ satisfies a **uniform Hölder condition** of order p iff there are constants M and ε such that

$$|f(x) - f(y)| \leq M|x - y|^p$$

for all $x, y \in [\alpha, \beta]$ with $|x - y| \leq \varepsilon$. Show that f satisfies a uniform Hölder condition of order p if and only if f satisfies a Hölder condition of order p.

The box dimension (and thus the Hausdorff dimension) of the graph of such a function is $\leq 2 - p$ [$MTFG$, Theorem 7.5.1]. And (since the projection of the graph onto the x-axis is an interval) the dimension is ≥ 1. So in particular, if a function satisfies a Hölder condition of order 1, then its graph has dimension 1.

(3.4.14) Proposition. *Let $1/2 < a < 1$. The Knopp function with parameter a satisfies a Hölder condition of order $-\log a / \log 2$.*

Proof. Write $p = -\log a / \log 2$, so that $a = 2^{-p}$. Let $x, x + h \in [0, 1]$. I must show that there is a constant M such that $|f(x+h) - f(x)| \leq M|h|^p$. This will certainly be true for $h = 0$, so assume $h \neq 0$. Now, first consider the sawtooth function $g(x)$. Since $|g(x)| \leq 1/2$ everywhere, we have $|g(2^n x + 2^n h) - g(2^n x)| \leq 1$. We will use this estimate for large n. On the other hand, g is continuous and $|g'(x)| = 1$ almost everywhere, so $|g(2^n x + 2^n h) - g(2^n x)| \leq 2^n h$. We will use this estimate for small values of n, where $2^n h < 1$. So, given h, let $N \in \mathbb{N}$ be such that

$$\frac{1}{2^{N-1}} > |h| \geq \frac{1}{2^N}.$$

Then $a^N = 2^{-pN} \leq |h|^p$ and $2^N = 2 \cdot 2^{N-1} \leq 2/|h|$. Now,

$$|f(x + h) - f(x)| \leq \sum_{n=0}^{\infty} a^n |g(2^n x + 2^n h) - g(2^n x)|$$

$$\leq \sum_{n=0}^{N-1} a^n 2^n |h| + \sum_{n=N}^{\infty} a^n$$

$$= \frac{2^N a^N - 1}{2a - 1} |h| + \frac{a^N}{1 - a}$$

$$\leq \frac{(2/|h|)|h|^p}{2a - 1} |h| + \frac{1}{1 - a}|h|^p$$

$$= \left(\frac{2}{2a - 1} + \frac{1}{1 - a} \right) |h|^p. \quad \text{☺}$$

(3.4.15) Exercise. What happens when $a \leq 1/2$ (so that $-\log a/\log 2 \geq 1$)?

The order $p = -\log a/\log 2$ is the best one for this function.

(3.4.16) Proposition. *Let $1/2 < a < 1$. The Knopp function f with parameter a satisfies no Hölder condition of order $> -\log a/\log 2$.*

Proof. let $p = -\log a/\log 2$, and suppose $p' > p$. If $x = 0$ and $h = 2^{-N}$, then (since $g(x) = x$ for $0 < x \leq 1/2$ and $g(x) = 0$ for integers x)

$$|f(x + h) - f(x)| = |f(h)| = \left| \sum_{n=0}^{\infty} a^n g(2^{n-N}) \right|$$

$$= \sum_{n=0}^{N-1} a^n 2^{n-N} = \frac{a^N - 2^{-N}}{2a - 1},$$

and thus

$$\frac{|f(x + h) - f(x)|}{h^{p'}} = \frac{2^{(-p+p')N} - 2^{(-1+p')N}}{2a - 1} \to \infty,$$

so f does not satisfy a Hölder condition of order p'.   ☺

(3.4.17) Exercise. Let $0 < a < 1$. Determine the fractal dimension of the graph of the Knopp function with parameter a. For example, the box dimension, the Hausdorff dimension, or the packing dimension.

(3.4.18) Exercise. Suppose the knots $x_0, x_1, \cdots, x_n$ are fixed, the horizontal scaling coefficients a_i are all positive, and the vertical scaling coefficients d_i are fixed. Show that the set of all continuous functions with self-affine graphs obeying these data forms a vector space. Compute the (linear) dimension of this vector space. Use a computer to graph a basis for this space in the simple case: $n = 2$; knots $0, 1/2, 1$; vertical scaling $1/2, 1/2$.

(3.4.19) Exercise. Let the function $G: [\alpha, \beta] \to \mathbb{R}$ have self-affine graph, where the affine transformations F_i have coefficients $(a_i, 0, c_i, d_i, e_i, f_i)$, as usual. Prove or disprove: If $|d_i| > |a_i|$ for all i, then G is nowhere differentiable. Is it enough that $|d_i| > |a_i|$ for some i, but not all?

(3.4.20) Exercise. Let the function $G: [\alpha, \beta] \to \mathbb{R}$ have self-affine graph, where the affine transformations F_i have coefficients $(a_i, 0, c_i, d_i, e_i, f_i)$. Suppose G is not affine on the whole interval. Let $p \in (0, 1)$ be given. Can we determine whether G satisfies a Hölder condition of order p by simple computations from the Barnsley coefficients?

Higher Dimensional Range. Let us consider the possibility of a function $f: [\alpha, \beta] \to \mathbb{R}^d$ for $d \geq 2$. A "self-affine graph" is possible also in that case. The graph lies in the space $\mathbb{R} \times \mathbb{R}^d$. The affine transformations on this space (mapping "vertical" lines to vertical lines) look like this in matrix form:

$$F \begin{bmatrix} x \\ \mathbf{y} \end{bmatrix} = \begin{bmatrix} a & \mathbf{0} \\ \mathbf{c} & D \end{bmatrix} \begin{bmatrix} x \\ \mathbf{y} \end{bmatrix} + \begin{bmatrix} e \\ \mathbf{f} \end{bmatrix}.$$

Here the argument consists of $x \in \mathbb{R}$ and $\mathbf{y} \in \mathbb{R}^d$. The coefficients $(a, \mathbf{0}, \mathbf{c}, D, e, \mathbf{f})$ involve $a, e \in \mathbb{R}$, $\mathbf{c}, \mathbf{f} \in \mathbb{R}^d$; and D is a $d \times d$ matrix. Written in terms of the components,

$$F(x, \mathbf{y}) = (ax + e, x\mathbf{c} + D\mathbf{y} + \mathbf{f}).$$

Many of the common examples of fractal sets in $\mathbb{R}^d$ are described together with a parametrization, so that a function $f: [\alpha, \beta] \to \mathbb{R}^d$ is defined and the fractal is the range $f[[\alpha, \beta]]$. The common space-filling curves are also described in this way.

Let us take the Hilbert curve as an example. It is a continuous function $f: [0, 1] \to \mathbb{R}^2$ with range equal to the square $[0, 1] \times [0, 1]$. Four affine transformations F_i describe it.

$$F_1 \begin{bmatrix} x \\ y_1 \\ y_2 \end{bmatrix} = \begin{bmatrix} 1/4 & 0 & 0 \\ 0 & 0 & 1/2 \\ 0 & 1/2 & 0 \end{bmatrix} \begin{bmatrix} x \\ y_1 \\ y_2 \end{bmatrix} + \begin{bmatrix} 0 \\ 0 \\ 0 \end{bmatrix},$$

$$F_2 \begin{bmatrix} x \\ y_1 \\ y_2 \end{bmatrix} = \begin{bmatrix} 1/4 & 0 & 0 \\ 0 & 1/2 & 0 \\ 0 & 0 & 1/2 \end{bmatrix} \begin{bmatrix} x \\ y_1 \\ y_2 \end{bmatrix} + \begin{bmatrix} 1/4 \\ 0 \\ 1/2 \end{bmatrix},$$

$$F_3 \begin{bmatrix} x \\ y_1 \\ y_2 \end{bmatrix} = \begin{bmatrix} 1/4 & 0 & 0 \\ 0 & 1/2 & 0 \\ 0 & 0 & 1/2 \end{bmatrix} \begin{bmatrix} x \\ y_1 \\ y_2 \end{bmatrix} + \begin{bmatrix} 1/2 \\ 1/2 \\ 1/2 \end{bmatrix},$$

$$F_4 \begin{bmatrix} x \\ y_1 \\ y_2 \end{bmatrix} = \begin{bmatrix} 1/4 & 0 & 0 \\ 0 & 0 & -1/2 \\ 0 & -1/2 & 0 \end{bmatrix} \begin{bmatrix} x \\ y_1 \\ y_2 \end{bmatrix} + \begin{bmatrix} 3/4 \\ 1 \\ 1/2 \end{bmatrix}.$$

If we think of (y_1, y_2) describing a point in the plane and x describing time, then as x ranges from 0 to 1, the point (y_1, y_2) fills the square by moving through a complex dance. Alternatively, we might think of (y_1, y_2) describing a position in a horizontal plane, and x describing an altitude. Then the self-affine set describes a complicated ramp from the point $(0, 0)$ at height 0 to the point $(1,0)$ at height 1. Some pictures are in [*MTFG*, p. 35]. Computer graphics fans might want to try drawing other pictures to aid the visualization of this curve.[5]

Cutting Corners. Start with a polygon. It consists of a sequence

$$\overline{P_0P_1}, \overline{P_1P_2}, \overline{P_2P_3}, \cdots, \overline{P_{n-1}P_n}$$

of adjoining line segments. In a closed polygon, $P_n = P_0$. Use this polygon to make a new one by "cutting off the corners" as follows. Trisect each of the segments: say $\overline{P_{k-1}P_k}$ is trisected as $\overline{P_{k-1}A_k}, \overline{A_kB_k}, \overline{B_kP_k}$. Thus (in vector notation), $A_k = (2/3)P_{k-1} + (1/3)P_k$ and $B_k = (1/3)P_{k-1} + (2/3)P_k$. The new polygon is

$$\overline{A_1B_1}, \overline{B_1A_2}, \overline{A_2B_2}, \overline{B_2A_3}, \cdots, \overline{B_{n-1}A_n}, \overline{A_nB_n}.$$

If the original polygon is closed, add the final edge $\overline{B_nA_1}$ so that the corner-cut polygon is again closed. What happens if we repeat this indefinitely?

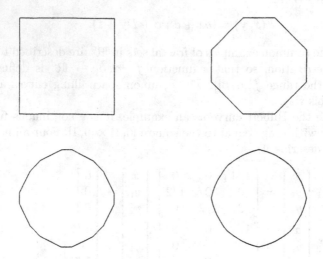

Figure 3.4.21. Cutting corners. Three steps and the limit.

For example, start with a square (Figure 3.4.21). What is the limit curve like? This is the second-derivative analogue of a singular function.

(3.4.22) Exercise. Describe this limit curve as the union of four sets, each the range of a function $\mathbb{R} \to \mathbb{R}^2$ with self-affine graph. The resulting curve is differentiable but has curvature 0 almost everywhere.

Higher-Dimensional Domain. Functions with domain contained in a Euclidean space $\mathbb{R}^d$ may be described with a self-affine graph. Most convenient

---

[5] In [74, Plate 2], I used colors to exhibit such parametrizations.

in $\mathbb{R}^2$ is to subdivide the domain into squares (rectangles, parallelograms) or triangles.

The affine transformations F will map $\mathbb{R}^d \times \mathbb{R}$ to itself. They should have the form

$$F\begin{bmatrix} \mathbf{x} \\ y \end{bmatrix} = \begin{bmatrix} A & \mathbf{0} \\ \mathbf{c} & d \end{bmatrix} \begin{bmatrix} \mathbf{x} \\ y \end{bmatrix} + \begin{bmatrix} \mathbf{e} \\ f \end{bmatrix}.$$

Here A is a square matrix, and $\mathbf{c}$ is a row vector (so that $\mathbf{cx}$ is the dot product). The iterated function system in this case will consist of transformations F_i with coefficients $(A_i, \mathbf{0}, \mathbf{c}_i, d_i, \mathbf{e}_i, f_i)$; the transformations $A_i\mathbf{x} + \mathbf{e}_i$ should make up an iterated function system in $\mathbb{R}^d$ that constructs the domain of our function. The other choices $(\mathbf{c}_i, d_i, f_i)$ should be made so that the overlaps match. The technical way to specify this matching could be complicated and is not discussed here.

One example will be described. Begin with an equilateral triangle T in the plane. A function $f\colon T \to \mathbb{R}$ is described using four affine transformations. The transformations map all of T onto four triangles with sides half the length of the sides of the original triangle. Let a, b, c be the three corners of T. Start with values at the corners, $f(a) = f(b) = f(c) = 0$. Then consider the midpoints of the edges of the triangle: Assign them values $1/2, 0, -1/2$ in a definite order. The vertical scaling is $1/4$. For the center triangle, rotate by $30°$ counterclockwise, and for the others rotate by $150°$ clockwise.

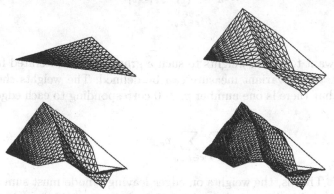

Figure 3.4.23. Midpoint displacement with 2-dimensional domain.

Fractal Functions. All of the types of functions with self-affine graph are special cases of a general definition sometimes called a **fractal function**. A fractal function $f\colon X \to Y$ is supposed to be defined as the solution of a functional equation of the form

$$(*) \qquad\qquad f(x) = v(x, f(b(x))),$$

where $b\colon X \to X$ and $v\colon X \times Y \to Y$ are given functions.

(3.4.24) Exercise. In the various examples of functions with self-affine graph described above, determine the appropriate functions v and b that will describe them in the form (∗).

*3.5 Graph Self-Similar Measures

For this section we will use (without explanation here) the notation and terminology concerning graph-directed iterated function system constructions (and in particular graph self-similar sets) from [MTFG, §§4.3, 6.4]. This combinatorial use of the term "graph" should not be confused with its use as the "graph of a function" elsewhere in the book.

We have a strongly connected directed multigraph (V, E); a family $(S_v)_{v \in V}$ of nonempty complete metric spaces; and a family $(f_e)_{e \in E}$ of functions with Lipschitz conditions

$$\rho\big(f_e(x), f_e(y)\big) \le r_e \rho(x, y)$$

for $x, y \in S_v$, $e \in E_{uv}$. We assume that the system of ratios (r_e) is contracting in the sense that all loops have ratio < 1. Then there exists a unique list $(K_v)_{v \in V}$ of nonempty compact sets satisfying the invariance condition

$$K_u = \bigcup_{\substack{v \in V \\ e \in E_{uv}}} f_e[K_v]$$

for all $u \in V$.

Now we want to assign weights to such a graph-directed iterated function system so that an invariant measure can be defined. The weights should be assigned so that there is one number $p_e > 0$ corresponding to each edge $e \in E$ and

$$(1) \qquad\qquad \sum_{v \in V} \sum_{e \in E_{uv}} p_e = 1$$

for all $u \in V$. That is, the weights on edges leaving a node must sum to 1.

Given all of this data, an **invariant list** of measures $(\mathfrak{I}_v)_{v \in V}$ is a list of probability measures, $\mathfrak{I}_v \in \mathfrak{P}(S_v)$, such that

$$\mathfrak{I}_u = \sum_{v \in V} \sum_{e \in E_{uv}} p_e f_{e*}(\mathfrak{I}_v)$$

for all $u \in V$.

(3.5.1) Exercise. Under all of the conditions enumerated, there is a unique invariant list of measures $(\mathfrak{I}_v)$.

---

* An optional section.

Now suppose the functions f_e are all similarities, with ratios r_e. A "similarity dimension" may be associated with the data $(V, E, (r_e)_{e \in E}, (p_e)_{e \in E})$. This is done as follows. Consider the matrix A, with rows and columns indexed by V, where the entry in row u column v is $\sum_{e \in E_{uv}} p_e$. Then (1) says that the row sums of A are all 1. Now, A is an irreducible nonnegative matrix, so by the Perron–Frobenius theory, it has spectral radius 1. So there is a unique normalized left eigenvector $(\lambda_v)_{v \in V}$, such that $\lambda_v > 0$, $\sum_{v \in V} \lambda_v = 1$, and

$$\sum_{u \in V} \sum_{e \in E_{uv}} \lambda_u p_e = \lambda_v \qquad \text{for all } v \in V.$$

The **similarity dimension** is

$$\frac{\sum_{u \in V} \sum_{v \in V} \sum_{e \in E_{uv}} \lambda_u p_e \log p_e}{\sum_{u \in V} \sum_{v \in V} \sum_{e \in E_{uv}} \lambda_u p_e \log r_e}.$$

There is a notion of "uniform" measure on a self-similar fractal. Let s be the similarity dimension associated with the Mauldin–Williams graph $(V, E, (r_e)_{e \in E})$. This means that 1 is the spectral radius of the matrix B formed with entry

$$\sum_{e \in E_{uv}} r_e^s$$

on row u, column v. Equivalently, it means that there are "Perron numbers" $q_v > 0$ such that

$$(2) \qquad \sum_{v \in V} \sum_{e \in E_{uv}} r_e^s q_v^s = q_u^s$$

for all $u \in V$. The "uniform" weights for this Mauldin–Williams graph are then

$$p_e = \left(\frac{q_v}{q_u} r_e \right)^s \qquad \text{for } e \in E_{uv}.$$

This definition in (2) should yield (1). So this is a legitimate system of weights.

(3.5.2) Proposition. *Let s be the similarity dimension for the Mauldin–Williams graph $(V, E, (r_e))$. Let p_e be the uniform weights defined above. Then the similarity dimension for the Mauldin–Williams graph with weights (p_e) is also s.*

Proof. Put $p_e = (q_v r_e / q_u)^s$ in the definitions. Decompose the logarithm $\log p_e = \log(q_v r_e / q_u)^s = s \log r_e + s \log q_v - s \log q_u$. Then the numerator (in the definition of the similarity dimension) is a sum of three terms:

$$T_1 = s \sum_{u \in V} \sum_{v \in V} \sum_{e \in E_{uv}} \lambda_u p_e \log r_e,$$

$$T_2 = s \sum_{u \in V} \sum_{v \in V} \sum_{e \in E_{uv}} \lambda_u p_e \log q_v,$$

$$T_3 = -s \sum_{u \in V} \sum_{v \in V} \sum_{e \in E_{uv}} \lambda_u p_e \log q_u.$$

Now, T_1 is s times the denominator, and

$$T_2 = s \sum_{v \in V} \left(\sum_{u \in V} \sum_{e \in E_{uv}} \lambda_u p_e \right) \log q_v = s \sum_{v \in V} \lambda_v \log q_v.$$

On the other hand,

$$T_3 = -s \sum_{u \in V} \lambda_u \left(\sum_{v \in V} \sum_{e \in E_{uv}} p_e \right) \log q_u = -s \sum_{u \in V} \lambda_u \cdot 1 \cdot \log q_u.$$

So in fact, $T_2 + T_3 = 0$, and our similarity dimension is s.   ☺

(3.5.3) Exercise. Verify that when the graph consists of a single node, the definitions of this section agree with those for ordinary self-similar measures.

As before, it is true that under proper separation conditions, the Hausdorff and packing dimensions of the graph self-similar measures $\mathcal{T}_v$ will all agree with the similarity dimension defined here. See [78], [203]. This fact will not be proved in this book.

*3.6 Remarks

The Szpilrajn proof that ind $S \le \dim S$ is also found in [141].

The approach to Hausdorff dimension using potential theory comes from Frostman [103]. Some of the density theorems in Chapter 1 that this is based on are also due to him.

Dimension of a measure is discussed in [50, 140, 78, 144]. For more on iterated function systems with weights, see Barnsley [13]. Existence and uniqueness of the invariant measure is (in some cases) in [142]. Evertsz and Mandelbrot [81] use self-similar measures as examples in their discussion of multifractals. Note the papers [246–249] by R. Strichartz on self-similar measures (and other iterated function system attractors); in [248] there is a computation of the Hausdorff dimension in certain iterated function systems with weights where the maps are not similarities. More references on self-similar measures are [11, 163].

Here is another useful result on pointwise dimension from [50, 52]. It is stated only for Borel sets in $\mathbb{R}^d$—what is required to obtain it for other metric spaces?

(3.6.1) Proposition. *Let $E \subseteq \mathbb{R}^d$ be a nonempty Borel set, and let s be a positive real number.*

(a) *If $\dim E > s$, then there exists a Borel measure $\mathcal{M}$ with $0 < \mathcal{M}(E) < \infty$ and $\underline{L}_{\mathcal{M}}(x) \ge s$ for all $x \in E$.*

---

* An optional section.

(b) *If* dim $E < s$, *then there exists a Borel measure* $\mathcal{M}$ *with* $0 < \mathcal{M}(\overline{E}) < \infty$
  and $\underline{L}_{\mathcal{M}}(x) \le s$ *for all* $x \in E$.
(c) *If* Dim $E > s$, *then there exists a Borel measure* $\mathcal{M}$ *with* $0 < \mathcal{M}(E) < \infty$
  and $\overline{L}_{\mathcal{M}}(x) \ge s$ *for* $\mathcal{M}$-*almost all* x.
(d) *If* Dim $E < s$, *then there exists a Borel measure* $\mathcal{M}$ *with* $0 < \mathcal{M}(\overline{E}) < \infty$
  and $\overline{L}_{\mathcal{M}}(x) \le s$ *for all* $x \in E$.

Recent work on *infinite* iterated function systems may be found in [187, 226]. For the noncompact case of self-similar fractal measures, see [3].

Arcsine Law. Consider the iterated function system consisting of these two maps on the interval $[-2, 2]$:

$$f_{\mathsf{L}}(x) = -\sqrt{x + 2}, \qquad f_{\mathsf{R}}(x) = \sqrt{x + 2}.$$

Use equal weights $(1/2, 1/2)$ for the two maps. These functions are not contractions. But nevertheless, there is a unique invariant probability measure $\mathcal{J}$ on $[-2, 2]$. This measure may be called the **arcsine law**, but I leave it to you to guess why it has that name.

Self-affine sets, as attractors of iterated function systems, are found in [13]. Evaluation of the Hausdorff dimension (and other fractal dimensions) of a self-affine set is a topic still under development. I regret that none of this material is included in this book. A few of the many references on self-affine sets are [178, 20, 21, 22, 56, 71, 75, 86, 191, 222, 265].

For functions with self-affine graph, I found useful the survey [99]. The survey [129] discusses nondifferentiable functions obtained in this way and contains a number of computer images.

The "cutting corners" example is one example of a class of functional equations studied by de Rham [222, 223, 224]. See [34] for an application to computer graphics. Do not miss [1], an elementary discussion of corner-cutting curves like this involving Farey fractions and interesting extrapolations of the idea.

Fractal functions are treated in [185, Chapters V, VI, VIII]. See also [65, 221].

Graph self-similar measures may be found implicitly in the literature on Markov chains. They are called "measures of Markov type" in [78].

On the Exercises. *(3.2.1)*: For (b), use spherical coordinates $\rho =$ radius, $\varphi =$ colatitude, $\theta =$ longitude. On the sphere $\rho = 1$, the element of surface area is $dA = \sin \varphi \, d\varphi \, d\theta$. The total surface area is $\int \int dA = 4\pi$, so the measure $\mathcal{M}$ is surface area times $1/(4\pi)$. By rotational invariance, $U_{\mathcal{M}}(\mathbf{x})$ is the same for all $\mathbf{x}$ with $|\mathbf{x}| = 1$. The distance from the north pole $(1, 0, 0)$ to a point $(1, \varphi, \theta)$ is $2\sin(\varphi/2)$. So if $\mathbf{x}$ is the north pole, compute

$$U_{\mathcal{M}}(\mathbf{x}) = \frac{1}{4\pi} \int_0^{2\pi} d\theta \int_0^{\pi} \frac{\sin \varphi \, d\varphi}{2\sin(\varphi/2)} = 1;$$

and thus $U_{\mathcal{M}}(\mathbf{x}) = 1$ for any $\mathbf{x}$ with $|\mathbf{x}| = 1$.

(3.2.2): For $0 < s < 1$,

$$U_{\mathcal{M}}^{s}(x) = \begin{cases} \dfrac{(1-x)^{1-s} - (-x)^{1-s}}{1-s}, & \text{for } x < 0, \\[2mm] \dfrac{(1-x)^{1-s} + x^{1-s}}{1-s}, & \text{for } 0 \le x \le 1, \\[2mm] \dfrac{x^{1-s} - (x-1)^{1-s}}{1-s}, & \text{for } x > 1. \end{cases}$$

(See Figure 3.6.2 for $s = 1/2$.) For $s = 1$,

$$U_{\mathcal{M}}^{1}(x) = \begin{cases} \log \dfrac{x-1}{x}, & \text{for } x < 0, \\[2mm] \infty, & \text{for } 0 \le x \le 1, \\[2mm] \log \dfrac{x}{x-1}, & \text{for } x > 1. \end{cases}$$

And for $s > 1$,

$$U_{\mathcal{M}}^{s}(x) = \begin{cases} \dfrac{(-x)^{1-s} - (1-x)^{1-s}}{s-1}, & \text{for } x < 0, \\[2mm] \infty, & \text{for } 0 \le x \le 1, \\[2mm] \dfrac{(x-1)^{1-s} - x^{1-s}}{s-1}, & \text{for } x > 1. \end{cases}$$

Recall that $[0, 1]$ has Hausdorff dimension 1, and $\mathcal{M}$ is the corresponding Hausdorff measure.

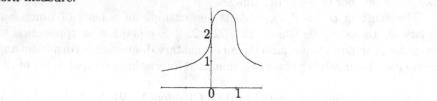

Figure 3.6.2. $U_{\mathcal{M}}^{1/2}(x)$.

(3.2.13): [138]

(3.3.4): Fix $\varepsilon, \eta > 0$. For each n, choose an $\mathcal{M}_n$-almost ε-cover $\{E_{in}\}_{i=1}^{\infty}$ with $\sum_{i}(\operatorname{diam} E_{in})^{s} \le \mathcal{H}_{\varepsilon}^{s}(\mathcal{M}_n) + \eta/2^{n}$. Then $\{E_{in}\}_{i=1}^{\infty}{}_{n=1}^{\infty}$ is an $\mathcal{M}$-almost ε-cover. And

$$\sum_{i,n}(\operatorname{diam} E_{in})^{s} \le \sum_{n} \mathcal{H}_{\varepsilon}^{s}(\mathcal{M}_n) + \eta,$$

so $\mathcal{H}_{\varepsilon}^{s}(\mathcal{M}) \le \sum \mathcal{H}_{\varepsilon}^{s}(\mathcal{M}_n) + \eta$. Now let $\eta \to 0$ to get $\mathcal{H}_{\varepsilon}^{s}(\mathcal{M}) \le \sum \mathcal{H}_{\varepsilon}^{s}(\mathcal{M}_n) \le \sum \mathcal{H}^{s}(\mathcal{M}_n)$. Then let $\varepsilon \to 0$ to get $\mathcal{H}^{s}(\mathcal{M}) \le \sum_{n} \mathcal{H}^{s}(\mathcal{M}_n)$.

(3.3.25): (a) $\log 3/\log 2$. (b) $(-p \log p - (1-p) \log(1-p))/\log 2$; the computation in [MTFG, p. 208] is actually showing that $\dim^{*} \mathcal{M}_p$ has this value.

(3.4.12): $(1/2, 0, 1/2, a, 0, 0), (-1/2, 0, 1/2, -a, 1, 0)$.

(3.4.17):   The paper [29] computes the Hausdorff dimension of many graphs of this type, but not this one in particular. The method of [12] will compute the box dimension as $2 - p = 2 + \log a / \log 2$, as expected.

(3.4.18): [99, Abb. 1/5].

(3.4.19): [99, Satz 2.2.1].

(3.4.20): [99, Satz 2.4.1]. In the special case where $d_1 = d_2 = \cdots = d_n = d$, the best Hölder order is $p = \log |d| / \log a$, where $a = \max\{a_1, a_2, \cdots, a_n\}$ (provided that $p < 1$).

(3.4.22): [222, 1].

4. Probability

This chapter concerns a branch of mathematics known as "probability theory."
It originated in connection with the description of random events, in particular, games of chance. Kolmogorov constructed a mathematical model for this study. Today, the mathematical model can be studied independently of any possible connection with random events in the real world. Probability theory has become a branch of mathematics in its own right.

J. E. Littlewood [173] notes that probability, applied in the real world, must be done simply by assumption; no "justification" or "proof" of such application can be made:

> Mathematics (by which I shall mean pure mathematics) has no grip on the real world; if probability is to deal with the real world it must contain elements outside mathematics; the *meaning* of "probability" must relate to the real world, and there must be one or more "primitive" propositions about the real world, from which we can then proceed deductively (i.e. mathematically) ... If he is consistent a man of the mathematical school washes his hands of applications. To some one who wants them he would say that the ideal system runs parallel to the usual theory: "If this is what you want, try it: it is not my business to justify application of the system; that can only be done by philosophizing; I am a mathematician."

As a mathematician, I will follow Littlewood here. The investigation of how (or to what extent) the mathematics applies to the real world is a question outside of mathematics itself. In this book, we do not consider such questions seriously—but we do frequently use familiar examples like coin tossing to illustrate the mathematics.

4.1 Events

The basic objects to be considered in probability theory are called **events**—think of a certain circumstance (or combination of circumstances) that may or may not occur, depending on the outcome of certain (known or unknown) random processes. There is to be a number associated with each event, called the **probability** of the event. It is a real number between 0 and 1; "likely" events are assigned high probabilities, and "unlikely" events are assigned low probabilities.

An example that will be used frequently concerns tossing a coin. This is an idealized, perfect coin. The result should be either "heads" or "tails" but not both. It should be "heads" with probability 1/2.

Interpretation. There are different ways to interpret a "probability" in applications outside mathematics. Two of the most common interpretations are the **frequentist** and **subjective** interpretations.

In the frequentist interpretation, when we say that the probability of heads is 1/2, we mean that if the coin is tossed repeatedly (say n times) the number of heads $H(n)$ should be about $(1/2)n$. Or, more mathematically but less practically,

$$\lim_{n \to \infty} \frac{H(n)}{n} = \frac{1}{2}.$$

The frequentist approach takes the Strong Law of Large Numbers (see 4.4.6) as the definition of probability.

Now, there are events that cannot be repeated: "It is 95% certain that President Lincoln was shot by John Wilkes Booth." This is a statement about the probability of an event. Except in science fiction stories, we cannot run the experiment many times to see how often our event occurs. And even in cases like tossing a coin, the experiment is never repeated exactly: whenever I touch the coin, it loses a few atoms and changes its balance slightly. Air currents and differences in muscular response may also change. So (like all mathematical models) the frequentist approach is only an approximation to reality.

Frequentists may postulate the existence of many universes in order to retain an objective interpretation of probability: Whenever there is a random occurrence, each of the possible outcomes occurs in some of the universes.

The "subjective" interpretation of probability involves "degree of belief." The probability of an event (for me) represents the degree to which I believe it will happen (or has happened). An advantage of this interpretation is that it makes sense even for events that cannot be repeated. On the other hand, the subjective interpretation denies that there is an objective probability associated with an event, independent of the observer. I have my opinion of the probability, you have your opinion, and it may seem pointless to do mathematics on such speculative numbers.

Language taken from each of these approaches will sometimes be used within the mathematics; but it should be remembered that the mathematics is the same whether or not I believe in multiple universes or objective probabilities. As noted above, discussion of how the mathematics applies to the real world is a question for philosophers.

Elementary Properties. Events may be combined to form other events. Let A and B be events. A new event, called "A or B," occurs iff A occurs or B occurs or both. Another new event, called "A and B," occurs iff A and B both occur. The event "not A" occurs iff A does not occur. Two events A, B are **mutually exclusive** iff they cannot both occur, that is, the event "A and B"

is impossible. Two events A, B are **exhaustive** iff at least one must occur, that is, the event "A or B" is certain to occur.

If A is an event, we will write $\mathbb{P}(A)$ for the probability of A. Probabilities are supposed to satisfy certain elementary properties. An event A that is impossible has probability zero, $\mathbb{P}(A) = 0$. (But we allow that possible events may have probability zero, as well.) A **certain** event A (that is, an event that necessarily must happen) has probability one. If A, B are mutually exclusive, then

$$\mathbb{P}(A \text{ or } B) = \mathbb{P}(A) + \mathbb{P}(B).$$

In particular, since A is mutually exclusive with "not A,"

$$\mathbb{P}(A) + \mathbb{P}(\text{not } A) = 1.$$

Sample Space. For mathematical discussion of probability, we will use the Kolmogorov model. There is a set Ω, called the **sample space**; events correspond to subsets of Ω. An element $\omega \in \Omega$ is sometimes called a **sample point**. There should be one sample point that corresponds to each possible (conceivable) combination of outcomes of the relevant random occurrences. Alternatively, each sample point corresponds to a possible universe (or state of the universe). It is important that every possible combination of outcomes is allowed. If we are too inclusive and actually include some impossible combinations as well, that is not a problem: such impossible events will be assigned probability 0.

Certain subsets of the sample space Ω are events. But (usually) other subsets do not correspond to events. Algebraic combinations of events are events: The impossible event is $\varnothing$. The certain event is Ω. The event "A or B" is the union $A \cup B$; the event "A and B" is the intersection $A \cap B$; the complementary event "not A" is the complementary set $\Omega \setminus A$. In fact, we will postulate that countable unions and intersections of events are also events. So the collection of all events constitutes a σ-algebra.

The probabilities $\mathbb{P}(A)$ of events A constitute a set-function $\mathbb{P}$. We postulate that it is a measure. Let Ω be the sample space and $\mathcal{F}$ the σ-algebra of events:

(a) $\mathbb{P}(\Omega) = 1$; $\mathbb{P}(\varnothing) = 0$.
(b) If $A_1, A_2, \cdots \in \mathcal{F}$ are disjoint, then $\mathbb{P}\left(\bigcup_{n=1}^{\infty} A_n\right) = \sum_{n=1}^{\infty} \mathbb{P}(A_n)$.

Thus, probability theory is interpreted as a branch of measure theory. For mathematical purposes, this is a very useful way to formulate probability theory. But the way in which probabilities and events are interpreted may be quite different from the way in which measures and measurable sets are interpreted. A measure $\mathcal{M}$ where the whole space has measure 1 may be called a **probability measure**, even when we are not thinking of a probabilistic interpretation.

In the mathematical sense, nothing is lost by postulating countable additivity:

(4.1.1) Exercise. Suppose Γ is a set, $\mathcal{G}$ is an algebra of subsets, and $\mathbb{P}'$ is a finitely additive measure on $(\Gamma, \mathcal{G})$; that is:

(a) $\Gamma \in \mathcal{G}$; $\mathbb{P}'(\Gamma) = 1$; $\varnothing \in \mathcal{G}$; $\mathbb{P}'(\varnothing) = 0$.
(b) If $A \in \mathcal{G}$, then $\Gamma \setminus A \in \mathcal{G}$ and $\mathbb{P}'(\Gamma \setminus A) = 1 - \mathbb{P}'(A)$.
(c) if $A, B \in \mathcal{G}$, then $A \cup B, A \cap B \in \mathcal{G}$.
(d) if $A, B \in \mathcal{G}$ are disjoint, then $\mathbb{P}'(A \cup B) = \mathbb{P}'(A) + \mathbb{P}'(B)$.

Prove that there is a countably additive model for $\mathbb{P}'$: There exist a set Ω, a σ-algebra $\mathcal{F}$ of subsets, a (countably additive) probability measure $\mathbb{P}$ on $\mathcal{F}$, and a one-to-one map $\Phi \colon \mathcal{G} \to \mathcal{F}$ such that:

(1) $\mathbb{P}(\Phi(A)) = \mathbb{P}'(A)$ for all $A \in \mathcal{G}$.
(2) $\Phi(A \cup B) = \Phi(A) \cup \Phi(B)$.
(3) $\Phi(A \cap B) = \Phi(A) \cap \Phi(B)$.
(4) $\Phi(\Gamma) = \Omega$, $\Phi(\varnothing) = \varnothing$.

Examples. We have already dealt with one of the most common examples of a probability space: the interval $[0, 1]$. More precisely, let $\Omega = [0, 1]$, let $\mathcal{F}$ be the σ-algebra of Lebesgue measurable sets, and let $\mathbb{P}$ be the restriction $\mathcal{L} \upharpoonright [0, 1]$ of Lebesgue measure to $[0, 1]$. Now, each single point $\{s\}$ has probability zero. This illustrates the importance of allowing probability zero even for an event that may be possible. Indeed, the entire sample space Ω is the union of these events $\{s\}$ of probability zero, so one of them is certain to occur. But since the interval is uncountable, this does not contradict the assumptions we made for probability.

We have also seen other common examples of probability spaces. Fix a positive integer n. To model the experiment of tossing a coin n times in succession, we may use a certain space of strings. Let us use a two-letter alphabet $E = \{\mathsf{H}, \mathsf{T}\}$, and let $\Omega = E^{(n)}$ be the set of strings of length n. Then each sample point $\omega \in \Omega$ corresponds to a sequence of outcomes, with H standing for "heads" and T for "tails." When we want to model a "fair" coin, that is, we want heads and tails each to have probability $1/2$ and the individual tosses to be independent of each other, then the probability measure $\mathbb{P}$ should be defined such that $\mathbb{P}(\{\omega\}) = 1/2^n$ for each $\omega \in \Omega$.

A related example would consist of the space of *infinite* strings of letters from the alphabet $E = \{\mathsf{H}, \mathsf{T}\}$. Or we can think of modeling an *infinite* sequence of tosses of a coin. The sample space is $\Omega = E^{(\omega)}$. Now we want to define a probability measure as we have done before. For each finite string $\alpha \in E^{(*)}$, the cylinder $[\alpha]$ should have probability $(1/2)^{|\alpha|}$.

Sequences of Events. Suppose an event A_n is given for each positive integer n. We may want to discuss the "limiting" behavior associated with such an infinite sequence.

As a simple example, consider the sample space $\Omega = E^{(\omega)}$ of infinite strings from the alphabet $E = \{\mathsf{H}, \mathsf{T}\}$. Let A_n be the event "the nth toss is heads." Other events may be defined in terms of these elementary events. For example,

$$\bigcap_{k=1}^{\infty} \bigcup_{n=k}^{\infty} A_n$$

is the event "there are infinitely many heads";

$$\bigcup_{k=1}^{\infty} \bigcap_{n=k}^{\infty} A_n$$

is the event "all but finitely many of the outcomes are heads."

In general, given a sequence A_n of events, we define

$$\limsup_{n \to \infty} A_n = \bigcap_{k=1}^{\infty} \bigcup_{n=k}^{\infty} A_n,$$

$$\liminf_{n \to \infty} A_n = \bigcup_{k=1}^{\infty} \bigcap_{n=k}^{\infty} A_n.$$

Thus, the event $\limsup A_n$ occurs iff A_n occurs for infinitely many n; and the event $\liminf A_n$ occurs iff A_n occurs for all except possibly finitely many n.

Next is the "easy direction" of the Borel–Cantelli Lemma. The converse "hard direction" will be discussed below in (4.3.10).

(4.1.2) Proposition. *Let A_n be a sequence of events. Suppose*

$$\sum_{n=1}^{\infty} \mathbb{P}(A_n) < \infty.$$

Then, with probability one, only finitely many of the events A_n occur. That is, $\mathbb{P}(\limsup A_n) = 0$.

Proof. Let $C = \Omega \setminus \limsup A_n$. We must show that $\mathbb{P}(C) = 1$. Fix $m \in N$. Then

$$C = \Omega \setminus \bigcap_{k} \bigcup_{n=k}^{\infty} A_n = \bigcup_{k} \left(\Omega \setminus \bigcup_{n=k}^{\infty} A_n \right) \supseteq \Omega \setminus \bigcup_{n=m}^{\infty} A_n,$$

so that

$$\mathbb{P}(C) \geq 1 - \mathbb{P}\left(\bigcup_{n=m}^{\infty} A_n \right) \geq 1 - \sum_{n=m}^{\infty} \mathbb{P}(A_n).$$

But because the series converges, the tail of the series goes to zero (as $m \to \infty$). Therefore, $\mathbb{P}(C) \geq 1 - 0 = 1$. ☺

A Technicality. A probability space $(\Omega, \mathcal{F}, \mathbb{P})$ is called **complete** iff every subset of an event of probability 0 is again an event; that is, if $A \subseteq B$, $B \in \mathcal{F}$, and $\mathbb{P}(B) = 0$ implies $A \in \mathcal{F}$. For technical reasons, it is customary to postulate that the sample space is complete.

4.2 Random Variables

A **random variable** is a randomly determined item. For example, a real random variable (also called a random real number) is a real number, but which real number it is may depend on the outcome of some random events. A spinner in some common games (see the figure) can be thought of as determining a random angle (which may be considered a random real number between 0 and 2π).

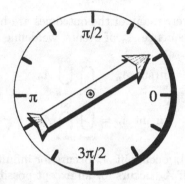

Figure 4.2.1. Random variable?

Technically, a random variable X is a function defined on the sample space Ω. For $\omega \in \Omega$, the value $X(\omega)$ is supposed to be the value the random variable has when the state of the universe is ω. On the other hand, a random variable determines some events. For example, if X is a real random variable, then for each real number t, the condition $X < t$ determines an event. In technical language, a real random variable is a measurable function $X \colon \Omega \to \mathbb{R}$.

Random variables with values other than real numbers are possible. In this book, we will normally consider only random elements of metric spaces. If (S, ρ) is a metric space, then an S-**valued random variable** is a measurable function $X \colon \Omega \to S$; measurability is determined by the given σ-algebra $\mathcal{F}$ of events on Ω and the Borel σ-algebra on the metric space S. An S-valued random variable may also be called a **random element of** S, or simply a **measurable function**.

Let us consider a few examples.

Random Real Number. The most often seen sort of random variable is one with values in the real line $\mathbb{R}$. If I measure the time between clicks of a Geiger counter, the result is a (random) positive real number. If I flip a coin 100 times and count the number of times it comes up heads, the result is a (random) number in the set $\{0, 1, 2, \cdots, 100\}$.

We will discuss below some of the more useful ways that real random variables may be classified.

Random Vector. The metric space $\mathbb{R}^d$ may be identified with d-dimensional Euclidean space or with the set of d-dimensional vectors. A random variable with values in $\mathbb{R}^d$ is then a measurable function $X \colon \Omega \to \mathbb{R}^d$. The Borel sets in $\mathbb{R}^d$ are generated by the coordinate half-spaces of the form

$$W(j,a) = \{\, x = (x_1, x_2, \cdots, x_d) : x_j < a \,\},$$

where $j \in \{1, 2, \cdots, d\}$ and $a \in \mathbb{R}$. We may write a function $X \colon \Omega \to \mathbb{R}^d$ in terms of its coordinates:

$$X(\omega) = \big(X_1(\omega), X_2(\omega), \cdots, X_d(\omega)\big).$$

The vector-valued function X is Borel measurable if and only if all of the components X_j are Borel measurable real-valued functions. Because of this identification, a random element of $\mathbb{R}^d$ is usually considered to be the same thing as a list $(X_1, X_2, \cdots, X_d)$ of d real random variables.

Let $G \subseteq \mathbb{R}^d$ be a Borel set with $0 < \mathcal{L}^d(G) < \infty$. A random element X of $\mathbb{R}^d$ is said to be **uniformly distributed** on G iff

$$\mathbb{P}\{X \in E\} = \frac{\mathcal{L}^d(E \cap G)}{\mathcal{L}^d(G)}$$

for all Borel sets $E \subseteq \mathbb{R}^d$. By a long-standing convention used by mathematicians, this is the property intended when we say "Let X be chosen at random in the set G" without any further explanation.[1]

Random Sequence. Let (S, ρ) be a metric space. Let $\Sigma = S^{\mathbb{N}}$ be the set of sequences in S. That is, an element $\mathbf{x} \in \Sigma$ is a sequence of elements of S; we will often write $\mathbf{x}(1), \mathbf{x}(2), \mathbf{x}(3), \cdots$ for the terms of the sequence $\mathbf{x}$. Or we may write $\mathbf{x} = (x_1, x_2, \cdots)$.

For a metric on Σ, we want to arrange the following condition for convergence: if $\mathbf{x}_n$ is a sequence of elements of Σ, and $\mathbf{x}_0 \in \Sigma$, then

$$\lim_{n \to \infty} \mathbf{x}_n = \mathbf{x}_0 \qquad \text{if and only if} \qquad \lim_{n \to \infty} \mathbf{x}_n(i) = \mathbf{x}_0(i) \text{ for each } i \in \mathbb{N}.$$

(This is sometimes known as "pointwise convergence.")

A metric with this convergence property may be specified in more than one way. For example,

$$\theta(\mathbf{x}, \mathbf{y}) = \sum_{i=1}^{\infty} \big(2^{-i} \wedge \rho(\mathbf{x}(i), \mathbf{y}(i))\big).$$

(4.2.2) Exercise. (a) The function $\theta \colon \Sigma \times \Sigma \to \mathbb{R}$ is a metric for the set Σ of sequences. (b) Convergence for the metric θ is pointwise convergence.

(4.2.3) Exercise. Let S be a separable metric space. Let τ be any metric on the sequence space $\Sigma = S^{\mathbb{N}}$ satisfying

---

[1] Beware: Choosing a point at random from a set G of infinite Lebesgue measure (or of 0 Lebesgue measure) is not governed by such a convention, so requires some further explanation.

$$\lim_{n\to\infty} \tau(\mathbf{x}_n, \mathbf{x}_0) = 0 \quad \text{if and only if} \quad \lim_{n\to\infty} \rho(\mathbf{x}_n(i), \mathbf{x}_0(i)) = 0 \text{ for } i \in \mathbb{N}.$$

Then the σ-algebra of Borel sets for the metric space (Σ, τ) is the σ-algebra generated by the sets

$$E(i, A) = \{\, \mathbf{x} \in \Sigma : \mathbf{x}(i) \in A \,\},$$

for integer $i \in \mathbb{N}$ and Borel set $A \subseteq S$.

A consequence of this exercise is that there is a one-to-one correspondence between (i) a random sequence in S and (ii) a sequence of random elements of S. For (i), we have a measurable function $X : \Omega \to S^{\mathbb{N}}$. For (ii), we have a sequence $(Y_n)_{n \in \mathbb{N}}$ where each $Y_n : \Omega \to S$ is measurable. The correspondence is specified by

$$X(\omega)(n) = Y_n(\omega).$$

Because of this correspondence, we will often not distinguish between (i) and (ii).

Random Function. A "random function" is a random element of a space of functions. In this book, we will use only a few metric spaces of functions, for example, $\mathfrak{C}(S, \mathbb{R})$ with the uniform metric, where S is a compact metric space, or $\mathrm{Lip}_b(S, R)$ with a metric like

$$\tilde{\rho}(f, 0) = \sup\{\, |f(x)| : x \in S \,\} + \sup\left\{\, \frac{|f(x) - f(y)|}{\rho(x, y)} : x, y \in S, x \neq y \,\right\}$$

and $\tilde{\rho}(f, g) = \tilde{\rho}(f - g, 0)$, where S is a metric space.

(4.2.4) Exercise. Verify that $\tilde{\rho}$ is a metric.

Let us consider the σ-algebra of Borel sets in a function space. Let S be a compact metric space, and consider the function space $\mathfrak{C}(S, \mathbb{R})$. If $s \in S$ and $a < b$ are real numbers, write

$$W(s, a, b) = \{\, f \in \mathfrak{C}(S, \mathbb{R}) : a < f(s) < b \,\}.$$

Then certainly, $W(s, a, b)$ is a Borel set (indeed, the "evaluation" $f \mapsto f(s)$ is continuous, so $W(s, a, b)$ is open).

(4.2.5) Theorem. *The sets $W(s, a, b)$ defined above generate the σ-algebra of Borel sets in $\mathfrak{C}(S, \mathbb{R})$.*

Proof. Let $\mathcal{B}$ be the σ-algebra of Borel sets in $\mathfrak{C}(S, \mathbb{R})$, and let $\mathcal{F}$ be the σ-algebra generated by the sets $W(s, a, b)$. As just noted, $\mathcal{F} \subseteq \mathcal{B}$. So we must prove the reverse inclusion. Since $\mathfrak{C}(S, \mathbb{R})$ is separable,[2] it is enough to show

---

[2] from the Weierstrass Approximation Theorem; see below, (4.4.11).

that any open ball in $\mathcal{C}(S, \mathbb{R})$ belongs to $\mathcal{F}$. So let $f_0 \in \mathcal{C}(S, \mathbb{R})$ and $r > 0$. Now let $V \subseteq S$ be a countable dense set. Then

$$B_r(f_0) = \bigcup_{n \in \mathbb{N}} \bigcap_{v \in V} W\left(v, f_0(v) - r + \frac{1}{n}, f_0(v) + r - \frac{1}{n}\right),$$

so $B_r(f_0) \in \mathcal{F}$ as required.   ☺

The collection of sets of the form

$$W(s_1, a_1, b_1) \cap W(s_2, a_2, b_2) \cap \cdots \cap W(s_n, a_n, b_n)$$

is a pi system that generates the Borel sets of $\mathcal{C}(S, \mathbb{R})$. Thus (by Proposition 1.1.8) a Borel probability measure on $\mathcal{C}(S, \mathbb{R})$ is uniquely determined by its values on sets of this form. (Such sets may be called "cylinders.")

Now consider the noncompact set $S = [0, \infty)$. Note that the functions in $\mathcal{C}([0, \infty), \mathbb{R})$ are allowed to be unbounded, so the supremum metric $\sup_x |f(x) - g(x)|$ is not a metric at all; and uniform convergence is not a useful topology. A reasonable topology for the space $\mathcal{C}([0, \infty), \mathbb{R})$ is "uniform convergence on compact sets"—a sequence f_n in $\mathcal{C}([0, \infty), \mathbb{R})$ converges to f iff f_n converges uniformly to f on every compact subset of $[0, \infty)$.

(4.2.6) Exercise. (a) Any compact set $K \subseteq [0, \infty)$ is a subset of some interval $[0, m]$ with $m \in \mathbb{N}$. (b) A sequence f_n converges to f uniformly on compact sets if and only if f_n converges uniformly to f on each interval $[0, m]$. (c) Formulate this with quantifiers and epsilons.

(4.2.7) Exercise. Define a metric on $\mathcal{C}([0, \infty), \mathbb{R})$ such that convergence according to your metric coincides with uniform convergence on compact sets. Can you make it also a complete metric?

(4.2.8) Exercise. The space $\mathcal{C}([0, \infty), \mathbb{R})$ is separable.

(4.2.9) Exercise. The Borel sets in $\mathcal{C}([0, \infty), \mathbb{R})$ are generated by the sets of the form $W(s, a, b) = \{ f \in \mathcal{C}([0, \infty), \mathbb{R}) : a < f(s) < b \}$.

Random Sets. Let S be a metric space. The set $\mathfrak{K}(S)$ of nonempty compact subsets of S may be made into a metric space using the **Hausdorff metric** [MTFG, §2.4]. Two sets $E, F \in \mathfrak{K}(S)$ are within distance ε iff each of the sets is contained in the ε-neighborhood of the other. When we talk about a random set, we will often refer to this metric (for the measurability required).

What are the measurability properties of a random set? Let S be a metric space, and let V be a random element of $\mathfrak{K}(S)$. Write D for the Hausdorff metric. For a given closed set F, the function "distance to F" is continuous in the Hausdorff metric: if $D(K_1, K_2) < \varepsilon$, then $|\operatorname{dist}(F, K_1) - \operatorname{dist}(F, K_2)| < \varepsilon$. Now, for F closed and K compact, $F \cap K \neq \varnothing$ if and only if $\operatorname{dist}(F, K) = 0$. Thus, if V is a random element of $\mathfrak{K}(S)$ and $F \subseteq S$ is closed, then

$$\{F \cap V \neq \varnothing\}$$

is measurable (it is an event). Next, if $A = \bigcup_n F_n$ is a countable union of closed sets (for example, any open set is a countable union of closed sets), then

$$\{A \cap V \neq \varnothing\} = \{F_n \cap V \neq \varnothing \text{ for some } n \in \mathbb{N}\} = \bigcup_n \{F_n \cap V \neq \varnothing\}$$

so it is measurable.

(4.2.10) Exercise. Let A be a Borel set. Does it follow that $\{A \cap V \neq \varnothing\}$ is measurable?

(4.2.11) Exercise. The "union" map (the function $(K_1, K_2) \mapsto K_1 \cup K_2$) is continuous. If V and W are random sets, then $V \cup W$ is also measurable (so it is a random set).

If we try to do the same for other cases, we may have certain problems. If V and W are random sets, does it follow that $V \cap W$ and $V \setminus W$ are random sets? If V_n is a random set for each $n \in \mathbb{N}$, then are $\bigcup_n V_n$ and $\bigcap_n V_n$ also random sets?

It is quite possible that $V \cap W$ is empty, thus not an element of $\mathfrak{K}(S)$. We can fix our definition to cover this case as follows. Let $(\Omega, \mathcal{F}, \mathbb{P})$ be our probability space. We say that a function $V : \Omega \to \mathfrak{K}(S) \cup \{\varnothing\}$ is measurable iff $\{V = \varnothing\} \in \mathcal{F}$ and $\{V \in A\} \in \mathcal{F}$ for all Borel sets $A \subseteq \mathfrak{K}(S)$.

But that fixed definition will still not cover cases like $V \setminus W$ and $\bigcup_n V_n$ where the sets are likely not even closed.

In some cases it is not hard to find a reasonable interpretation, for example, for a random closed set in $\mathbb{R}$. For each $\omega \in \Omega$, suppose $V(\omega)$ is a closed subset of $\mathbb{R}$. We will say that V is measurable iff $V \cap [-n, n]$ is measurable for all $n \in \mathbb{N}$. For each n, the function $V \cap [-n, n]$ has values in $\mathfrak{K}(\mathbb{R}) \cup \{\varnothing\}$. We may define a random closed set in $\mathbb{R}^d$ in a similar way. A random open set is the complement of a random closed set. This should be enough for our use in this book.

*[Still more general notions of random set (not necessarily compact) may be defined. A random analytic set (see the end of §1.7) may be specified as

$$E = \bigcup_{(n_1, n_2, \cdots) \in \mathbb{N}^{\mathbb{N}}} \bigcap_{k \in \mathbb{N}} C(n_1, n_2, \cdots, n_k),$$

where, $C(n_1, n_2, \cdots, n_k)$ is a random compact set for each finite sequence $(n_1, n_2, \cdots, n_k)$. All of the random sets in this book should be analytic, but we will often not stop to check.]

Random Measures. Let S be a metric space. The set $\mathfrak{P}(S)$ of tight probability measures on S will be given the narrow topology. The metrics ρ_γ of

---

* Optional

§2.5 all produce the same open sets, and the same Borel sets, for $\mathfrak{P}(S)$. By the definition of the narrow topology, for any bounded continuous function h, the map $\mathcal{M} \mapsto \int h(x)\,\mathcal{M}(dx)$ is a continuous function on $\mathfrak{P}(S)$. This should be enough for you to complete the following exercise. (Of course, parts (b), (d), (e) can be done only after you know the definitions of "upper semicontinuous," "mean," and "variance.")

(4.2.12) Exercise. Let S be a metric space. Then:

(a) If h is a bounded Borel function on S, then $\mathcal{M} \mapsto \int h(x)\,\mathcal{M}(dx)$ is a bounded Borel function on $\mathfrak{P}(S)$.

(b) If h is a bounded upper semicontinuous function on S, then $\mathcal{M} \mapsto \int h(x)\,\mathcal{M}(dx)$ is a bounded upper semicontinuous function on $\mathfrak{P}(S)$.

(c) If $A \subseteq S$ is a Borel set, then $\mathcal{M} \mapsto \mathcal{M}(A)$ is a Borel function on $\mathfrak{P}(S)$.

(d) "Mean" is continuous on $\mathfrak{P}(\mathbb{R})$; that is, $\mathcal{M} \mapsto$ (the mean of $\mathcal{M}$) is a continuous function on the set $\{\mathcal{M} \in \mathfrak{P}(\mathbb{R}) :$ the mean of $\mathcal{M}$ exists $\}$.

(e) "Variance" is continuous on $\mathfrak{P}(\mathbb{R})$.

(4.2.13) Exercise. The functions $\mathcal{M} \mapsto \mathcal{M}(A)$ generate the Borel sets. That is, a function $V \colon \Omega \to \mathfrak{P}(S)$ is measurable if and only if for all Borel sets A, the function $V(A) \colon \Omega \to \mathbb{R}$ is measurable.

Interpretation of this exercise needs a little care. Let V be a random measure on S, that is, a random element of $\mathfrak{P}(S)$. Each element of $\mathfrak{P}(S)$ is a tight measure; but this does not mean that the set of all values of V is a uniformly tight set. There need not be a single (nonrandom) compact set K such that $V(K) > 1 - \varepsilon$ with probability one.

On the other hand, for each Borel set $E \subseteq S$, we have a real random variable $V(E)$. These real random variables satisfy a family of identities: if (A_n) is a disjoint sequence of Borel sets, then

$$(*) \qquad V\left(\bigcup_n A_n\right) = \sum_n V(A_n).$$

But in order to define a random measure, it is not enough to specify a collection $(V(E))_{E \in \mathcal{B}(S)}$ of nonnegative random variables that satisfy each identity $(*)$ with probability one. There are uncountably many sequences (A_n), and even if each identity $(*)$ holds with probability one, it need not follow that all identities $(*)$ hold simultaneously for any sample point.

Mathematical Expectation. If X is a real random variable, we will often need to measure the "average" or "mean" value of X. This is known as the **mathematical expectation** of X. Technically, it is the same thing as the integral of X; it is important to recall that we are dealing with a probability measure, where $\mathbb{P}(\Omega) = 1$.

Let $(\Omega, \mathcal{F}, \mathbb{P})$ be a probability space. Let $X \colon \Omega \to \mathbb{R}$ be a real random variable. The mathematical expectation of X is

$$\mathbb{E}\left[X\right] = \int_{\Omega} X(\omega)\,\mathbb{P}(d\omega),$$

provided that it exists.[3] Recall that this means $\int |X(\omega)|\,\mathbb{P}(d\omega) < \infty$. A real random variable X is sometimes called **integrable** iff $\mathbb{E}\left[|X|\right] < \infty$. Probabilists may say that X has "finite first moment."

As a simple example, consider the toss of a fair coin $\mathbb{P}\{X = 1\} = \mathbb{P}\{X = 0\} = 1/2$. Then $\mathbb{E}\left[X\right] = (1/2) \cdot 1 + (1/2) \cdot 0 = 1/2$.

(4.2.14) Proposition. *Let X be a real random variable. Then, for every $\lambda > 0$,*

$$\mathbb{P}\left\{|X| \geq \lambda\right\} \leq \frac{1}{\lambda}\mathbb{E}\left[|X|\right].$$

Proof. Let $E = \{|X| \geq \lambda\}$. Then, since $|X| \geq \lambda$ on E,

$$\mathbb{P}(E) = \frac{1}{\lambda}\mathbb{E}\left[\lambda\,\mathbb{1}_E\right] \leq \frac{1}{\lambda}\mathbb{E}\left[|X|\,\mathbb{1}_E\right] \leq \frac{1}{\lambda}\mathbb{E}\left[|X|\right],$$

as required.   ☺

(4.2.15) Exercise. A real random variable X belongs to **weak L_1** iff there is a constant C such that $\mathbb{P}\{|X| \geq \lambda\} \leq C/\lambda$ for all $\lambda > 0$. The preceding proposition shows that if X is integrable, then X belongs to weak L_1. Is the converse true: if X belongs to weak L_1, does it necessarily follow that X is integrable?

(4.2.16) Exercise. Let X be a nonnegative random variable. Then the expected value may be computed as a (possibly improper) Riemann integral:

$$\mathbb{E}\left[X\right] = \int_0^{\infty} \mathbb{P}\{X > \lambda\}\,d\lambda.$$

Variance. The expectation of a random variable is a useful characteristic. But it does not tell the whole story. For example, a random variable X that is identically equal to 0 has expectation zero. But other random variables may also have expectation zero because the positive and negative values balance out.

A second useful characteristic of a real random variable is one that measures (in a single number) the amount that X tends to "spread out" above and below its mean. If X is a real random variable with mean $\mathbb{E}\left[X\right] = \mu$, then the **variance** of X is

$$\mathrm{Var}\left[X\right] = \mathbb{E}\left[|X - \mu|^2\right].$$

It is nonnegative, but it may be finite or infinite.

---

[3] In physics texts you will often see the notation $\langle X \rangle$ for the expectation of a random variable X.

Suppose X is a simple random variable describing the toss of a fair coin, $\mathbb{P}\{X = 1\} = \mathbb{P}\{X = 0\} = 1/2$. Then the variance is

$$\text{Var}\,[X] = |1 - \mu|^2 \cdot \mathbb{P}\{X = 1\} + |0 - \mu|^2 \mathbb{P}\{X = 0\}$$
$$= (1/4) \cdot (1/2) + (1/4) \cdot (1/2) = 1/4.$$

(4.2.17) Exercise. Let X be the random variable describing a "biased coin": Fix a number p with $0 < p < 1$ as the probability of heads. Then let X satisfy $\mathbb{P}\{X = 1\} = p, \mathbb{P}\{X = 0\} = 1 - p$. Compute the mean and variance.

(4.2.18) Exercise. Let X be the random variable describing a die: that is, let $\mathbb{P}\{X = k\} = 1/6$ for $k = 1, 2, \cdots, 6$. Compute the mean and variance.

A real random variable is called **square-integrable** iff $\mathbb{E}\left[|X|^2\right] < \infty$.[4] From this it follows that the variance exists and is finite. First, the mean exists, since $\mathbb{E}\,[|X|] \leq \mathbb{E}\left[|X|^2\right]^{1/2}$. Second, the integral defining the variance converges. In fact,

(4.2.19) Exercise. $\text{Var}\,[X] = \mathbb{E}\left[X^2\right] - \mathbb{E}\,[X]^2$.

A related concept, the **standard deviation**, is simply the square root of the variance. The Greek letter σ is often used for the standard deviation. In the fair-coin example, $\text{Var}\,[X] = 1/4$, so the standard deviation is $\sigma = 1/2$.

(4.2.20) Exercise. Let X be a real random variable. For every $\lambda > 0$, show that

$$\mathbb{P}\{|X| \geq \lambda\} \leq \frac{1}{\lambda^2} \mathbb{E}\left[X^2\right].$$

This is sometimes called "Chebyshev's inequality."

(4.2.21) Exercise. Let us say that a random variable X belongs to **weak** L_2 iff there is a constant C such that for all $\lambda > 0$, we have $\mathbb{P}\{|X| \geq \lambda\} \leq C/\lambda^2$. Chebyshev's inequality shows that if X is square-integrable, then it belongs to weak L_2. What about the converse: if X belongs to weak L_2, does it follow that X is square-integrable?

The following simple convergence theorem (based on Chebyshev's inequality and the Borel–Cantelli Lemma) will be used in the next chapter.

(4.2.22) Proposition. *For each $n \in \mathbb{N}$, let X_n be a real random variable with mean μ_n and variance σ_n^2. Suppose $\lim_n \mu_n = \mu$ and $\sum_n \sigma_n^2 < \infty$. Then with probability one, the sequence X_n converges to μ.*

---

[4] You may think it strange that I wrote $|X|^2$ and not just X^2, which is equal to it. Well, I guess I have the tendency to write $|X|^2$ from situations where I dealt with complex functions.

Proof. For $p \in \mathbb{N}$, there is m_p so large that

$$\sum_{n=m_p}^{\infty} \sigma_n^2 < 2^{-3p} \quad \text{and} \quad |\mu_n - \mu| < 2^{-p} \text{ for all } n \geq m_p.$$

Let $E_p = \{|X_n - \mu_n| \leq 2^{-p} \text{ for all } n \geq m_p\}$. Now,

$$\mathbb{P}(E_p) \geq 1 - \sum_{n=m_p}^{\infty} \mathbb{P}\{|X_n - \mu_n| > 2^{-p}\} \geq 1 - \sum_{n=m_p}^{\infty} \frac{1}{2^{-2p}} \sigma_n^2 \geq 1 - 2^{-p}.$$

The series $\sum_p 2^{-p}$ converges, so by the Borel–Cantelli Lemma 4.1.2, with probability one, all but finitely many of the events E_p must occur. So for large enough p,

$$|X_n - \mu| \leq |X_n - \mu_n| + |\mu_n - \mu| \leq 2^{-p} + 2^{-p}$$

for all $n \geq m_p$. Therefore, X_n converges to μ.   ☺

Probability Distributions. Suppose $X: \Omega \to S$ is a random variable with values in the metric space S. Then X determines a measure $\mathcal{D}_X$ on S:

$$\mathcal{D}_X(E) = \mathbb{P}\{\omega \in \Omega : X(\omega) \in E\} \quad \text{for all Borel sets } E \subseteq S.$$

In practice, the sample point ω is almost never mentioned:

$$\mathcal{D}_X(E) = \mathbb{P}\{X \in E\}.$$

The measure $\mathcal{D}_X$ determined in this way is called the **distribution** (or **law**) of the random variable X.

Two random variables defined in completely different ways, perhaps even defined on two different sample spaces, may have the same distribution. We write

$$X \overset{\mathcal{D}}{=} Y$$

when X and Y have the same distribution; that is, $\mathcal{D}_X(A) = \mathcal{D}_Y(A)$ for all Borel sets A. It follows that X and Y have many other properties in common. For example, two real random variables with the same distribution have the same expectation and the same variance. Indeed,

$$\mathbb{E}[X] = \int_{\mathbb{R}} x \, \mathcal{D}_X(dx) = \mu, \quad \text{Var}[X] = \int |x - \mu|^2 \, \mathcal{D}_X(dx),$$

and in general, if f is a Borel function,

$$\mathbb{E}[f(X)] = \int f(x) \, \mathcal{D}_X(dx).$$

This is merely the change of variables formula 2.2.4.

Two (or more) random variables that have the same distribution are said to be **identically distributed**.

Suppose X is a random variable with values in a metric space S, and Y is a random variable with values in T. Then the ordered pair (X, Y) will be a random variable with values in the Cartesian product $S \times T$. The distribution of the ordered pair (X, Y) is a measure on $S \times T$; it is sometimes known as the **joint distribution** of X and Y. It contains more information than just the individual distributions of X and Y.

(4.2.23) Exercise. Give two examples of joint distributions for an ordered pair (X, Y) on $[0, 1] \times [0, 1]$. In both examples, the random variables X and Y should be uniformly distributed on $[0, 1]$. But the joint distributions should be different in the two examples.

Certain distributions occur so often that they have special names. Only a few of the more important ones will be mentioned here.

Uniform Distribution. Let $a < b$ be real numbers. The **uniform distribution** on the interval $[a, b]$ is the probability measure

$$\frac{1}{b - a} \, \mathcal{L} \!\restriction\! [a, b];$$

that is, Lebesgue measure restricted and normalized to the interval. Equivalently, this is the measure $\mathcal{M}$ on $\mathbb{R}$ defined by

$$\mathcal{M}(E) = \frac{\mathcal{L}(E)}{b - a}.$$

Thus, a real random variable X is said to be **distributed uniformly on** $[a, b]$ iff

$$\mathbb{P}\{X \in E\} = \frac{\mathcal{L}(E)}{b - a}$$

for all Borel sets E. Equivalently, for all real t,

$$\mathbb{P}\{X \le t\} = \begin{cases} 0, & \text{if } t < a, \\ \dfrac{t - a}{b - a}, & \text{if } a \le t \le b, \\ 1, & \text{if } b < t. \end{cases}$$

Dirac Distribution. Let a be a real number. The **Dirac distribution** at a is the measure $\mathcal{E}_a$ defined by

$$\mathcal{E}_a(E) = \begin{cases} 0, & \text{if } a \notin E, \\ 1, & \text{if } a \in E. \end{cases}$$

Sometimes this measure is called the **unit point mass** at a. A random variable X with this distribution satisfies $X = a$ with probability one, so we may say that X is **deterministic**, or **nonrandom**.

Note that Dirac distributions may be defined in the same way for random variables in other spaces, as well.

Bernoulli and Binomial Distributions. Let $0 < p < 1$ be a fixed number. A real random variable X has **Bernoulli distribution** with probability p of success iff

$$\mathbb{P}\{X = 1\} = p, \qquad \mathbb{P}\{X = 0\} = 1 - p$$

(and thus, necessarily, $\mathbb{P}\{X \notin \{0,1\}\} = 0$). This distribution is a linear combination of two Dirac measures:

$$p\mathcal{E}_1 + (1 - p)\mathcal{E}_0.$$

We think of such a random variable as an indicator function. If A is an event, then the random variable $\mathbb{1}_A$, which is 1 if A occurs and 0 if A does not occur, is called the **indicator function** of the event A. When probability is formulated as measure theory (via the Kolmogorov model, as we have done here), an indicator function is the same thing as what is known as a characteristic function in certain other branches of mathematics.[5]

Now, in addition to the parameter p, let n be a positive integer. The **binomial distribution** with parameters p and n is a combination of Dirac measures where the coefficients are the terms of the binomial expansion:

$$\sum_{k=0}^{n} \binom{n}{k} p^k (1 - p)^{n-k} \, \mathcal{E}_k.$$

Equivalently, a random variable X has binomial distribution iff

$$\mathbb{P}\{X = k\} = \binom{n}{k} p^k (1 - p)^{n-k}$$

for $k = 0, 1, 2, \cdots, n$ and $\mathbb{P}\{X = t\} = 0$ for other values of t.

Here is an exercise that will test your understanding of independence. If you cannot do it now, return and try again after our discussion of independence in §4.3.[6]

(4.2.24) Exercise. Let $Y_1, Y_2, \cdots, Y_n$ be independent random variables, each with Bernoulli distribution

$$\mathbb{P}\{Y_k = 1\} = p, \qquad \mathbb{P}\{Y_k = 0\} = (1 - p).$$

Then the sum $X = Y_1 + Y_2 + \cdots + Y_n$ has binomial distribution with parameters p, n.

(4.2.25) Exercise. Let $(Y_k)_{k=1}^{\infty}$ be an infinite sequence of random variables, independent, and all with Bernoulli distribution

$$\mathbb{P}\{Y_k = 1\} = \frac{1}{2}, \qquad \mathbb{P}\{Y_k = 0\} = \frac{1}{2}.$$

---

[5] In probability theory, accordingly, there is *another* concept known as "characteristic function."

[6] We will refer to independence many times in this section, despite the fact that independence is not defined until §4.3.

Show that the combination

$$X = \sum_{k=1}^{\infty} 2^{-k} Y_k$$

is uniformly distributed on $[0,1]$.

Phrased in a more colorful way, in order to generate a uniform distribution on $[0,1]$, generate the digits for the binary expansion by tossing a fair coin independently for each place.

Cantor–Lebesgue Distribution. Suppose you use a fair coin to generate digits, but interpret them in base 3, and let heads mean digit 0 and tails mean digit 2. More precisely, let $(Y_k)_{k=1}^{\infty}$ be an infinite sequence of random variables, independent, all with Bernoulli distribution

$$\mathbb{P}\{Y_k = 1\} = \frac{1}{2}, \qquad \mathbb{P}\{Y_k = 0\} = \frac{1}{2}.$$

Then

$$X = 2 \sum_{k=1}^{\infty} 3^{-k} Y_k$$

has the **Cantor–Lebesgue distribution.** Because of the description of the Cantor ternary set C in terms of base 3 expansions $[MTFG$, Proposition 1.1.5$]$, we clearly have $\mathbb{P}\{X \in C\} = 1$. But recall that $\mathcal{L}(C) = 0$.

Two Borel measures $\mathcal{M}_1, \mathcal{M}_2$ defined in a metric space X are said to be **singular** to each other iff there is a Borel set E with $\mathcal{M}_1(E) = 0$ and $\mathcal{M}_2(X \setminus E) = 0$. We write $\mathcal{M}_1 \perp \mathcal{M}_2$. A measure $\mathcal{M}$ in Euclidean space $\mathbb{R}^d$ is simply called **singular** iff it is singular to Lebesgue measure $\mathcal{L}^d$. Thus, the Cantor–Lebesgue distribution in $\mathbb{R}$ is a singular distribution.

Of course, the Dirac distributions and the binomial distributions are also singular. But the Cantor–Lebesgue distribution has another interesting property. A measure $\mathcal{M}$ on a metric space X is called **continuous** iff $\mathcal{M}(\{x\}) = 0$ for every $x \in X$.

(4.2.26) Exercise. The Cantor–Lebesgue distribution is continuous.

The other examples (Dirac, binomial) are singular, but for the trivial reason that their mass is concentrated on just a finite number of points. They are not continuous distributions. But the Cantor–Lebesgue distribution is continuous and singular. It is not concentrated on a finite (or even countable) number of points.

(4.2.27) Exercise. Let $\mathcal{M}$ be a finite measure on $\mathbb{R}$. Its **cumulative distribution function** (c.d.f.) is the function $F \colon \mathbb{R} \to \mathbb{R}$ defined by

$$F(x) = \mathcal{M}\{t \in \mathbb{R} : t \le x\}, \qquad \text{for } x \in \mathbb{R}.$$

Show that the measure $\mathcal{M}$ is continuous (in the sense just defined) if and only if the function F is continuous (in the conventional sense).

(4.2.28) Exercise. Describe the c.d.f. of the Dirac and binomial distributions.

Exponential Distribution. Let $\lambda > 0$ be given. The **exponential distribution** with parameter λ is the measure $\mathcal{M}$ on $\mathbb{R}$ defined by

$$\mathcal{M}(A) = \int_{A \cap (0,\infty)} \lambda e^{-\lambda x}\, dx.$$

Or equivalently (since $\int_0^x \lambda e^{-\lambda t}\, dt = 1 - e^{-\lambda x}$), a random variable X has exponential distribution with parameter λ iff

$$\mathbb{P}\{X \le x\} = \begin{cases} 0, & \text{if } x \le 0, \\ 1 - e^{-\lambda x}, & \text{if } x > 0. \end{cases}$$

This is an example of an **absolutely continuous** distribution. In general, a measure $\mathcal{M}$ in $\mathbb{R}^d$ is said to be **absolutely continuous** iff it is absolutely continuous with respect to Lebesgue measure $\mathcal{L}^d$. (In this example, the distribution has density $\lambda e^{-\lambda x}$ on $(0,\infty)$.)

An exponential random variable is often used to model a "waiting time" for an event to happen. Say we wait for the elevator to come. When I arrive and press the button I start the clock, and when the elevator door opens I stop the clock. The time elapsed is the value of X. But suppose you also arrive there while I am waiting. You start your clock when you arrive and stop it when the elevator door opens. The exponential distribution for the waiting time has the interesting feature that both you and I can imagine the same probability distribution for our waiting times, even though you arrived after I did. This "memoryless" property of the distribution will be discussed more precisely after our discussion of conditional probabilities.

(4.2.29) Exercise. Compute the mean and variance of the exponential distribution with parameter λ.

Normal Distribution. Let $\mu \in \mathbb{R}$ and $\sigma > 0$ be given. The **normal distribution** (or Gaussian distribution) with parameters μ and σ is the real distribution with density

$$\frac{1}{\sqrt{2\pi}\sigma} \exp\left(\frac{-(x-\mu)^2}{2\sigma^2}\right).$$

(4.2.30) Exercise. Review some integration facts to verify that the constant has been chosen correctly to make the total mass equal to 1:

$$\int_{-\infty}^{\infty} \exp\left(\frac{-(x-\mu)^2}{2\sigma^2}\right) dx = \sqrt{2\pi}\sigma.$$

Then verify that the mean is μ and the variance is σ^2:

$$\frac{1}{\sqrt{2\pi}\sigma} \int_{-\infty}^{\infty} x \exp\left(\frac{-(x-\mu)^2}{2\sigma^2}\right) dx = \mu;$$

$$\frac{1}{\sqrt{2\pi}\sigma} \int_{-\infty}^{\infty} (x-\mu)^2 \exp\left(\frac{-(x-\mu)^2}{2\sigma^2}\right) dx = \sigma^2.$$

As the name suggests, the "normal" distribution is thought to model many of the random phenomena in the real world. For mathematical purposes, it is used in many contexts. We will see it below in connection with Brownian motion.

(4.2.31) Exercise. Suppose the random variable X has normal distribution with mean μ and variance σ^2. If z is a fixed real number, what is the distribution of the translated random variable $X - z$? If u is a fixed positive number, what the distribution of the random variable X/u? Combine these two principles to describe the distribution of the random variable $aX + b$, for given constants a, b.

The **standard** normal distribution is the normal distribution with mean 0 and variance 1. If Y has standard normal distribution, then $\sigma Y + \mu$ is normal with mean μ and variance σ^2.

Poisson Distribution. Let $\lambda > 0$ be given. The **Poisson distribution** with parameter λ is a distribution $\mathcal{M}$ concentrated on the nonnegative integers, with

$$\mathcal{M}(\{k\}) = \frac{e^{-\lambda}\lambda^k}{k!} \qquad \text{for } k = 0, 1, 2, \cdots.$$

Of course, the factor $e^{-\lambda}$ was chosen so that the sum is 1.

(4.2.32) Exercise. Compute the mean and variance for the Poisson distribution.

Cauchy Distribution. The standard **Cauchy distribution** in $\mathbb{R}$ has density

$$\frac{1}{\pi(1+x^2)}.$$

The graph is shown in Figure 4.2.33. So if a random variable X has this distribution, then for any Borel set $E \subseteq \mathbb{R}$,

$$\mathbb{P}\{X \in E\} = \int_E \frac{1}{\pi(1+x^2)} \, dx.$$

The corresponding c.d.f. is

$$F(x) = \frac{1}{2} + \frac{1}{\pi} \arctan x.$$

A random variable X with this distribution may be obtained starting with a random variable U, uniformly distributed on $[0, 1]$, and applying the inverse of this c.d.f.: $X = \tan(\pi(U - 1/2))$. This distribution is not integrable: $\mathbb{E}[|X|] = \infty$. The expected magnitude is infinite, but of course each instance is finite.[7] Asymptotic estimates for the size of the tails of the distribution are

$$\mathbb{P}\{X > \lambda\} \sim \frac{1}{\pi\lambda},$$

$$\mathbb{P}\{X < -\lambda\} \sim \frac{1}{\pi\lambda}, \qquad \text{as } \lambda \to \infty.$$

Thus X is in weak L_1 but not L_1 (Exercise 4.2.15).

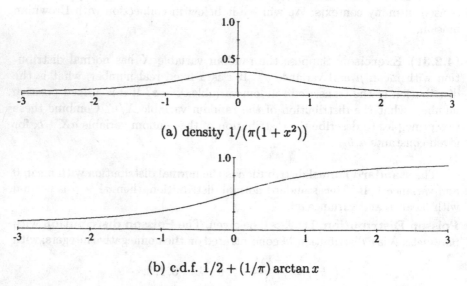

(a) density $1/(\pi(1 + x^2))$

(b) c.d.f. $1/2 + (1/\pi)\arctan x$

Figure 4.2.33. The Cauchy distribution.

(4.2.34) Exercise. Check the following integrals to justify the statements made above:

(a) The constant $1/\pi$ has been chosen correctly:

$$\int_{-\infty}^{\infty} \frac{1}{\pi(1 + x^2)} \, dx = 1.$$

(b) The c.d.f.:

$$\int_{-\infty}^{x} \frac{1}{\pi(1 + t^2)} \, dt = \frac{1}{2} + \frac{1}{\pi} \arctan x.$$

(c) X is not integrable:

$$\int_{-\infty}^{\infty} |x| \, \frac{1}{\pi(1 + x^2)} \, dx = \infty.$$

---

[7] Let us say it once more. With probability one, $|X| < \infty$, but $\mathbb{E}[|X|] = \infty$. This distinction between $\mathbb{E}[|X|] = \infty$ and $|X| = \infty$ is sometimes overlooked.

(d) The tail estimate:

$$\lim_{\lambda \to \infty} \pi\lambda \int_\lambda^\infty \frac{1}{\pi(1+x^2)}\, dx = 1.$$

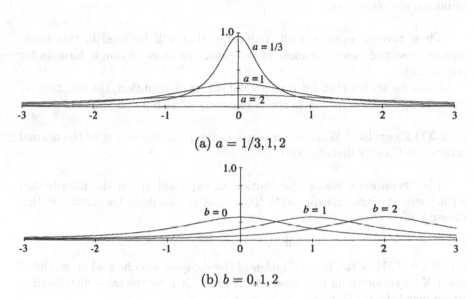

(a) $a = 1/3, 1, 2$

(b) $b = 0, 1, 2$

Figure 4.2.35. Parameters.

Let $a > 0, b$ be real numbers. The general **Cauchy distribution** with parameters b, a has density

$$\frac{a}{\pi(a^2 + (x-b)^2)}.$$

This is related to the standard Cauchy distribution by a simple change of variables: if X has standard Cauchy distribution, then $Y = aX + b$ has Cauchy distribution with parameters b, a. Now, b is a "location" parameter: it shows the center of the distribution. Since a Cauchy distribution is not integrable, this center is not a "mean" in the strict sense. It is a **median** in the sense that

$$\mathbb{P}\{Y \le b\} = \frac{1}{2} = \mathbb{P}\{Y \ge b\}.$$

It is also a **mode**, since the high point of the density $a/\big(\pi(a^2 + (x-b)^2)\big)$ occurs at $x = b$. The parameter a is a "concentration" parameter, indicating how spread out the distribution is. Since Y is not integrable, a is not a standard deviation, but it plays a similar role in some circumstances.

Stable Distributions.* A probability measure $\mathcal{M}$ on the line $\mathbb{R}$ is called **Lévy stable** iff for every $n \in \mathbb{N}$, there are constants a_n and b_n such that if

---

* Optional topic.

$X_1, X_2, \cdots, X_n$ are i.i.d. with distribution $\mathcal{M}$, then $a_n(X_1+X_2+\cdots+X_n)+b_n$ also has distribution $\mathcal{M}$.

(4.2.36) Exercise. Show that the normal distribution and the Cauchy distribution are stable.

These two examples are the only ones that will be used in this book. Unlike these two cases, for most of the others there is no simple formula for the density.

It can be shown that for a nontrivial stable distribution, the constant a_n must be of the form n^α for a certain exponent α with $0 < \alpha \leq 2$.

(4.2.37) Exercise. What are the exponents α corresponding to the normal and to the Cauchy distribution?

The "symmetric stable distribution of exponent α" is the distribution with Fourier transform $\exp(-|t|^\alpha)$. That is, X is distributed according to this measure iff

$$(*) \qquad\qquad \mathbb{E}\left[e^{itX}\right] = e^{-|t|^\alpha}$$

for all $t \in \mathbb{R}$. Here the interpretation of the complex exponential is simplified since X is symmetric in the sense that X and $-X$ have the same distribution, so we may take $(*)$ to mean

$$\mathbb{E}\left[\cos(tX)\right] = e^{-|t|^\alpha}.$$

(4.2.38) Exercise. Show that the standard normal and Cauchy distributions are symmetric stable distributions for their appropriate exponents.

Multidimensional stable random variables are also possible. If we choose a unit vector $u \in \mathbb{R}^d$ (perhaps at random) and then choose a stable random variable X, the product Xu will be stable in $\mathbb{R}^d$. If the unit vector u is chosen uniformly distributed on the unit sphere of $\mathbb{R}^d$, then Xu will have "isotropic" distribution in the sense that all directions look the same.

Multidimensional Normal Distribution. A random vector X in $\mathbb{R}^d$ is said to be normally distributed iff the dot product $c \cdot X$ is normally distributed for every vector $c \in \mathbb{R}^d$. If we write X in terms of its scalar components $X = (X_1, X_2, \cdots, X_d)$, then this means that $c_1X_2 + c_2X_2 + \cdots c_dX_d$ has normal distribution for all constants c_i.

It may be shown [32, p. 335] that such a normal random variable in $\mathbb{R}^d$ has a density with respect to $\mathcal{L}^d$ of the form

$$(2\pi)^{-d/2}(\det A)^{-1/2} \exp\left(-\frac{x^T A x}{2}\right).$$

Here A is a certain symmetric positive definite $d \times d$ matrix, $x \in \mathbb{R}^d$ is considered a column vector, and x^T is the transpose of x.

(4.2.39) Exercise. Practice some linear algebra and multidimensional integration by showing that this is a density for a d-dimensional normal random variable.

The standard d-dimensional normal random vector is the one where A is the identity matrix. Then $x^T A x = |x|^2$, so the density is

$$(2\pi)^{-d/2} \exp\left(-\frac{|x|^2}{2}\right).$$

If we write this using the components $x = (x_1, x_2, \cdots, x_n)$, the density is

$$(2\pi)^{-d/2} \exp\left(-\frac{\sum_{i=1}^d x_i^2}{2}\right) = \prod_{i=1}^d (2\pi)^{-1/2} \exp\left(-\frac{x_i^2}{2}\right),$$

so this random vector X may be constructed using d independent standard normal real random variables as the coordinates.

(4.2.40) Exercise. Let X be a standard d-dimensional normal random vector. For any $c \in \mathbb{R}^d$ with $|c| = 1$, the dot product $c \cdot X$ is a standard normal real random variable.

4.3 Dependence

Let K be an event with positive probability. If we know that K occurs, how does this change our ideas about the probabilities of other events? Given that K occurs, the **conditional probability** that A occurs is

$$\mathbb{P}[A \mid K] = \frac{\mathbb{P}(A \cap K)}{\mathbb{P}(K)}.$$

This may be read "the probability of A given K." (We will see later that conditional probabilities with respect to events K of probability 0 are also sometimes useful, but of course they cannot be defined by this formula.)

As an example, consider an experiment consisting of three successive tosses of a fair coin. We can think of the sample space as the set of all sequences of length three from the set $\{\mathbf{H}, \mathbf{T}\}$; all sequences have the same probability, $1/8$. Let A be the event "exactly two of the three tosses are heads," and K the event "the first coin is heads." Then $\mathbb{P}(A) = 3/8$, since there are three outcomes: $A = \{\mathbf{HHT}, \mathbf{HTH}, \mathbf{THH}\}$. Similarly, $\mathbb{P}(K) = 1/2$. And $\mathbb{P}(A \cap K) = 1/4$. So the conditional probability is

$$\mathbb{P}[A \mid K] = \frac{\mathbb{P}(A \cap K)}{\mathbb{P}(K)} = 1/2.$$

Comparing $\mathbb{P}(A) = 3/8$ and $\mathbb{P}[A \mid K] = 1/2 > 3/8$, before the experiment starts, we might say that we believe the probability of getting exactly 2 heads in 3 tosses is 3/8. But when we see that the outcome of the first toss is heads, we may revise our estimate upward to 1/2.

Now let us consider the "memoryless" property of the exponential distribution. Let X be a random variable with exponential distribution with parameter λ. That is,

$$\mathbb{P}\{X \leq x\} = 1 - e^{-\lambda x} \qquad \text{for } x > 0.$$

The distribution has density $\lambda e^{-\lambda x}$ on the positive real axis. Let us think of X as a "waiting time." Suppose we have already waited for a length of time α, and X has not been reached yet; what new information will we have about the waiting time? Surprisingly, conditioned on the event that $X > \alpha$, the distribution for the *remaining* time $X - \alpha$ is the same as the original distribution of X:

$$\mathbb{P}[X - \alpha \leq x \mid X > \alpha] = \frac{\mathbb{P}\{\alpha < X \leq x + \alpha\}}{\mathbb{P}\{X > \alpha\}}$$

$$= \frac{e^{-\lambda \alpha} - e^{-\lambda(x+\alpha)}}{e^{-\lambda \alpha}} = 1 - e^{-\lambda x}.$$

(4.3.1) Exercise. Are there other memoryless distributions? If X is a real random variable concentrated on $(0, \infty)$ and

$$\mathbb{P}[X - \alpha \leq x \mid X > \alpha] = \mathbb{P}\{X \leq x\}$$

for all $x, \alpha > 0$, must X be exponentially distributed?

Independence. The event A is **independent** of K iff its probability is not changed by whether K occurs:

$$\mathbb{P}[A \mid K] = \mathbb{P}(A),$$

or, equivalently,

$$\mathbb{P}(A \cap K) = \mathbb{P}(A)\mathbb{P}(K).$$

This equation is symmetric in A and K, so we may simply say that the events A, K are independent.[8]

As an example, consider again the experiment consisting of three successive tosses of a fair coin. If A is the event "the first toss is heads," then

$$A = \{\mathbf{HHH}, \mathbf{HHT}, \mathbf{HTH}, \mathbf{HTT}\}$$

consists of 4 of the 8 elements of Ω, so $\mathbb{P}(A) = 1/2$. If B is the event "the second toss is heads," then similarly, $\mathbb{P}(B) = 1/2$. Finally, the event $A \cap B$

---

[8] The word "independent" is used in other ways by mathematicians; so if necessary, we may say "statistically independent" for this kind of independence.

consists of 2 of the 8 elements, so $\mathbb{P}(A \cap B) = 1/4$. But then $\mathbb{P}(A \cap B) = \mathbb{P}(A)\mathbb{P}(B)$, so these two events are independent.

Let X, Y be random variables. We say that X and Y are independent iff any event determined by X is independent of any event determined by Y. That is, if A and B are Borel sets, then

$$\mathbb{P}\{X \in A \text{ and } Y \in B\} = \mathbb{P}\{X \in A\}\mathbb{P}\{Y \in B\}.$$

(4.3.2) Exercise. Let X and Y be real random variables. Show that X and Y are independent iff

$$\mathbb{P}\{X < s, Y < t\} = \mathbb{P}\{X < s\}\mathbb{P}\{Y < t\}$$

for all rational numbers s, t.

The next theorem says that if random variables X, Y are independent, then the distribution of the pair (X, Y) is the product measure obtained from the distribution of X and the distribution of Y.

(4.3.3) Theorem. *Let S and T be separable metric spaces, let X be a random element of S, and let Y be a random element of T. Suppose X and Y are independent. Then*

(a) $\mathcal{B}(S \times T) = \mathcal{B}(S) \otimes \mathcal{B}(T)$.
(b) $\mathcal{D}_{(X,Y)} = \mathcal{D}_X \otimes \mathcal{D}_Y$.
(c) *If $A \subseteq S$ and $B \subseteq T$ are Borel sets, then*

$$\mathbb{P}(X \in A \text{ and } Y \in B) = \mathbb{P}(X \in A)\mathbb{P}(Y \in B).$$

(d) *If $C \subseteq S \times T$ is a Borel set, then the cross-sections*

$$C_x = \{ y \in T : (x,y) \in C \}$$

are all Borel, and

$$\mathbb{P}((X,Y) \in C) = \int_S \mathcal{D}_Y(C_x)\, \mathcal{D}_X(dx).$$

Proof. (a) is from (2.1.3). (c) is the definition of independence. (b) follows from (a) and (c). (d) is a consequence of (b). ☺

(4.3.4) Exercise. Suppose X and Y are independent real random variables. Show that $\mathbb{E}[XY] = \mathbb{E}[X]\,\mathbb{E}[Y]$.

Suppose we have more than two random variables, $X_1, X_2, \cdots, X_n$. We say they are **independent** iff for any sequence $B_1, \cdots, B_n$ of Borel sets,

$$\mathbb{P}\{X_i \in B_i \text{ for } i = 1, 2, \cdots n\} = \prod_{i=1}^{n} \mathbb{P}\{X_i \in B_i\}.$$

(4.3.5) Exercise. We say that a set $\{X_1, X_2, \cdots, X_n\}$ is **pairwise independent** iff X_i and X_j are independent for every pair $i \neq j$. Give an example showing that "pairwise independent" is not equivalent to "independent" as defined above.

(4.3.6) Exercise. A set $\{A_1, A_2, \cdots, A_n\}$ of events is **independent** iff the corresponding set $\{\mathbb{1}_{A_1}, \mathbb{1}_{A_2}, \cdots, \mathbb{1}_{A_n}\}$ of indicator functions is independent. Show that this is equivalent to

$$\mathbb{P}\left(\bigcap_{i \in I} A_i\right) = \prod_{i \in I} \mathbb{P}(A_i)$$

for every subset $I \subseteq \{1, 2, \cdots, n\}$.

Next is a useful formula for the variance of a sum.

(4.3.7) Proposition. *Let* $X_1, X_2, \cdots, X_n$ *be independent square-integrable real random variables. Then* $\mathrm{Var}\,[X_1 + X_2 + \cdots + X_n] = \mathrm{Var}\,[X_1] + \mathrm{Var}\,[X_2] + \cdots + \mathrm{Var}\,[X_n]$.

Proof. Since the random variables are square-integrable, they are also integrable. Write $\mu_i = \mathbb{E}\,[X_i]$ for their means and $\sigma_i^2 = \mathrm{Var}\,[X_i] = \mathbb{E}\left[(X_i - \mu_i)^2\right]$ for their variances. Now by independence, $\mathbb{E}\,[X_i X_j] = \mu_i \mu_j$ when $i \neq j$. As a consequence, $\mathbb{E}\left[(X_i - \mu_i)(X_j - \mu_j)\right] = \mathbb{E}\,[X_i X_j] - \mu_i \mathbb{E}\,[X_j] - \mu_j \mathbb{E}\,[X_i] + \mu_i \mu_j = \mu_i \mu_j - \mu_i \mu_j - \mu_j \mu_i + \mu_i \mu_j = 0$. If we write $S = X_1 + X_2 + \cdots + X_n$, its mean μ is

$$\mu = \mathbb{E}\,[S] = \mathbb{E}\left[\sum_i X_i\right] = \sum_i \mathbb{E}\,[X_i] = \sum_i \mu_i.$$

Now we may compute

$$\mathrm{Var}\,[S] = \mathbb{E}\left[(S - \mu)^2\right] = \mathbb{E}\left[\left(\sum_i (X_i - \mu_i)\right)^2\right]$$

$$= \mathbb{E}\left[\sum_i (X_i - \mu_i)^2 + 2\sum_{i<j}(X_i - \mu_i)(X_j - \mu_j)\right]$$

$$= \mathbb{E}\left[\sum_i (X_i - \mu_i)^2\right] + 0$$

$$= \sum_i \mathbb{E}\left[(X_i - \mu_i)^2\right] = \sum_i \sigma_i^2. \qquad \smiley$$

(4.3.8) Exercise. Give an example showing that the equation

$$\mathrm{Var}\,[X_1 + X_2 + \cdots + X_n] = \mathrm{Var}\,[X_1] + \mathrm{Var}\,[X_2] + \cdots + \mathrm{Var}\,[X_n]$$

may fail when the random variables are not independent. But show that it is true when they are pairwise independent.

Now let $\{X_i\}_{i \in I}$ be a family of random variables, where the index set I is infinite (possibly even uncountable). We say that the family $\{X_i\}_{i \in I}$ is **independent** iff every finite subfamily is independent as defined above.

(4.3.9) Exercise. A sequence $\{X_n\}_{n \in \mathbb{N}}$ is independent iff for every sequence $\{B_i\}_{i \in \mathbb{N}}$ of Borel sets,

$$\mathbb{P}\{X_i \in B_i \text{ for } i \in \mathbb{N}\} = \prod_{i=1}^{\infty} \mathbb{P}\{X_i \in B_i\}.$$

A family $\{X_i\}_{i \in I}$ of random variables is called **identically distributed** iff the random variables all have the same distribution:

$$\mathbb{P}\{X_i \in B\} = \mathbb{P}\{X_j \in B\}$$

for all $i, j \in I$ and all Borel sets B. The abbreviation i.i.d. ("independent and identically distributed") will sometimes be used.

Here is the "converse" part of the Borel–Cantelli Lemma. (The easy part was done in (4.1.2).)

(4.3.10) Theorem. *Let A_n be an independent sequence of events. Suppose*

$$\sum_{n=1}^{\infty} \mathbb{P}(A_n) = \infty.$$

Then with probability one, infinitely many of the events A_n occur. That is, $\mathbb{P}(\limsup A_n) = 1$.

Proof. We must show $\mathbb{P}(\Omega \setminus \limsup A_n) = 0$. Now, $\Omega \setminus \limsup A_n$ is the increasing union of the events $B_m = \bigcap_{n=m}^{\infty}(\Omega \setminus A_n)$, so it is enough to show $\mathbb{P}(B_m) = 0$ for all m. But if $M > m$, we have by the independence

$$\mathbb{P}\left(\bigcap_{n=m}^{M}(\Omega \setminus A_n)\right) = \prod_{n=m}^{M} \mathbb{P}(\Omega \setminus A_n) = \prod_{n=m}^{M}(1 - \mathbb{P}(A_n))$$
$$\leq \prod_{n=m}^{M} \exp(-\mathbb{P}(A_n)) = \exp\left(-\sum_{n=m}^{M} \mathbb{P}(A_n)\right).$$

Now, by the divergence of the series $\sum \mathbb{P}(A_n)$, for any fixed m, we have $\lim_M \sum_{n=m}^{M} \mathbb{P}(A_n) = \infty$. Therefore, $\lim_M \mathbb{P}(\bigcap_{n=m}^{M}(\Omega \setminus A_n)) = 0$, so that $\mathbb{P}(\bigcap_{n=m}^{\infty}(\Omega \setminus A_n)) = 0$, as required. ☺

The two parts of the Borel–Cantelli Lemma state that if A_n are independent events, then $\limsup A_n$ has probability 0 or 1 according as the series $\sum \mathbb{P}(A_n)$ converges or diverges.

The next result is a "maximal inequality" due to Kolmogorov.

(4.3.11) Proposition. *Let* $X_1, X_2, \cdots, X_n$ *be independent real random variables with mean* 0 *and variances* $\sigma_1^2, \sigma_2^2, \cdots, \sigma_n^2$. *Write* $S_k = X_1 + \cdots + X_k$. *Then for any* $\lambda > 0$,

$$\mathbb{P}\left\{ \max_{1 \leq k \leq n} |S_k| \geq \lambda \right\} \leq \frac{1}{\lambda^2} \sum_{k=1}^{n} \sigma_k^2.$$

Proof. Define events[9]

$$A_1 = \{ |S_1| \geq \lambda \},$$
$$A_2 = \{ |S_1| < \lambda, |S_2| \geq \lambda \},$$
$$\cdots$$
$$A_k = \{ |S_1| < \lambda, \cdots, |S_{k-1}| < \lambda, |S_k| \geq \lambda \},$$
$$\cdots$$

Then the A_k are disjoint, and

$$\left\{ \max_{1 \leq k \leq n} |S_k| \geq \lambda \right\} = \bigcup_{k=1}^{n} A_k.$$

Next note that for $1 \leq k \leq n$, the random variable $S_n - S_k = X_{k+1} + \cdots + X_n$ is independent of the random variables $X_1, \cdots, X_k$ and therefore independent of $S_k \mathbb{1}_{A_k}$. So $\mathbb{E}\left[(S_n - S_k) S_k \mathbb{1}_{A_k} \right] = 0$. Then we have

$$\mathbb{E}\left[(S_n^2 - S_k^2) \mathbb{1}_{A_k} \right] = \mathbb{E}\left[\left((S_n - S_k)^2 + 2 S_k (S_n - S_k) \right) \mathbb{1}_{A_k} \right]$$
$$= \mathbb{E}\left[(S_n - S_k)^2 \mathbb{1}_{A_k} \right] \geq 0.$$

Now note that $|S_k| \geq \lambda$ on the set A_k, so $\mathbb{E}\left[S_k^2 \mathbb{1}_{A_k} \right] \geq \lambda^2 \mathbb{P}(A_k)$, and we may compute

$$\mathbb{P}\left\{ \max_{1 \leq k \leq n} |S_k| \geq \lambda \right\} = \sum_{k=1}^{n} \mathbb{P}(A_k) \leq \frac{1}{\lambda^2} \sum_{k=1}^{n} \mathbb{E}\left[S_k^2 \mathbb{1}_{A_k} \right]$$
$$\leq \frac{1}{\lambda^2} \sum_{k=1}^{n} \mathbb{E}\left[S_n^2 \mathbb{1}_{A_k} \right] \leq \frac{1}{\lambda^2} \mathbb{E}[S_n^2] = \frac{1}{\lambda^2} \sum_{k=1}^{n} \sigma_k^2. \quad \text{☺}$$

As a corollary, we will prove a simple criterion for convergence of a random series.

---

[9] The knowledgeable reader may note that we are defining here a "stopping time." More on stopping times will be seen below.

(4.3.12) Corollary. *Let $(X_i)_{i \in \mathbb{N}}$ be independent real random variables. Suppose X_i has mean 0 and variance σ_i^2 and the series $\sum \sigma_i^2$ converges. Then with probability one, the series $\sum X_i$ converges.*

Proof. Let $S_n = X_1 + X_2 + \cdots + X_n$ be the partial sums. We will show that the sequence (S_n) satisfies the Cauchy criterion with probability one. For a positive integer p, choose m_p so large that

$$\sum_{i=m_p}^{\infty} \sigma_i^2 < 2^{-3p}.$$

Now let

$$E_p = \left\{ |S_n - S_{m_p}| \leq 2^{-p} \text{ for all } n \geq m_p \right\}.$$

By the maximal inequality just proved,

$$\mathbb{P}(E_p) = 1 - \mathbb{P}\{ \sup_{n \geq m_p} |S_n - S_{m_p}| > 2^{-p} \}$$

$$\geq 1 - \frac{1}{2^{-2p}} \sum_{i=m_p}^{\infty} \sigma_i^2 > 1 - 2^{-p}.$$

Now the series $\sum_p 2^{-p}$ converges, so by the Borel–Cantelli Lemma 4.1.2, with probability one all but finitely many of the events E_p must occur. But then the sequence (S_n) satisfies the Cauchy criterion. ☺

(4.3.13) Exercise. Suppose we have independent random variables with mean μ_i and variance σ_i^2. State and prove a convergence theorem for the series $\sum X_i$ that will apply even when the means are not 0.

Conditioning. The conditional probability formula

$$\mathbb{P}[A \mid K] = \frac{\mathbb{P}(A \cap K)}{\mathbb{P}(K)}$$

does not make sense when K is an event of probability zero. Now normally, events of probability zero may be ignored, so this is not a problem. But in some cases it is useful to consider events of probability zero. For example, let X be a random variable with standard normal distribution. Then all of the events $\{X = t\}$ have probability zero. But it still may make sense to say things like this:

$$\mathbb{P}[X > t \mid X = t] = 0, \qquad \mathbb{P}[X \geq t \mid X = t] = 1.$$

Certainly, these are not computed as ratios $\mathbb{P}[A \mid K] = \mathbb{P}(A \cap K)/\mathbb{P}(K)$. But there are sometimes useful things that can be said concerning such conditional probabilities.

Another example. Suppose X and Y are independent, both with exponential distribution

$$\mathbb{P}\{X \leq t\} = \mathbb{P}\{Y \leq t\} = 1 - e^{-t}, \qquad t > 0.$$

Suppose we know the value of the sum $X + Y$. What can we then say about the value of X? Note that X and $X + Y$ are not independent: for example, we know that $X + Y > X$. We will see that

$$\mathbb{P}\left[X \leq x \mid X + Y = z\right] = \begin{cases} 0, & \text{if } x \leq 0, \\ x/z, & \text{if } 0 < x < z, \\ 1, & \text{if } z \leq x. \end{cases}$$

That is, given that the value of $X + Y$ is z, the conditional distribution of X is uniform on the interval $(0, z)$.

Technically, we will consider an event A and a random variable X. Then for $t \in \mathbb{R}$, we wish to deal with the conditional probability $\mathbb{P}\left[A \mid X = t\right]$. This should be something that depends on t, of course. By definition, $\mathbb{P}\left[A \mid X = t\right]$ is a function $f(t)$ such that

$$\int_E f(t)\, \mathcal{D}_X(dt) = \mathbb{P}\left(A \cap \{X \in E\}\right)$$

for all Borel sets $E \subseteq \mathbb{R}$.

(4.3.14) Exercise. Let A be an event, and X a real random variable. Show that $\mathcal{M}(E) = \mathbb{P}\left(A \cap \{X \in E\}\right)$ defines a finite Borel measure on $\mathbb{R}$, absolutely continuous with respect to $\mathcal{D}_X$. Conclude (using the Radon–Nikodým Theorem) that a function $f(t)$ as above exists.

The function $f(t) = \mathbb{P}\left[A \mid X = t\right]$ is unique only up to sets of $\mathcal{D}_X$-measure zero. So if X is a continuous random variable, the expression $\mathbb{P}\left[A \mid X = t\right]$ is undefined for any given value t; but it is defined (as a function of t) up to sets of measure zero. Many of the common probability distributions have continuous density functions, and it may therefore often happen that the conditional probability function $\mathbb{P}\left[A \mid X = t\right]$ may be chosen to be a continuous function of t. If so, we will declare the continuous function $f(t)$ to be the value of $\mathbb{P}\left[A \mid X = t\right]$, even for one fixed value of t.

(4.3.15) Exercise. Suppose the random variable X has continuous density h. That is, $\mathbb{P}\{X \in E\} = \int_E h(t)\, dt$ for all Borel sets E. Investigate the possibility of computing the conditional probability $\mathbb{P}\left[A \mid X = t\right]$, for individual points t, as a density (see 2.3.9(e)).

Conditional Distribution. Let X and Y be two random variables. If they are not independent, then observing the value of X will give us information about the distribution of Y. But if Y is not totally dependent on X, that observation will not tell us the precise value of Y. Instead, we get a new probability distribution that depends on the observation.

The **conditional distribution** of Y given X will be a distribution $\mathcal{M}_x$ that depends on the value x of X as follows:

$$\mathcal{M}_x(E) = \mathbb{P}\left[Y \in E \mid X = x\right].$$

If Y is a real random variable, the distribution may be described by its c.d.f. F_x like this:

$$F_x(y) = \mathbb{P}\left[Y \leq y \mid X = x\right].$$

If $\mathcal{M}_x$ is absolutely continuous, then it will be described by a conditional density h_x, so that

$$\mathbb{P}\left[Y \in E \mid X = x\right] = \int_E h_x(y)\, dy.$$

There does not seem to be a standard notation for a conditional distribution (similar to the notations $\mathbb{E}\left[Y \mid X = t\right]$, $\mathbb{P}\left[A \mid X = t\right]$ for the conditional expectation and conditional probability). In this book we will write

$$\mathcal{D}_{Y \mid X = t}$$

for the conditional distribution of Y given $X = t$. More formally, let S and T be metric spaces. Let X be a random element of T and let Y be a random element of S. For Borel sets $E \subseteq S$ and points $t \in T$,

$$\mathcal{D}_{Y \mid X = t}(E) = \mathbb{P}\left[Y \in E \mid X = t\right].$$

Certainly, for each given E, this is defined for almost all t (in the sense of the measure $\mathcal{D}_X$). In many special cases (and all the cases we will deal with), for almost all t this is defined for all E at once, despite the fact that there are uncountably many sets E. For example, if the right-hand side can be chosen to be a continuous function of t, we will make that choice.

Perhaps a computation will help in understanding. Suppose X and Y are independent, both with exponential distribution

$$\mathbb{P}\{X \leq t\} = \mathbb{P}\{Y \leq t\} = 1 - e^{-t}, \qquad t > 0.$$

Let $Z = X + Y$. How can we compute the conditional distribution of X given Z, that is, $\mathbb{P}\left[X \leq x \mid Z = z\right]$? First, since $X, Y > 0$ with probability one, we certainly have $\mathbb{P}\left[X \leq x \mid Z = z\right] = 0$ when $x \leq 0$ and $\mathbb{P}\left[X \leq x \mid Z = z\right] = 1$ when $x \geq z$. So the interesting case is $0 < x < z$.

Let's do this as a density calculation. Fix x and z with $0 < x < z$. Then for $\varepsilon > 0$ small enough, $x < z - \varepsilon$, so

$$
\begin{aligned}
&\mathbb{P}\left[X \leq x \mid z - \varepsilon < Z < z + \varepsilon\right] \\
&= \frac{\mathbb{P}\{X \leq x, z - \varepsilon < Z < z + \varepsilon\}}{\mathbb{P}\{z - \varepsilon < Z < z + \varepsilon\}} \\
&= \frac{\mathbb{P}\{X \leq x, z - \varepsilon - X < Y < z + \varepsilon - X\}}{\mathbb{P}\{z - \varepsilon - X < Y < z + \varepsilon - X\}} \\
&= \frac{x\left(e^{-z+\varepsilon} - e^{-z-\varepsilon}\right)}{(z+1)\left(e^{-z+\varepsilon} - e^{-z-\varepsilon}\right) - \varepsilon\left(e^{-z+\varepsilon} + e^{-z-\varepsilon}\right)},
\end{aligned}
$$

and therefore

$$\mathbb{P}\left[X \le x \,|\, Z = z\right] = \lim_{\varepsilon \to 0} \frac{x\left(e^{-z+\varepsilon} - e^{-z-\varepsilon}\right)}{(z+1)\left(e^{-z+\varepsilon} - e^{-z-\varepsilon}\right) - \varepsilon\left(e^{-z+\varepsilon} + e^{-z-\varepsilon}\right)} = \frac{x}{z}.$$

The conditional distribution $\mathcal{D}_{X|Z=z}$ is uniform on the interval $(0, z)$.

There is another way to do this computation. The pair (X, Y) has joint density e^{-x-y} on the set $(0, \infty) \times (0, \infty)$. The pair (X, Y) is related to (X, Z) by the change of variables $Z = X + Y, X = X$. The Jacobian determinant for the change of variables is 1, so the joint density for the pair (X, Z) is $e^{-x-(z-x)} = e^{-z}$ on the set $\{(x, z) : 0 < x < z\}$. A density for the conditional distribution of X given $Z = z$ is

$$\frac{e^{-z}}{\int_0^z e^{-z}\, ds} = \frac{1}{z}$$

on the set $\{x : 0 < x < z\}$. So again, this is the uniform distribution on the interval $(0, z)$.

The justification of this method is left to the reader:

(4.3.16) Exercise. Let real random variables X, Y have joint density $h(x, y)$. Then the conditional distribution of Y given $X = x$ has density

$$h_x(y) = \frac{h(x, y)}{\int_{\mathbb{R}} h(x, s)\, ds}$$

for almost all x.

Conditional Expectation. The expectation of a conditional distribution is used often enough that it has its own notation and terminology. For an event K with positive probability, the **conditional expectation** of a random variable Y given K is

$$\mathbb{E}\left[Y \,|\, K\right] = \frac{\mathbb{E}\left[Y\, \mathbb{1}_K\right]}{\mathbb{P}(K)}.$$

But for conditioning events K of probability 0 we use a definition similar to that for conditional probability.

Let X and Y be random variables, Y real and integrable. The **conditional expectation** of Y given $X = t$ is a function of t

$$f(t) = \mathbb{E}\left[Y \,|\, X = t\right]$$

such that for all Borel sets E,

$$\int_E f(t)\, \mathcal{D}_X(dt) = \mathbb{E}\left[Y\, \mathbb{1}_{\{X \in E\}}\right].$$

The right-hand side of this equation defines a measure, absolutely continuous with respect to $\mathcal{D}_X$, so such a function f exists by the Radon–Nikodým theorem. (Thus, this definition will make sense even in strange situations where conditional distributions may not exist.)

(4.3.17) Exercise. The conditional expectation is the expectation of the conditional distribution (when it exists). More precisely, let X and Y be as above. Suppose $\mathcal{D}_{Y|X=t}$ exists. Show that

$$\mathbb{E}\left[Y \mid X = t\right] = \int y\, \mathcal{D}_{Y|X=t}(dy).$$

(4.3.18) Exercise. Let X and Y be integrable real random variables. Show that if X and Y are independent, then

$$\mathbb{E}\left[Y \mid X = t\right] = \mathbb{E}\left[Y\right], \qquad \text{for all } t \in \mathbb{R}.$$

Is the converse true?

The notion of a conditional expectation is useful for certain computations. If a random variable X has its values in a countable set Q, then of course

$$\sum_{a \in Q} \mathbb{P}\{X = a\} = 1.$$

If Y is another random variable, then

$$\mathbb{E}\left[Y\right] = \sum_{a \in Q} \mathbb{E}\left[Y \mid X = a\right] \mathbb{P}\{X = a\}.$$

There is an analogue of this formula that works even when X is a continuous random variable.

(4.3.19) Exercise. Let X and Y be random variables. Suppose Y is an integrable real random variable. Prove that

$$\mathbb{E}\left[Y\right] = \int \mathbb{E}\left[Y \mid X = x\right] \mathcal{D}_X(dx).$$

[Hint. Approximate Y by simple functions.]

If X is also a real random variable and has a density $h(x)$ with respect to Lebesgue measure, then this becomes

$$\mathbb{E}\left[Y\right] = \int \mathbb{E}\left[Y \mid X = x\right] h(x)\, dx.$$

Suppose we have two real random variables, X and Y. How is the distribution of their sum related to their individual distributions? Well,

$$\mathcal{D}_{X+Y}(E) = \mathbb{P}\{X + Y \in E\}$$
$$= \int \mathbb{P}\left[x + Y \in E \mid X = x\right] \mathcal{D}_X(dx)$$
$$= \int \mathbb{P}\left[Y \in E - x \mid X = x\right] \mathcal{D}_X(dx).$$

Now let us see what further simplifications will occur if X and Y are independent.

$$\int \mathbb{P}\left[Y \in E - x \mid X = x\right] \mathcal{D}_X(dx) = \int \mathbb{P}\{Y \in E - x\} \mathcal{D}_X(dx)$$

$$= \int \mathcal{D}_Y(E - x) \mathcal{D}_X(dx).$$

Finally, suppose further that $\mathcal{D}_X$ and $\mathcal{D}_Y$ are absolutely continuous, with densities h_X, h_Y.

$$\int \mathcal{D}_Y(E - x) \mathcal{D}_X(dx) = \int \int \mathbb{1}_{E-x}(y) \, h_Y(y) \, dy \, h_X(x) \, dx$$

$$= \int \int \mathbb{1}_E(u) \, h_Y(u - x) \, du \, h_X(x) \, dx$$

$$= \int \int_E h_Y(u - x) \, du \, h_X(x) \, dx$$

$$= \int_E \int h_Y(u - x) \, h_X(x) \, dx \, du.$$

(Tonelli's Theorem applies since the integrand is nonnegative.) So, combining everything, we get

$$\mathcal{D}_{X+Y}(E) = \int_E h(u) \, du,$$

where $h(u) = \int h_Y(u - x) \, h_X(x) \, dx$. That is, $h(u)$ is the density for $\mathcal{D}_{X+Y}$. This combination

$$h(u) = \int h_Y(u - x) \, h_X(x) \, dx$$

is known as the **convolution** of the functions h_X and h_Y.

Sum of Independent Normal Random Variables. Let X and Y be independent random variables with normal distributions; let X have mean μ and variance σ^2; let Y have mean μ' and variance σ'^2. Thus X and Y have densities

$$h_X(x) = \frac{1}{\sqrt{2\pi}\sigma} \exp\left(-\frac{(x - \mu)^2}{2\sigma^2}\right), \quad h_Y(x) = \frac{1}{\sqrt{2\pi}\sigma'} \exp\left(-\frac{(y - \mu')^2}{2\sigma'^2}\right).$$

The density for the sum $X + Y$ is the convolution

$$h_{X+Y}(u) = \int_{-\infty}^{\infty} h_Y(u - x) h_X(x) \, dx$$

$$= \int_{-\infty}^{\infty} \frac{1}{\sqrt{2\pi}\sigma'} \exp\left(-\frac{(u - x - \mu')^2}{2\sigma'^2}\right) \frac{1}{\sqrt{2\pi}\sigma'} \exp\left(-\frac{(x - \mu)^2}{2\sigma^2}\right) dx.$$

This integral may be computed by combining the exponentials, completing the square inside, and then comparing with known integrals (done above in Exercise 4.2.30). The result is

$$\frac{1}{\sqrt{2\pi}\sigma''} \exp\left(-\frac{(u - \mu'')^2}{2\sigma''^2}\right),$$

where $\mu'' = \mu + \mu'$ and $\sigma'' = \sqrt{\sigma^2 + \sigma'^2}$. So in fact, the sum of two independent normal random variables is again a normal random variable. The means and variances add.

A computation in reverse will be useful later. Suppose X and Y are independent normal random variables with mean 0 and variance $1/2$. Then $Z = X + Y$ is a standard normal random variable. What is the conditional distribution of X given the sum Z? To find this distribution, recall that X and Y have continuous densities

$$\frac{1}{\sqrt{\pi}} e^{-x^2}.$$

But X and Y are independent, so the joint distribution for the pair (X, Y) has density

$$\frac{1}{\sqrt{\pi}} e^{-x^2} \frac{1}{\sqrt{\pi}} e^{-y^2} = \frac{1}{\pi} e^{-(x^2 + y^2)}.$$

This is the density with respect to 2-dimensional Lebesgue measure. The pair (X, Z) is related to the pair (X, Y) by the linear transformation $X = X, Z = X + Y$; the Jacobian determinant for this transformation is 1. So the density for the pair (X, Z) is

$$\frac{1}{\pi} e^{-(x^2 + (z-x)^2)} = \frac{1}{\pi} e^{-(2x^2 - 2xz + z^2)}.$$

Now we may compute

$$\frac{1}{\pi} \int_{-\infty}^{\infty} e^{-(2x^2 - 2xz + z^2)} \, dx = \frac{1}{\sqrt{2\pi}} e^{-z^2/2}$$

and conclude that the conditional distribution $\mathcal{D}_{X|Z=z}$ has density

$$\frac{(1/\pi) \exp\left(-(2x^2 - 2xz + z^2)\right)}{(1/\sqrt{2\pi}) \exp\left(-z^2/2\right)} = \frac{\sqrt{2}}{\sqrt{\pi}} \exp\left(-2(x - z/2)^2\right).$$

So the conditional distribution $\mathcal{D}_{X|Z=z}$ is normal, with mean $z/2$ and variance $1/4$.

Sum of Independent Poisson Random Variables. Let X and Y be independent random variables. Suppose X has Poisson distribution with mean λ, and Y has Poisson distribution with mean λ'. What is the distribution of $X + Y$? Compute as follows:

$$\mathbb{P}\{X + Y = n\} = \sum_{k=0}^{n} \mathbb{P}\{X = k, Y = n - k\}$$

$$= \sum_{k=0}^{n} \mathbb{P}\{X = k\}\mathbb{P}\{Y = n - k\}$$

$$= \sum_{k=0}^{n} \frac{e^{-\lambda}\lambda^k}{k!} \frac{e^{-\lambda'}\lambda'^{n-k}}{(n - k)!}$$

$$= \frac{e^{-\lambda-\lambda'}}{n!} \sum_{k=0}^{n} \binom{n}{k} \lambda^k \lambda'^{n-k}$$

$$= \frac{e^{-\lambda-\lambda'}(\lambda + \lambda')^n}{n!}.$$

So the sum $X + Y$ has Poisson distribution with mean $\lambda + \lambda'$.

Now we consider the distribution of the independent summand, given the sum. Suppose X and Y are as before, so that $Z = X + Y$ has Poisson distribution. If we know the value of Z, what will be the conditional distribution of X?

$$\mathbb{P}[X = k \mid Z = n] = \frac{\mathbb{P}\{X = k, Z = n\}}{\mathbb{P}\{Z = n\}} = \frac{\mathbb{P}\{X = k, Y = n - k\}}{\mathbb{P}\{Z = n\}}$$

$$= \frac{\left(e^{-\lambda}\lambda^k/k!\right)\left(e^{-\lambda'}\lambda'^{n-k}/(n - k)!\right)}{e^{-\lambda-\lambda'}(\lambda + \lambda')^n/n!} = \binom{n}{k} \frac{\lambda^k \lambda'^{n-k}}{(\lambda + \lambda')^n}.$$

The conditional distribution $\mathcal{D}_{X|Z=n}$ is a binomial distribution with parameters $\lambda/(\lambda + \lambda')$ and n. Symmetrically, the conditional distribution $\mathcal{D}_{Y|Z=n}$ for the other summand is a binomial distribution with parameters $\lambda'/(\lambda + \lambda')$ and n.

Sum of Independent Exponential Random Variables. Let X and Y be independent exponential random variables with parameter λ and λ', respectively.

(4.3.20) Exercise. What is the distribution of $Z = X + Y$? (We computed the conditional distribution $\mathcal{D}_{X|Z=z}$ before.)

Sum of Independent Cauchy Random Variables. * Let X and Y be independent random variables, where X has Cauchy distribution with parameters b, a and Y has Cauchy distribution with parameters b', a'. Then the sum $Z = X + Y$ has Cauchy distribution with parameters $b + b', a + a'$.

---

* Optional topic.

(4.3.21) Exercise. Verify the integral

$$\int_{-\infty}^{\infty} \frac{a}{\pi(a^2 + (x-b)^2)} \frac{a'}{\pi(a'^2 + (z-x-b')^2)}\, dx$$

$$= \frac{a+a'}{\pi((a+a')^2 + (z-b-b')^2)}.$$

Suppose X and Y both have Cauchy distribution with parameters $0, 1/2$, so that their sum $Z = X + Y$ has (standard) Cauchy distribution with parameters $0, 1$.

(4.3.22) Exercise. Verify that the conditional distribution $\mathcal{D}_{X|Z=z}$ has density

$$\theta_z(x) = \frac{4(1+z^2)}{\pi(1+4x^2)(1+4(z-x)^2)}.$$

We may also expand in partial fractions with respect to the variable x:

$$\theta_z(x) = \frac{z+2x}{\pi z(1+4x^2)} + \frac{z+2(z-x)}{\pi z(1+4(z-x)^2)}.$$

But it is *not* another Cauchy distribution. When $|z| \leq 1$ it is a unimodal distribution, but when $|z| > 1$ it is bimodal. The two peaks are concentrated about 0 and z.

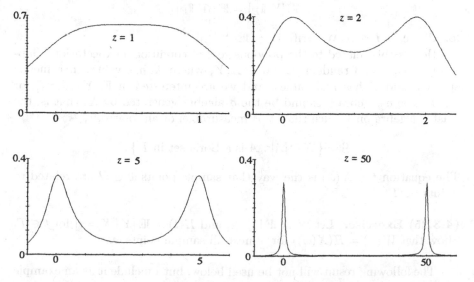

Figure 4.3.23. Density of Cauchy summand.

The Cauchy distribution behaves differently than the normal distribution in this regard. If we know the value z of the sum $Z = X + Y$, and that value is relatively large, then we may expect that one of the summands is nearly equal to 0 and the other summand is nearly equal to z.

Note that the limit as $z \to \infty$ is

$$\lim_{z \to \infty} \theta_z(x) = \frac{1}{\pi(1 + 4x^2)},$$

half the density of a Cauchy distribution. So each of the peaks nearly has a Cauchy distribution. So each of the peaks should satisfy a tail estimate as in Exercise 4.2.34(d). This shows how little of the distribution lies between the two peaks.

(4.3.24) Exercise. Prove an estimate like this:

$$\int_{\lambda}^{z-\lambda} \theta_z(x) \, dx < \frac{1}{2\pi\lambda}$$

for z large enough and $0 < \lambda < z/2$.

Conditioning on a σ-Algebra. A slightly more general notion of conditioning is sometimes used. Recall that when we discuss probability in the mathematical sense, we always have in the background a "sample space" $(\Omega, \mathcal{F}, \mathbb{P})$. Let $\mathcal{G} \subseteq \mathcal{F}$ be a sub-σ-algebra. Let $Y \colon \Omega \to \mathbb{R}$ be an integrable random variable. Then a random variable W is called the **conditional expectation** of Y given $\mathcal{G}$ iff W is $\mathcal{G}$-measurable and

$$\mathbb{E}[Y \, \mathbb{1}_E] = \mathbb{E}[W \, \mathbb{1}_E]$$

for all events $E \in \mathcal{G}$. We write $W = \mathbb{E}[Y \mid \mathcal{G}]$.

How is this related to the previous sort of conditional expectation? Suppose we have two random variables X, Y, where X has values in a metric space T and Y has real values, and we are interested in $\mathbb{E}[Y \mid X = t]$ for $t \in T$. The σ-algebra $\mathcal{G}$ should be the σ-algebra generated by X, that is, the least σ-algebra on Ω such that X is measurable; or alternatively,

$$\mathcal{G} = \left\{ X^{-1}[A] : A \text{ is a Borel set in } T \right\}.$$

The equation $t = X(\omega)$ is the way that sample points $\omega \in \Omega$ are related to points $t \in T$.

(4.3.25) Exercise. Let $W = \mathbb{E}[Y \mid \mathcal{G}]$ and $H(t) = \mathbb{E}[Y \mid X = t]$ for $t \in T$. Show that $W(\omega) = H(X(\omega))$ for almost all sample points ω.

The following result will not be used below, but I include it as an example of computations with conditional expectations.

(4.3.26) Theorem. *A conditional expectation is a "best approximation" in the mean-square sense. That is, let Y be a square-integrable random variable; let $\mathcal{G} \subseteq \mathcal{F}$ be a sub-σ-algebra. Then among all $\mathcal{G}$-measurable square-integrable random variables Z, the mean square error $\mathbb{E}\left[|Y - Z|^2\right]$ is minimum when $Z = \mathbb{E}[Y \mid \mathcal{G}]$.*

Proof. (a) I claim first that if U and Y are square-integrable random variables and U is $\mathcal{G}$-measurable, then $\mathbb{E}[U\mathbb{E}[Y \mid \mathcal{G}]] = \mathbb{E}[UY]$. This is proved in stages: it is true when $U = \mathbb{1}_E$ for $E \in \mathcal{G}$ by the definition of $\mathbb{E}[Y \mid \mathcal{G}]$. Then, it is true for $\mathcal{G}$-measurable simple functions by linearity. Then, it is true for nonnegative $\mathcal{G}$-measurable functions U by approximating U with an increasing sequence of $\mathcal{G}$-measurable simple functions and using the Dominated Convergence Theorem. Finally, it is true for a general $\mathcal{G}$-measurable function U, since the positive and negative parts are also $\mathcal{G}$-measurable.

(b) Next I claim that if Y and Z are square-integrable random variables and Z is $\mathcal{G}$-measurable, then

$$\mathbb{E}\left[(Z - \mathbb{E}[Y \mid \mathcal{G}])(Y - \mathbb{E}[Y \mid \mathcal{G}])\right] = 0.$$

Now Z is $\mathcal{G}$-measurable, so $\mathbb{E}[Z\mathbb{E}[Y \mid \mathcal{G}]] = \mathbb{E}[ZY]$; and $\mathbb{E}[Y \mid \mathcal{G}]$ is $\mathcal{G}$-measurable, so $\mathbb{E}\left[\mathbb{E}[Y \mid \mathcal{G}]^2\right] = \mathbb{E}[Y\mathbb{E}[Y \mid \mathcal{G}]]$. Therefore,

$$\mathbb{E}\left[(Z - \mathbb{E}[Y \mid \mathcal{G}])(Y - \mathbb{E}[Y \mid \mathcal{G}])\right]$$
$$= \mathbb{E}[YZ] - \mathbb{E}[Z\mathbb{E}[Y \mid \mathcal{G}]] - \mathbb{E}[Y\mathbb{E}[Y \mid \mathcal{G}]] + \mathbb{E}\left[\mathbb{E}[Y \mid \mathcal{G}]^2\right]$$
$$= 0.$$

(c) To see that $\mathbb{E}[Y \mid \mathcal{G}]$ is the best approximation among $\mathcal{G}$-measurable functions, let Z be any $\mathcal{G}$-measurable function, and compute:

$$\mathbb{E}\left[(Y - Z)^2\right] = \mathbb{E}\left[\left((Y - \mathbb{E}[Y \mid \mathcal{G}]) + (\mathbb{E}[Y \mid \mathcal{G}] - Z)\right)^2\right]$$
$$= \mathbb{E}\left[(Y - \mathbb{E}[Y \mid \mathcal{G}])^2\right] - 2\mathbb{E}\left[(Z - \mathbb{E}[Y \mid \mathcal{G}])(Y - \mathbb{E}[Y \mid \mathcal{G}])\right]$$
$$\quad + \mathbb{E}\left[(\mathbb{E}[Y \mid \mathcal{G}] - Z)^2\right]$$
$$= \mathbb{E}\left[(Y - \mathbb{E}[Y \mid \mathcal{G}])^2\right] + \mathbb{E}\left[(\mathbb{E}[Y \mid \mathcal{G}] - Z)^2\right]$$
$$\leq \mathbb{E}\left[(Y - \mathbb{E}[Y \mid \mathcal{G}])^2\right],$$

and there is equality only if $Z = \mathbb{E}[Y \mid \mathcal{G}]$ almost everywhere.   ☺

Martingales. A sequence $X_1, X_2, \cdots$ of integrable real random variables is called a **martingale** iff for all n and almost all $(t_1, t_2, \cdots, t_n) \in \mathbb{R}^n$,

$$t_n = \mathbb{E}[X_{n+1} \mid (X_1, X_2, \cdots, X_n) = (t_1, t_2, \cdots, t_n)].$$

This equation is often written

(*) $$X_n = \mathbb{E}[X_{n+1} \mid (X_1, X_2, \cdots, X_n)].$$

An example is the "double-or-nothing" game. You start with a bankroll of one dollar to bet. For each play of the game, you bet your entire bankroll on a flip of a fair coin. If the coin comes up heads, you win, and the house matches your bet (so you double your money). But if the coin comes up tails, you lose, and the house gets the money (so you have nothing). To analyze this game, let us write S_n for your bankroll after n plays. The sample space is the set of infinite strings from the alphabet $E = \{\mathsf{H}, \mathsf{T}\}$. The value of S_n depends only on the first n tosses. We start with $S_0 = 1$. For $n \geq 1$, if the first n tosses were all H, then you won n times, so you have doubled your money n times, and $S_n = 2^n$. On the other hand, if there is at least one T among the first n tosses, you lost all your money and so you now have nothing left; $S_n = 0$.

We claim that this sequence S_n is a martingale. Consider a sequence $(t_1, t_2, \cdots, t_n)$ of values. (We need only consider sequences that occur as values of $(S_1, S_2, \cdots, S_n)$ with positive probability.) If $t_n = 0$, then $S_{n+1} = 0$ on the set $\{(S_1, S_2, \cdots, S_n) = (t_1, t_2, \cdots, t_n)\}$ so that

$$\mathbb{E}[S_{n+1} \mid (S_1, S_2, \cdots, S_n) = (t_1, t_2, \cdots, t_n)] = 0 = t_n.$$

On the other hand, if $t_n = 2^n$, then

$$\mathbb{P}\{(S_1, S_2, \cdots, S_n) = (1, 2, \cdots, 2^n)\} = \mathbb{P}(\{\mathsf{HH}\cdots\mathsf{H}\})$$
$$= 2^{-n},$$
$$\mathbb{P}\{(S_1, S_2, \cdots, S_n, S_{n+1}) = (1, 2, \cdots, 2^n, 2^{n+1})\} = \mathbb{P}(\{\mathsf{HH}\cdots\mathsf{HH}\})$$
$$= 2^{-n-1},$$
$$\mathbb{P}\{(S_1, S_2, \cdots, S_n, S_{n+1}) = (1, 2, \cdots, 2^n, 0)\} = \mathbb{P}(\{\mathsf{HH}\cdots\mathsf{HT}\})$$
$$= 2^{-n-1}.$$

Therefore,

$$\mathbb{E}[S_{n+1} \mid (S_1, S_2, \cdots, S_n) = (t_1, t_2, \cdots, t_n)] = \frac{0 \cdot 2^{-n-1} + 2^{n+1} \cdot 2^{-n-1}}{2^{-n}}$$
$$= 2^n = t_n.$$

Thus, in all cases,

$$\mathbb{E}[S_{n+1} \mid (S_1, S_2, \cdots, S_n)] = S_n,$$

as required. Given the amount you have now, the expected amount after one more play is equal to that present amount.

Because of this interpretation, a martingale is often called a "fair" game, since it is neither advantageous nor disadvantageous to the player (on the average).

A slightly more abstract version of the definition may be formulated.

A **stochastic basis** in the sample space $(\Omega, \mathcal{F}, \mathbb{P})$ is a sequence $(\mathcal{F}_n)$ of σ-algebras satisfying $\mathcal{F}_n \subseteq \mathcal{F}$ and $\mathcal{F}_n \subseteq \mathcal{F}_{n+1}$ for all $n \in \mathbb{N}$. When we think of $\mathbb{N}$ as "time," we may call $\mathcal{F}_n$ the σ-algebra of "events prior to time n." We might think of events $E \in \mathcal{F}_n$ as events that depend only on random occurrences up to time n.

A sequence (X_n) of integrable real random variables is a **martingale** adapted to the stochastic basis $(\mathcal{F}_n)$ iff X_n is $\mathcal{F}_n$-measurable for all n and

$$(**) \qquad\qquad \mathbb{E}\left[X_{n+1} \mid \mathcal{F}_n\right] = X_n$$

for all n.

If X_n is a martingale in the original sense and $\mathcal{F}_n$ is defined as the σ-algebra generated by $\{X_1, X_2, \cdots, X_n\}$, then $(**)$ becomes exactly $(*)$, so X_n is also a martingale with respect to this stochastic basis.

Let a stochastic basis $(\mathcal{F}_n)$ be given. A **stopping time** with respect to $(\mathcal{F}_n)$ is a random natural number T such that $\{T = n\} \in \mathcal{F}_n$ for all n. So a stopping time is a random time such that we can tell when it occurs. At time n we know whether or not $T = n$. We do not need to wait for future events to determine when T happens.

Of course, a gambler must make decisions on what he will do based only on information he has at that time. He is not allowed to use future events (such as the outcome of the race) when placing his bet! Or (to explain the terminology), the gambler is not allowed to use future information to decide when to stop playing a game.[10]

(4.3.27) Proposition. *Let $(\mathcal{F}_n)$ be a stochastic basis. Let (X_n) be a sequence of real integrable random variables such that X_n is $\mathcal{F}_n$-measurable for all n. Then (X_n) is a martingale with respect to $(\mathcal{F}_n)$ if and only if $\mathbb{E}[X_T]$ is constant (as T runs through all bounded stopping times with respect to $(\mathcal{F}_n)$).*

Proof. Suppose $\mathbb{E}[X_T]$ is constant. Let $n \in \mathbb{N}$. We must show that

$$\mathbb{E}\left[X_{n+1} \mid \mathcal{F}_n\right] = X_n.$$

To do this, since X_n is $\mathcal{F}_n$-measurable by assumption, it is enough to show that $\mathbb{E}\left[X_{n+1}\, \mathbb{1}_A\right] = \mathbb{E}\left[X_n\, \mathbb{1}_A\right]$ for all $A \in \mathcal{F}_n$. So let $A \in \mathcal{F}_n$. Now, the constant n is a bounded stopping time, and so is

$$T = \begin{cases} n + 1, & \text{on } A, \\ n, & \text{otherwise.} \end{cases}$$

Therefore, $\mathbb{E}[X_T] = \mathbb{E}[X_n]$. That is,

$$\mathbb{E}\left[X_{n+1}\, \mathbb{1}_A + X_n\, \mathbb{1}_{\Omega \setminus A}\right] = \mathbb{E}\left[X_n\, \mathbb{1}_A + X_n\, \mathbb{1}_{\Omega \setminus A}\right].$$

Subtracting $\mathbb{E}\left[X_n\, \mathbb{1}_{\Omega \setminus A}\right]$ from both sides, we get $\mathbb{E}\left[X_{n+1}\, \mathbb{1}_A\right] = \mathbb{E}\left[X_n\, \mathbb{1}_A\right]$, as required.

Conversely, suppose (X_n) is a martingale. Because $\mathbb{E}\left[X_{n+1} \mid \mathcal{F}_n\right] = X_n$, we may also conclude that $\mathbb{E}\left[X_m \mid \mathcal{F}_n\right] = X_n$ for $n < m$ (by induction on the difference $m - n$). Now, for constant stopping times n we have

---

[10]There is an old recipe for making perfect toast: Heat the bread until it starts to smoke, then thirty seconds less.

$$\mathbb{E}\left[X_1\right] = \mathbb{E}\left[\mathbb{E}\left[X_n \mid \mathcal{F}_1\right]\right] = \mathbb{E}\left[X_n\right].$$

For a nonconstant bounded stopping time T, choose a constant $n > T$, and then write

$$\mathbb{E}\left[X_T\right] = \sum_{k=1}^{n-1} \mathbb{E}\left[X_k \mathbb{1}_{\{T=k\}}\right] = \sum_{k=1}^{n-1} \mathbb{E}\left[\mathbb{E}\left[X_n \mid \mathcal{F}_k\right] \mathbb{1}_{\{T=k\}}\right]$$

$$= \sum_{k=1}^{n-1} \mathbb{E}\left[X_n \mathbb{1}_{\{T=k\}}\right] = \mathbb{E}\left[X_n\right] = \mathbb{E}\left[X_1\right].$$

So $\mathbb{E}\left[X_T\right]$ has the same value for all bounded stopping times T.   ☺

This proposition illustrates the "fair game" description of the martingale. No matter how we choose to stop (as long as we do not use future information), our expected fortune is the same.

Random Walk. As another example, consider the "symmetric simple random walk." Define a random sequence (U_n) recursively as follows. First, $U_0 = 0$. Given U_n, flip a fair coin, and let $U_{n+1} = U_n + 1$ if it comes out heads and $U_{n+1} = U_n - 1$ if it comes out tails. Now the conditional expectations are easy:

$$\mathbb{E}\left[U_{n+1} \mid (U_0, U_1, \cdots, U_n)\right] = \frac{1}{2}(U_n + 1) + \frac{1}{2}(U_n - 1) = U_n.$$

So it is a martingale.

Notice a few other items here. The distribution of U_n is essentially a binomial distribution with parameters $1/2$ and n; the only difference is that the values are $-n, -n+2, \cdots, n-2, n$. So in fact, $(U_n + n)/2$ is exactly the binomial random variable.

We may write $U_n = X_1 + X_2 + \cdots + X_n$, where the X_i are independent with $\mathbb{P}\{X_i = 1\} = \mathbb{P}\{X_i = -1\} = 1/2$.

This random walk may be viewed as a simple model of two gamblers playing successive fair games against each other with unit bets at even money. If a player wins, his holdings increase by one dollar. If he loses, his holdings decrease by one dollar. Whatever one player wins, the other loses. Suppose player A begins with an amount a, and player B begins with an amount b, and they play until one of them loses everything. Proposition 4.3.27 may be used to evaluate the probability that each player wins.

Let U_n be the random walk discussed above. It represents the amount that A has won from B after n plays. The question is, What is the probability that U_n reaches $+b$ before it reaches $-a$? This is the probability that player B loses everything.

Now first I claim that with probability one, there is an n with $U_n = b$ or $U_n = -a$. That is, our series of games eventually ends. We may draw this conclusion from the Borel–Cantelli Lemma. Disjoint strings of $a + b$ plays are independent of each other. With positive probability $(1/2)^{a+b}$, any one of those strings is a run of $a + b$ heads in a row. If that happens, and we are still between b and $-a$ at the beginning of the run, then certainly we pass b sometime during the run. The infinite series with all terms equal to $(1/2)^{a+b}$ diverges, so by the Borel–Cantelli Lemma at least one (indeed, infinitely many) such runs occur.

Define a stopping time T as the first time we reach either b or $-a$:

$$T = \inf \{ n : U_n = b \text{ or } U_n = -a \}.$$

In order to apply Proposition 4.3.27 we need a *bounded* stopping time. For $k \in \mathbb{N}$, certainly $T \wedge k$ is a bounded stopping time. Now, as $k \to \infty$, we have $\mathbb{P}\{U_{T\wedge k} = b\} \nearrow \mathbb{P}\{U_T = b\}$, $\mathbb{P}\{U_{T\wedge k} = -a\} \nearrow \mathbb{P}\{U_T = -a\}$, and $\mathbb{P}\{T > k\} \searrow 0$. But U_n is a martingale, so $\mathbb{E}[U_{T\wedge k}] = \mathbb{E}[U_0] = 0$. Thus (by the Dominated Convergence Theorem),

$$\left| b\mathbb{P}\{U_T = b\} - a\mathbb{P}\{U_T = -a\} \right| = \lim_k \left| b\mathbb{P}\{U_{T\wedge k} = b\} - a\mathbb{P}\{U_{T\wedge k} = -a\} \right|$$

$$\leq \lim_k (a \vee b)\mathbb{P}\{T > k\} = 0.$$

So we get

$$\mathbb{P}\{B \text{ loses everything first}\} = \mathbb{P}\{U_T = b\} = \frac{a}{a+b}.$$

As you might expect, the one with the smaller stake is the one most likely to lose everything. But this is a quantitative version of that conclusion.

4.4 Limit Theorems

In this section we will consider a few convergence theorems in probability theory. We will treat only a few such theorems for use below. There is a much more extensive theory than we can possibly cover here.

Convergence of Nonnegative Martingales. First we discuss the convergence of martingales. Many martingales converge with probability one. (But not all of them. The random walk U_n at the end of the previous section is an example of a martingale that does not converge.) For our purposes it will be enough to know that a nonnegative martingale converges with probability one. The first example (double-or-nothing) is an example here. In that case, with probability one $S_n = 0$ for large enough n, so it certainly converges.

(4.4.1) Theorem. *Let $(\mathcal{F}_n)$ be a stochastic basis, and let (X_n) be a martingale adapted to it. Suppose $X_n \geq 0$. Then with probability one, (X_n) converges.*

Proof. First I claim that $\mathbb{P}\{X_n \to \infty\} = 0$. If not, say $\mathbb{P}\{X_n \to \infty\} = \varepsilon > 0$; then for any $M > 0$ we would have $\mathbb{P}\{X_n > M\} > \varepsilon/2$ for large enough n, so $\mathbb{E}[X_n] \geq M\varepsilon/2$. This is true for any $M > 0$, so $\mathbb{E}[X_n]$ is certainly not constant.

Assume (for purposes of contradiction) that $\mathbb{P}\{X_n$ does not converge$\} > 0$. Then there exist $\alpha < \beta$ such that $\mathbb{P}(A) > 0$, where

$$A = \left\{ \liminf_{n \to \infty} X_n < \alpha < \beta < \limsup_{n \to \infty} X_n \right\}.$$

We will use this to construct two bounded stopping times S, T with $\mathbb{E}\,[X_S] \neq \mathbb{E}\,[X_T]$, contradicting the fact that (X_n) is a martingale. We will carry out this construction carefully for the benefit of readers not familiar with such measurability considerations.

Write $\varepsilon = \mathbb{P}(A)$. Let $\delta > 0$ be so small that $(\beta - \alpha)(\varepsilon - 2\delta) - \alpha\delta > 0$. Now, $\mathcal{R} = \bigcup_{n=1}^{\infty} \mathcal{F}_n$ is an algebra of subsets of Ω. Let $\mathcal{G}$ be the σ-algebra generated by $\mathcal{R}$. Each of the random variables X_n is $\mathcal{G}$-measurable, so we may conclude that $A \in \mathcal{G}$. Then by the Algebra Approximation Theorem 1.1.9, there is $B \in \mathcal{R}$ with $\mathbb{P}(A \Delta B) < \delta$. So there is $N \in \mathbb{N}$ with $B \in \mathcal{F}_N$. Now,

$$(*) \qquad B \cap \{\text{there exists } n \geq N, X_n < \alpha\} \supseteq B \cap A,$$

so the set on the left has probability at least $\mathbb{P}(A) - \mathbb{P}(A \Delta B) > \varepsilon - \delta$. Now, there is $N' > N$ such that $\mathbb{P}(C) > \varepsilon - \delta$, where

$$C = B \cap \{\text{there exists } n, N \leq n \leq N', X_n < \alpha\},$$

because this set increases to the set on the left in $(*)$ as $N' \to \infty$. Note that $C \in \mathcal{F}_{N'}$. Next, in the same way,

$$C \cap \{\text{there exists } n \geq N', X_n > \beta\} \supseteq C \cap A,$$

so the set on the left has probability at least $\mathbb{P}(C) - \mathbb{P}(A \Delta B) > \varepsilon - 2\delta$. There is $N'' > N'$ such that $\mathbb{P}(D) > \varepsilon - 2\delta$, where

$$D = C \cap \{\text{there exists } n, N' \leq n \leq N'', X_n > \beta\}.$$

Note that $D \in \mathcal{F}_{N''}$ and $\mathbb{P}(C \setminus D) < \delta$.

We are ready to define two bounded stopping times, adapted to the stochastic basis $(\mathcal{F}_n)$.

$$T = \begin{cases} N & \text{outside } B, \\ \min\{n : N \leq n \leq N', X_n < \alpha\} & \text{on } C, \\ N' & \text{on } B \setminus C. \end{cases}$$

$$S = \begin{cases} N & \text{outside } B, \\ N' & \text{on } B \setminus C, \\ \min\{n : N' \leq n \leq N'', X_n > \beta\} & \text{on } D, \\ N'' & \text{on } C \setminus D. \end{cases}$$

(Convince yourself that these are, indeed, stopping times.) We must show that $\mathbb{E}\,[X_S] \neq \mathbb{E}\,[X_T]$. To estimate $\mathbb{E}\,[X_S - X_T]$, we note that

$$X_S - X_T \begin{cases} = X_N - X_N = 0 & \text{outside } B, \\ = X_{N'} - X_{N'} = 0 & \text{on } B \setminus C, \\ > \beta - \alpha & \text{on } D, \\ > X_{N''} - \alpha > -\alpha & \text{on } C \setminus D. \end{cases}$$

Thus $\mathbb{E}[X_S - X_T] > (\beta - \alpha)\mathbb{P}(D) - \alpha\mathbb{P}(C \setminus D) > (\beta - \alpha)(\varepsilon - 2\delta) - \alpha\delta > 0$, as claimed. ☺

Laws of Large Numbers. A **law of large numbers** says that a sequence of independent identically distributed random variables has a long run average that approximates the expectation. That is, if X_n are i.i.d., then

$$\lim_{n \to \infty} \frac{1}{n} \sum_{k=1}^{n} X_k$$

converges to the constant $\mathbb{E}[X_1]$. As the hypotheses are varied, the mode of convergence also may be varied. Convergence in probability and convergence with probability one are the ones we will discuss here.

We begin with the "weak law of large numbers." A sequence X_n of random variables is said to **converge in probability** to a random variable X iff for every $\varepsilon > 0$,

$$\lim_n \mathbb{P}\{|X_n - X| > \varepsilon\} = 0.$$

This is a "weaker" form of convergence than "$X_n \to X$ with probability one."

(4.4.2) Exercise. Let $I_1, I_2, \cdots$ be the set of all dyadic subintervals of $[0, 1]$ arranged in some order. On the probability space $[0, 1]$ with Lebesgue measure, the sequence $X_n = \mathbb{1}_{I_n}$ converges to 0 in probability but diverges for every sample point $\omega \in [0, 1]$.

(4.4.3) Weak Law of Large Numbers. *Let X_n be independent random variables with means μ_n and variances σ_n^2. Suppose*

$$\lim_{n \to \infty} \frac{1}{n^2} \sum_{k=1}^{n} \sigma_k^2 = 0.$$

Then

$$\frac{1}{n} \sum_{k=1}^{n} (X_k - \mu_k)$$

converges to 0 in probability as $n \to \infty$.

Proof. Consider the random variable $Y_n = (1/n) \sum_{k=1}^{n} (X_k - \mu_k)$. Compute $\mathbb{E}[Y_n] = 0$ and (by independence) $\text{Var}[Y_n] = (1/n^2) \sum_{k=1}^{n} \sigma_k^2$. Thus, by Chebyshev's inequality 4.2.20, for any $\varepsilon > 0$,

$$\mathbb{P}\{Y_n \ge \varepsilon\} \le \frac{1}{\varepsilon^2} \frac{1}{n^2} \sum_{k=1}^{n} \sigma_k^2.$$

For each fixed ε, this goes to 0 as n goes to infinity. Thus Y_n converges to 0 in probability.   ☺

(4.4.4) Corollary. *Let X_n be i.i.d. random variables with $\mathbb{E}\left[X_n^2\right] < \infty$. Then $(1/n)\sum_{k=1}^{n} X_k$ converges in probability to the constant $\mathbb{E}[X_1]$.*

Proof. Since the X_n are identically distributed, all μ_n are the same, $\mathbb{E}[X_1]$, and all σ_n^2 are the same, $\mathrm{Var}[X_1] < \infty$. Thus $(1/n^2)\sum_{k=1}^{n}\sigma_k^2 = (1/n)\,\mathrm{Var}[X_1] \to 0$.   ☺

A "strong law of large numbers" is one where the convergence occurs pointwise (with probability one).

(4.4.5) Theorem. *Let X_n be independent random variables with means μ_n and variances σ_n^2. Suppose*

$$\sum_{n=1}^{\infty}\frac{\sigma_n^2}{n^2} < \infty.$$

Then with probability one,

$$\frac{1}{n}\sum_{k=1}^{\infty}(X_k - \mu_k)$$

converges to the constant 0.

Proof. Write $S_n = \sum_{k=1}^{n}(X_k - \mu_k)$. Let $\varepsilon > 0$ be given. For $j \in \mathbb{N}$, let

$$E_j = \left\{\frac{|S_n|}{n} \geq \varepsilon \text{ for some } n \text{ such that } 2^{j-1} \leq n < 2^j\right\}.$$

Then on the set E_j we have $|S_n| \geq \varepsilon 2^{j-1}$ for some $n < 2^j$, so by Kolmogorov's maximal inequality 4.3.11,

$$\mathbb{P}(E_j) \leq \mathbb{P}\left\{\max_{1 \leq n \leq 2^j}|S_n| \geq \varepsilon 2^{j-1}\right\} \leq \frac{1}{(\varepsilon 2^{j-1})^2}\sum_{n=1}^{2^j}\sigma_n^2.$$

Therefore,

$$\sum_{j=1}^{\infty}\mathbb{P}(E_j) \leq \frac{4}{\varepsilon^2}\sum_{j=1}^{\infty}\sum_{n=1}^{2^j}2^{-2j}\sigma_n^2$$

$$= \frac{4}{\varepsilon^2}\sum_{n=1}^{\infty}\left(\sum_{j \geq \log_2 n}2^{-2j}\right)\sigma_n^2 \leq \frac{8}{\varepsilon^2}\sum_{n=1}^{\infty}\frac{\sigma_n^2}{n^2} < \infty.$$

By the Borel–Cantelli Lemma (easy direction) 4.1.2, with probability one, E_j occurs for only finitely many j. That is, $|S_n|/n \geq \varepsilon$ for only finitely many n.

So, with probability one, $\limsup_n |S_n|/n \le \varepsilon$. This is true for every positive (rational) ε, so $\limsup |S_n|/n = 0$ with probability one.   ☺

(4.4.6) Corollary. *Let X_n be i.i.d. random variables with $\mathbb{E}\left[X_1^2\right] < \infty$. Then with probability one, $(1/n)\sum_{k=1}^{n} X_k$ converges to the constant $\mathbb{E}[X_1]$.*

Proof. Since the X_n are identically distributed, all μ_n are the same, $\mathbb{E}[X_1]$, and all σ_n^2 are the same, $\mathrm{Var}[X_1] < \infty$. Since the series $\sum_{n=1}^{\infty}(1/n^2)$ converges, we have $\sum_{n=1}^{\infty} \sigma_n^2/n^2 < \infty$.   ☺

Next is a slight variant, where independence is relaxed slightly. Let k be a positive integer. A sequence (X_n) of random variables is called k-**independent** iff the subsequences $X_i, X_{i+k}, X_{i+2k}, \cdots$ are independent sequences. For example, imagine tossing a fair coin repeatedly to generate an infinite sequence (Y_n) of 0s and 1s. The sequence $X_n = Y_n + Y_{n-1}$ is not an independent sequence (for example, if $X_4 = 2$, then $X_5 \ne 0$). But it is a 2-independent sequence. The dependence occurs only for random variables that are close to each other in the sequence.

(4.4.7) Corollary. *Let X_n be k-independent identically distributed random variables with $\mathbb{E}\left[X_1^2\right] < \infty$. Then with probability one, $(1/n)\sum_{k-1}^{n} X_k$ converges to the constant $\mathbb{E}[X_1]$.*

Proof. Write $\mu = \mathbb{E}[X_1] = \mathbb{E}[X_n]$ for all n. For fixed value $i \in \{1, 2, 3, \cdots, k\}$ the sequence

$$X_i, X_{i+k}, X_{i+2k}, \cdots$$

is i.i.d. So by the usual Law of Large Numbers, $\lim_{m\to\infty}(1/m)Y_{i,m} = \mu$, where

$$Y_{i,m} = \sum_{j=0}^{m-1} X_{i+jm}.$$

Now consider the full sequence. For $n \in N$, we have

$$\sum_{j=1}^{n} X_j = \sum_{i=1}^{k} Y_{i,m(i,n)},$$

where $m(i,n)$ is an integer that differs from n/k by at most 2. In particular, as $n \to \infty$ we have $m(i,n)/n \to 1/k$ for all i. So

$$\lim_{n\to\infty} \frac{1}{n}\sum_{j=1}^{n} X_j = \sum_{i=1}^{k} \frac{1}{k} \lim_{m\to\infty}\frac{1}{m}Y_{i,m} = \sum_{i=1}^{k} \frac{1}{k}\mu = \mu.   ☺$$

The Weierstrass Approximation Theorem. We will now exhibit a clever use of the Law of Large Numbers (actually Chebyshev's inequality) to prove the Weierstrass Approximation Theorem. Often, the Weierstrass theorem is considered more basic than the law of large numbers and is proved at an earlier stage of a mathematician's education. But since we have not proved it here, we may do it the other way around, using the probabilistic ideas presented here to prove the Weierstrass theorem.

(4.4.8) Weierstrass Approximation Theorem. *Let $a < b$ be real numbers, and let $f: [a, b] \to \mathbb{R}$ be continuous. There is a sequence p_n of polynomials such that p_n converges to f uniformly on $[a, b]$.*

Making a linear change of variables, we may reduce the proof to the special case of the interval $[0, 1]$. Then the theorem will be a consequence of this version (due to S. Bernstein):

(4.4.9) Theorem. *Let $f: [0, 1] \to \mathbb{R}$ be continuous. Define polynomials $B_n(x)$ by*

$$B_n(x) = \sum_{k=1}^{n} f\left(\frac{k}{n}\right) \binom{n}{k} x^k (1-x)^{n-k}.$$

Then $B_n(x)$ converges to $f(x)$ uniformly on $[0, 1]$.

Proof. Let $\varepsilon > 0$ be given. Since f is uniformly continuous, there is $\eta > 0$ such that $|f(x) - f(y)| < \varepsilon/2$ whenever $|x - y| < \eta$. Also, f is bounded on $[0, 1]$, say $M = \sup |f(x)|$. Let $N \in \mathbb{N}$ be so large that $4M/N < \eta^2 \varepsilon$. We claim that $|B_n(x) - f(x)| \le \varepsilon$ for all $n \ge N$ and all $x \in [0, 1]$.

So fix $n \ge N$ and $x \in [0, 1]$. Let X_i be an i.i.d. sequence of Bernoulli random variables with $\mathbb{P}\{X_i = 1\} = x$ and $\mathbb{P}\{X_i = 0\} = 1 - x$. The means are $\mathbb{E}[X_i] = x$; the variances are $\operatorname{Var}[X_i] = \mathbb{E}\left[|X_i - x|^2\right] = x(1 - x) \le 1$. Now, if $S = X_1 + X_2 + \cdots + X_n$,[11] then $\mathbb{E}[f(S/n)] = B_n(x)$. Now S/n should (by the Law of Large Numbers) be close to x with large probability. To make this precise, note that

$$\left| f\left(\frac{S}{n}\right) - f(x) \right| \le \begin{cases} \varepsilon/2 & \text{on } \{|S/n - x| < \eta\}, \\ 2M & \text{on } \{|S/n - x| \ge \eta\}. \end{cases}$$

Now, $S/n - x$ has variance $\le 1/n$, so $\mathbb{P}\{|S/n - x| \ge \eta\} \le 1/(n\eta^2)$ by Chebyshev's inequality. Thus

$$|B_n(x) - f(x)| \le \mathbb{E}\left[\left| f\left(\frac{S}{n}\right) - f(x) \right| \right]$$

$$\le \frac{\varepsilon}{2} + \frac{2M}{n\eta^2} \le \frac{\varepsilon}{2} + \frac{\varepsilon}{2}. \qquad \smiley$$

---

[11] S has a binomial distribution with parameters x, n.

(4.4.10) Corollary. *The function space* $\mathfrak{C}([0,1], \mathbb{R})$ *is separable.*

Proof. Polynomials with rational coefficients are dense in $\mathfrak{C}([0,1], \mathbb{R})$.   ☺

(4.4.11) Exercise. Let K be a compact metric space. Then $\mathfrak{C}(K, \mathbb{R})$ is separable. Prove it along these lines:

(a) First, $\mathfrak{C}([0,1]^{\mathbb{N}}, \mathbb{R})$ is separable, where $[0,1]^{\mathbb{N}}$ is the countable product topological space, since functions that depend on finitely many coordinates are dense.

(b) There is a sequence of continuous functions $s_n : K \to [0,1]$ that separates points of K; that is, if $x \neq y$ in K, then $s_n(x) \neq s_n(y)$ for some n.

(c) If $s : K \to [0,1]^{\mathbb{N}}$ is defined with components s_n, then s is a homeomorphism of K onto a subset of $[0,1]^{\mathbb{N}}$.

(d) If $s : K \to [0,1]^{\mathbb{N}}$ is a homeomorphism of K onto a subset of $[0,1]^{\mathbb{N}}$, then $f \mapsto f \circ s$ defines a continuous function of $\mathfrak{C}([0,1]^{\mathbb{N}}, \mathbb{R})$ onto $\mathfrak{C}(K, \mathbb{R})$. The continuous image of a separable space is separable.

*4.5 Remarks

See an interesting article by J. Doob [62] on the history of probability theory and how it became more mathematical over the years.

There are many texts dealing with probability theory as a mathematical discipline. Generally speaking, these are the texts that presuppose a knowledge of measure theory. A few examples are [32, 46, 69, 97, 168]. Most, if not all, of the material in Sections 4.1–4.3 is contained in each of those books.

Our use of conditional probabilities of the form $\mathbb{P}[E \mid X = x]$ rather than the more general $\mathbb{P}[E \mid \mathcal{G}]$ may be unusual (compare [32, Exercise 33.5]). I hope it is more intuitive; most of the applications here are of the form $\mathbb{P}[E \mid X = x]$.

As noted in §4.3, there does not seem to be a standard notation for a conditional probability distribution $\mathcal{D}_{Y \mid X = t}$ or $\mathcal{D}_{Y \mid \mathcal{G}}$. See [32, p. 390] for the existence of conditional distributions and [32, Exercise 33.13] for a description of a case where $\mathcal{D}_{Y \mid \mathcal{G}}$ does not exist as a measure for almost all sample points.

Martingales[12] are discussed in [32, §35]. (My own work on a generalization is described in [79].) L. Snell [240] wrote an entertaining account dealing with martingales together with popular topics like gambling.

Stopping times are also known as optional times. Proposition 4.3.27 may be called "optional stopping" of the martingale (X_n).

---

* An optional section.

[12] The third meaning of the word "martingale" listed by the Oxford English Dictionary: "3. A system in gambling which consists in doubling the stake when losing in the hope of eventually recouping oneself." They trace the word back to Rabelais but seem unsure of the etymology.

On the Exercises. *(4.2.7)*: One possible metric:

$$\rho(f,g) = \sum_{m=1}^{\infty} \frac{1}{2^m} \left(1 \wedge \sup_{x \in [0,m]} |f(x) - g(x)| \right).$$

This metric is complete.

(4.2.8): Use Corollary 4.4.10, below.

(4.2.29): Mean $1/\lambda$, variance $1/\lambda^2$.

(4.2.32): Mean λ, variance λ.

(4.3.2): Hint. You may use the Pi–Lambda Theorem.

(4.3.14): [32, Theorem 33.3].

(4.3.15): [32, (33.19)].

(4.3.16): [32, (33.12), (33.15)].

(4.3.17): [32, Theorem 34.5].

(4.3.19): See [32, (33.16)].

5. Probability and Fractals

In this chapter, we will consider some of the applications of probability theory to the study of fractals. The most important of these is the possibility of *random fractals,* that is, random sets (or random measures) that exhibit the irregularities (and regularities) characteristic of the sets and measures that we normally associate with the word "fractal." The basic idea of self-similarity can be modified slightly to statistical self-similarity. A random set is said to be statistically self-similar if the set is made up of smaller parts, and each part is similar to an instance of the same random set. The parts are not actually similar to the whole, but they are similar to sets that might have been obtained for the whole if the random events had turned out differently. (More technically, the part is similar to a random set whose probability distribution is the same as the distribution of the whole set.)

We will also consider other ways that probability theory may be used in the study of fractal geometry. Some of the deterministic (nonrandom) fractal sets (and measures) may be constructed using random methods. Probabilistic theorems (such as the Borel–Cantelli Lemma or the Strong Law of Large Numbers) may be used in proofs involved with the study of the properties of fractal sets, even when the sets are deterministic.

5.1 The Chaos Game

Begin with three points in the plane, corners of an equilateral triangle. (We call them a_L, a_R, a_U.) Pick an arbitrary starting point x_0 in the plane. From these data we will define a random sequence $(x_n)_{n \in \mathbb{N}}$ of points in the plane. Suppose x_n has been defined. For the next stage, select at random one of the three points a_e. The next term x_{n+1} in our sequence is the midpoint of the line segment joining x_n and a_e; that is, $x_{n+1} = (a_e + x_n)/2$. A few steps are illustrated in Figure 5.1.1.

This construction is called "the chaos game" by Barnsley [13]. We get a random sequence of points in the plane. But when many steps are actually carried out by a computer, the resulting picture looks less and less "random" as time passes. See Figure 5.1.2.

This random sequence construction converges to the "Sierpiński gasket" fractal, the self-similar set in the plane determined by the iterated function system consisting of three transformations,

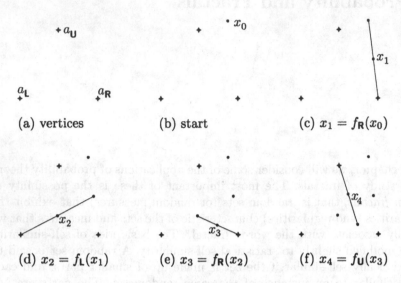

(a) vertices   (b) start   (c) $x_1 = f_\mathsf{R}(x_0)$

(d) $x_2 = f_\mathsf{L}(x_1)$ (e) $x_3 = f_\mathsf{R}(x_2)$ (f) $x_4 = f_\mathsf{U}(x_3)$

Figure 5.1.1. The chaos game.

$$f_\mathsf{L}(x) = \frac{a_\mathsf{L} + x}{2},$$
$$f_\mathsf{R}(x) = \frac{a_\mathsf{R} + x}{2},$$
$$f_\mathsf{U}(x) = \frac{a_\mathsf{U} + x}{2}.$$

We will now consider the sense in which it converges.

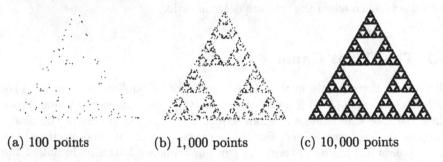

(a) 100 points   (b) 1,000 points   (c) 10,000 points

Figure 5.1.2. The chaos game played out.

Convergence. In fact, let us consider a more general iterated function system. Let S be a nonempty complete metric space, let E be a finite alphabet, and for each $e \in E$ let $f_e \colon S \to S$ be a function. These data define an **iterated function system**. We assume that there is a constant $r < 1$ such that

$$\rho\big(f_e(x), f_e(y)\big) \le r\rho(x, y)$$

for all $x, y \in S$ and all $e \in E$; that is, the iterated function system is contracting. The attractor K is the unique nonempty compact set such that

$$K = \bigcup_{e \in E} f_e[K].$$

Now let us consider a random sequence defined as follows. We specify a set $(p_e)_{e \in E}$ of weights (or probabilities), where $p_e > 0$ for all $e \in E$ and $\sum_{e \in E} p_e = 1$. This, in turn, specifies a natural product measure $\mathcal{M}$ on the space $E^{(\omega)}$ of infinite strings. For our probability sample space, we take $E^{(\omega)}$ and let $\mathbb{P} = \mathcal{M}$ be the probability measure. This product measure construction may be stated in another way: If σ is an infinite string, write σ_n for its nth letter. We have specified a way to choose such a string σ at random, so that the individual letters σ_n are independent and identically distributed, with distribution

$$\sum_{e \in E} p_e \mathcal{E}_e.$$

Now choose an arbitrary point $x_0 \in S$ as starting point. When x_{n-1} is defined, let $x_n = f_{\sigma_n}(x_{n-1})$. We may simplify our notation by writing a composition

$$f_{e_1 e_2 \cdots e_n} = f_{e_1} \circ f_{e_2} \circ \cdots \circ f_{e_n}.$$

In this notation, our random sequence may be written $x_n = f_{\sigma_n \sigma_{n-1} \cdots \sigma_2 \sigma_1}(x_0)$. A similar convention will be used to abbreviate a product of the p's:

$$p(e_1 e_2 \cdots e_n) = p_{e_1} p_{e_2} \cdots p_{e_n}.$$

(5.1.3) Theorem. *Let a contracting iterated function system (f_e), weights (p_e), and starting point x_0 be specified. Let K be the attractor of the iterated function system. Let the random sequence (x_n) be defined as above. Then, with probability one, the set of all cluster points of the sequence (x_n) is exactly the set K.*

Proof. If y_0 is any point of K, we have

$$\rho\big(f_{\sigma_n \sigma_{n-1} \cdots \sigma_2 \sigma_1}(x_0), f_{\sigma_n \sigma_{n-1} \cdots \sigma_2 \sigma_1}(y_0)\big) \leq r^n \rho(x_0, y_0) \to 0,$$

so our random sequence starting at x_0 has the same cluster points as the one starting at y_0. So we may assume that $x_0 \in K$. Each function f_e maps K into K, and K is a closed set. This makes it clear that every cluster point is in K.

Conversely, we will show that every point of K is a cluster point of the sequence (x_n) (with probability one).

If α is a finite string, we say that α **occurs in** a string σ iff $\sigma = \beta \alpha \tau$ for some strings β, τ. Now, given a finite string $\alpha \in E^{(m)}$, consider the events

$$S_{km}(\alpha) = \Big\{ \beta \alpha \tau : \beta \in E^{(km)}, \tau \in E^{(\omega)} \Big\}.$$

The events S_{km} ($k = 0, 1, 2, \cdots$) depend on disjoint blocks of letters in the string σ, so they are independent events. The probability of each S_{km} is $p(\alpha) >$

0. By the Borel–Cantelli Lemma (or by the Law of Large Numbers), with probability one, infinitely many of the S_{km} happen. In particular, at least one of the S_{km} happens. So (with probability one) α occurs in the string σ.

There are countably many finite strings α, so with probability one, all finite strings occur in σ. We will use this to show that every element of K is a cluster point of the sequence (x_n).

Fix $x \in K$. Let $\theta \in E^{(\omega)}$ be an address of x, so that

$$\{x\} = \bigcap_{n \to \infty} f_{\theta \restriction n}[K].$$

Given n, let $\alpha = \theta_n \theta_{n-1} \cdots \theta_2 \theta_1$ be the reversal of the first n letters of θ. We know that α occurs in σ; say $\sigma = \beta \alpha \tau$, where $|\beta| = m$. Then $x_{n+m} = f_{\theta_1 \theta_2 \cdots \theta_n \beta_m \cdots \beta_1}(x_0) \in f_{\theta \restriction n}[K]$. Thus these points x_{n+m} converge to the point x, so x is a cluster point.   ☺

The limit set K is not a "random fractal" even though it may be (with probability one) constructed in this random manner.

(5.1.4) Exercise. Formulate (and prove) the corresponding result where the iterated function system is governed by a directed multigraph [MTFG, §4.3].

Self-Similar Measures. The random sequence (x_n) constructed here can be used to construct not only a fractal set, but also a fractal measure. If $m \in \mathbb{N}$, define a discrete random measure by

$$\mathcal{T}_m = \frac{1}{m} \sum_{n=1}^{m} \varepsilon_{x_n}.$$

That is, $\mathcal{T}_m(E)$ is the proportion of the points $\{x_1, x_2, \cdots, x_m\}$ that fall in the set E. We will see that this sequence of measures converges to the invariant measure $\mathcal{T}$ for the iterated function system (f_e) with weights (p_e). Recall that $\mathcal{T}$ is the unique probability measure that satisfies

$$\int h(y)\, \mathcal{T}(dy) = \sum_{e \in E} p_e \int h(f_e(y))\, \mathcal{T}(dy)$$

for all bounded continuous functions h.

(5.1.5) Theorem. *Let a contracting iterated function system (f_e), weights (p_e), and starting point x_0 be specified. Let $\mathcal{T}$ be the invariant measure of the iterated function system (f_e) with weights (p_e). Let the random sequence of $(\mathcal{T}_m)$ of measures be defined as above. Then, with probability one, the sequence $\mathcal{T}_m$ converges to $\mathcal{T}$ narrowly.*

Proof. This is just a more careful version of the proof of Theorem 5.1.3—we need to know not only that every string α occurs in σ, but that it occurs with the proper frequency. To do this, we will apply the Strong Law of Large Numbers.

We will continue to use (x_n) and K with the same meaning. First, note that if $y_0 \in K$, then for any $h \in V_1$ we have

$$\left| \frac{1}{m} \sum_{n=1}^{m} h\big(f_{\sigma_n \cdots \sigma_1}(x_0)\big) - \frac{1}{m} \sum_{n=1}^{m} h\big(f_{\sigma_n \cdots \sigma_1}(y_0)\big) \right|$$

$$\leq \frac{1}{m} \sum_{n=1}^{m} \left| h\big(f_{\sigma_n \cdots \sigma_1}(x_0)\big) - h\big(f_{\sigma_n \cdots \sigma_1}(y_0)\big) \right|$$

$$\leq \frac{1}{m} \sum_{n=1}^{m} r^n \rho(x_0, y_0)$$

$$= \frac{\rho(x_0, y_0)}{m} \frac{r - r^{m+1}}{1 - r}.$$

This shows that the ρ_1-distance between the measure $\mathcal{T}_m$ defined with starting point x_0 and the same thing defined with starting point y_0 is at most $\rho(x_0, y_0) r / (m(1 - r))$. This goes to 0 as $m \to \infty$. So we may assume in this proof that $x_0 \in K$.

Now let $h \in \mathfrak{C}(K, \mathbb{R})$. We will show that $\int h(y) \, \mathcal{T}_m(dy) \to \int h(y) \, \mathcal{T}(dy)$. This will show that $\mathcal{T}_m \to \mathcal{T}$ by the definition of narrow convergence. Fix $\varepsilon > 0$. Now, h is continuous and K is compact, so h is uniformly continuous. Thus, there is $\delta > 0$ such that $|h(x) - h(y)| < \varepsilon/6$ whenever $x, y \in K$ with $\rho(x, y) < \delta$. Let $k \in \mathbb{N}$ be so large that $r^k \operatorname{diam} K < \delta$. Now for $n \geq k$,

(1) $$\left| h\big(f_{\sigma_n \cdots \sigma_1}(x_0)\big) - h\big(f_{\sigma_n \cdots \sigma_{n-k+1}}(x_0)\big) \right| < \frac{\varepsilon}{6}.$$

If α is a string of length k, then for any $y \in K$ we have $|h(f_\alpha(y)) - h(f_\alpha(x_0))| < \varepsilon/6$. So

$$\left| \sum_{|\alpha|=k} p(\alpha) h\big(f_\alpha(y)\big) - \sum_{|\alpha|=k} p(\alpha) h\big(f_\alpha(x_0)\big) \right| < \frac{\varepsilon}{6}.$$

Integrate with respect to y to obtain

$$\left| \int \sum_{|\alpha|=k} p(\alpha) h\big(f_\alpha(y)\big) \, \mathcal{T}(dy) - \sum_{|\alpha|=k} p(\alpha) h\big(f_\alpha(x_0)\big) \right| < \frac{\varepsilon}{6}.$$

Or, by the invariance of $\mathcal{T}$,

(2) $$\left| \int h(y) \, \mathcal{T}(dy) - \sum_{|\alpha|=k} p(\alpha) h\big(f_\alpha(x_0)\big) \right| < \frac{\varepsilon}{6}.$$

Now, h is continuous and K is compact, so h is bounded on K. Choose $m_0 \in \mathbb{N}$ so large that $(k/m_0) \sup_{y \in K} |h(y)| < \varepsilon/6$. Using this inequality for the first k terms and (1) for the rest of the terms, we have for $m \geq m_0$,

$$(3) \qquad \left| \frac{1}{m} \sum_{n=1}^{m} h\big(f_{\sigma_n \cdots \sigma_1}(x_0)\big) - \frac{1}{m} \sum_{n=k}^{m} h\big(f_{\sigma_n \cdots \sigma_{n-k+1}}(x_0)\big) \right| < \frac{\varepsilon}{6} + \frac{\varepsilon}{6} = \frac{\varepsilon}{3}.$$

Now we are ready for our application of the Strong Law of Large Numbers. Disjoint blocks of length k in the string σ are independent; the terms

$$h\big(f_{\sigma_n \cdots \sigma_{n-k+1}}(x_0)\big)$$

are k-independent and identically distributed with mean

$$\sum_{|\alpha|=k} p(\alpha) h\big(f_\alpha(x_0)\big).$$

But k is fixed, so we have by the Strong Law (with probability one)

$$\lim_{m \to \infty} \frac{1}{m} \sum_{n=k}^{m} h\big(f_{\sigma_n \cdots \sigma_{n-k+1}}(x_0)\big) = \sum_{|\alpha|=k} p(\alpha) h\big(f_\alpha(x_0)\big).$$

So there is a (random) integer $m_1 > m_0$ so large that for all $m \geq m_1$,

$$(4) \qquad \left| \frac{1}{m} \sum_{n=k}^{m} h\big(f_{\sigma_n \cdots \sigma_{n-k+1}}(x_0)\big) - \sum_{|\alpha|=k} p(\alpha) h\big(f_\alpha(x_0)\big) \right| < \frac{\varepsilon}{3}.$$

Adding $(2), (3),$ and (4) we get for all $m \geq m_1$,

$$\left| \frac{1}{m} \sum_{n=1}^{m} h\big(f_{\sigma_n \cdots \sigma_1}(x_0)\big) - \int h(y) \, \mathcal{J}(dy) \right| < \frac{\varepsilon}{3} + \frac{\varepsilon}{3} + \frac{\varepsilon}{3} = \varepsilon.$$

This completes the proof of convergence. (Since $\mathcal{C}(K, \mathbb{R})$ is separable, it is enough to use countably many functions h, so the union of the exceptional events of probability 0 is still of probability 0.)   ☺

Again, the invariant measure $\mathcal{J}$ is not a *random* fractal, even though it may be constructed by this random method.

5.2 Dimension of Self-Similar Measures

Recall the construction of the self-similar measure given in §3.3. We will use the same notation and terminology here. In a complete metric space S the iterated function system (f_e) consisting of similarities with ratios r_e together with weights p_e defines the self-similar measure $\mathcal{J}$. Write $r_{\min} = \min\{r_e : e \in E\}$ and $r_{\max} = \max\{r_e : e \in E\}$, so that $0 < r_{\min} \leq r_{\max} < 1$.

String Space. First let us consider the string space (or model space) that corresponds to the ratio list (r_e). The set $E^{(\omega)}$ of infinite strings has on it a metric such that $\operatorname{diam}[\alpha] = r(\alpha)$ for every finite string α. (We assume there are at least 2 letters.) For each letter $e \in E$ there is a "right shift"

$\theta_e \colon E^{(\omega)} \to E^{(\omega)}$ defined by $\theta_e(\sigma) = e\sigma$. Then θ_e is a similarity with ratio r_e. The whole space $E^{(\omega)} = [\Lambda]$ is the disjoint union of the parts $[e] = \theta_e([\Lambda])$. So $E^{(\omega)}$ is the attractor of the iterated function system (θ_e).

Now add the information of the weights (p_e). A natural measure $\mathcal{M}$ on $E^{(\omega)}$ is defined if we specify $\mathcal{M}([\alpha]) = p(\alpha)$ for all finite strings α and extend by method I. When we do this, we get

$$(*) \qquad \mathcal{M} = \sum_{e \in E} p_e \theta_{e*}(\mathcal{M}),$$

so that $\mathcal{M}$ is the invariant measure defined by this iterated function system with weights. To prove $(*)$, first prove it for cylinders (which constitute a pi system), and then conclude that the two sides agree on all Borel sets by Proposition 1.1.8.

Recall that the similarity dimension for these data is defined as

$$\frac{\sum p_e \log p_e}{\sum p_e \log r_e} = s_0.$$

(5.2.1) Theorem. *Let s_0 be the similarity dimension for the ratio system (r_e) with weights (p_e). Then the self-similar measure $\mathcal{M}$ on the string space $E^{(\omega)}$ satisfies* $\mathrm{Dim}^* \mathcal{M} = \mathrm{Dim}_* \mathcal{M} = \dim^* \mathcal{M} = \dim_* \mathcal{M} = s_0$.

Proof. We will use a probabilistic argument here. Think of choosing an infinite string $\sigma \in E^{(\omega)}$ at random, so that the letters σ_i are independent and distributed according to the measure

$$\sum_{e \in E} p_e \mathcal{E}_e.$$

This will mean that $\mathcal{M}$ is the distribution of the random string σ.

For any function $g \colon E \to \mathbb{R}$, the random real numbers $g(\sigma_i)$ are independent and have distribution

$$\sum_{e \in E} p_e \mathcal{E}_{g(e)}.$$

Since E is finite, these random variables are bounded and integrable. So by the Strong Law of Large Numbers 4.4.6,

$$\lim_{k \to \infty} \frac{1}{k} \sum_{i=1}^{k} g(\sigma_i) = \sum_e p_e g(e)$$

with probability one. If we apply this to the function $g(e) = \log p_e$, we get

$$\frac{\log p(\sigma \restriction k)}{k} \to \sum p_e \log p_e$$

with probability one. If we apply it to the function $g(e) = \log r_e$, we get

$$\frac{\log r(\sigma \restriction k)}{k} \to \sum p_e \log r_e$$

with probability one. Therefore,

$$\frac{\log p(\sigma \restriction k)}{\log r(\sigma \restriction k)} \to \frac{\sum p_e \log p_e}{\sum p_e \log r_e} = s_0$$

with probability one.

Now suppose $\tau \in E^{(\omega)}$ and $\varepsilon > 0$ are given. Let k be such that $r(\tau \restriction k) \leq \varepsilon < r(\tau \restriction (k-1))$. Then $\overline{B}_\varepsilon(\tau) = [\tau \restriction k]$. Also, $\varepsilon < r(\tau \restriction (k-1)) \leq r(\tau \restriction k)/r_{\min}$. Thus

$$\frac{\log \mathcal{M}(\overline{B}_\varepsilon(\tau))}{\log \varepsilon} \leq \frac{\log p(\tau \restriction k)}{\log r(\tau \restriction k) - \log r_{\min}}.$$

Now as $\varepsilon \to 0$ we will have $k \to \infty$, $p(\tau \restriction k) \to 0$, and $r(\tau \restriction k) \to 0$, so the $r_{\min}$ term is negligible compared to the others. Therefore,

$$\limsup_{\varepsilon \to 0} \frac{\log \mathcal{M}(\overline{B}_\varepsilon(\tau))}{\log \varepsilon} \leq \limsup_{k \to \infty} \frac{\log p(\tau \restriction k)}{\log r(\tau \restriction k)}.$$

But for almost every τ, this converges to s_0. Similarly,

$$\frac{\log \mathcal{M}(\overline{B}_\varepsilon(\tau))}{\log \varepsilon} \geq \frac{\log p(\tau \restriction k)}{\log r(\tau \restriction k) - \log r_{\max}},$$

and thus

$$\liminf_{\varepsilon \to 0} \frac{\log \mathcal{M}(\overline{B}_\varepsilon(\tau))}{\log \varepsilon} \geq s_0$$

for almost every τ. We conclude that the pointwise dimension is

$$\lim_{\varepsilon \to 0} \frac{\log \mathcal{M}(\overline{B}_\varepsilon(\tau))}{\log \varepsilon} = s_0$$

for $\mathcal{M}$-almost every $\tau \in E^{(\omega)}$. By Theorem 3.3.14, we conclude that $\mathrm{Dim}^* \mathcal{M} = \mathrm{Dim}_* \mathcal{M} = \dim^* \mathcal{M} = \dim_* \mathcal{M} = s_0$. (Since $E^{(\omega)}$ is ultrametric, every finite measure has the strong Vitali property (by Exercise 1.3.4).)   ☺

Upper Bound. Now we will consider the more general case. Let S be a nonempty complete metric space, and let (f_e) be a realization of the ratio list (r_e) in S. That is, for each $e \in E$,

$$\rho\big(f(x), f(y)\big) = r_e \rho(x, y)$$

for all $x, y \in S$. The attractor $K \in \mathfrak{K}(S)$ is the unique nonempty compact set K that satisfies

$$K = \bigcup_{e \in E} f_e[K].$$

Corresponding to the weights (p_e), the attractor $\mathcal{T} \in \mathfrak{P}(S)$ is the unique tight Borel probability measure that satisfies

$$\mathcal{T} = \sum_{e \in E} p_e f_{e*}(\mathcal{T}).$$

There is a "model map" relating the string space to the general attractor. This is a function $h\colon E^{(\omega)} \to S$ with $h \circ \theta_e = f_e \circ h$ for all e. The image of $E^{(\omega)}$ is the self-similar set K, and the image of the measure $\mathcal{M}$ is the self-similar measure $\mathcal{J}$. The map satisfies the Lipschitz condition $\rho(h(\sigma), h(\tau)) \leq \rho(\sigma, \tau)$.

This information will allow an easy proof of the upper bound estimates.

(5.2.2) Exercise.   Suppose $f\colon S \to T$ satisfies a Lipschitz condition. Then $\dim f[S] \leq \dim S$ and $\operatorname{Dim} f[S] \leq \operatorname{Dim} S$. Let $\mathcal{M} \in \mathfrak{P}(S)$. Then $\dim^* f_*(\mathcal{M}) \leq \dim^* \mathcal{M}$, $\dim_* f_*(\mathcal{M}) \leq \dim_* \mathcal{M}$, $\operatorname{Dim}^* f_*(\mathcal{M}) \leq \operatorname{Dim}^* \mathcal{M}$, and $\operatorname{Dim}_* f_*(\mathcal{M}) \leq \operatorname{Dim}_* \mathcal{M}$.

(5.2.3) Corollary. *Let S be a nonempty complete metric space. Let $\mathcal{J}$ be the self-similar measure defined by an iterated function system (f_e) on S with ratio list (r_e) and weights (p_e). Let s_0 be the similarity dimension. Then $\dim^* \mathcal{J} \leq s_0$, $\dim_* \mathcal{J} \leq s_0$, $\operatorname{Dim}^* \mathcal{J} \leq s_0$, and $\operatorname{Dim}_* \mathcal{J} \leq s_0$.*

Lower Bound. For a lower bound estimate on the dimension, we will of course require a restriction on the "overlap" in the iterated function system. The case that will be discussed here requires a geometric estimate in Euclidean space that follows from the open set condition (for example, [11, 234]). (For a general metric space, the open set condition is not enough: there is a "strong open set condition" in the literature that may be used instead [209, 235].)

Let (r_e) be a ratio list and let (f_e) be a realization of it in Euclidean space $\mathbb{R}^d$. We say that (f_e) satisfies the **open set condition** iff there is a nonempty bounded open set $U \subseteq \mathbb{R}^d$ such that $f_e[U] \subseteq U$ for all U and $f_e[U] \cap f_{e'} U = \varnothing$ for all $e \neq s' \in E$.

The following lemma is contained in [MTFG, Theorem 6.3.12], but we spell it out carefully here for reference. Some convenient notation will be used: $U_\alpha = f_{\alpha_1} \circ f_{\alpha_2} \circ \cdots \circ f_{\alpha_k}[U]$ if $\alpha \in E^{(k)}$; $\alpha^- = \alpha{\restriction}(k-1)$ if $\alpha \in E^{(k)}$.

(5.2.4) Lemma. *Let (f_e) satisfy the open set condition in $\mathbb{R}^d$ using open set U. There is a constant C such that for every nonempty bounded set A,*

$$\left\{ \alpha \in E^{(*)} : r(\alpha) < \operatorname{diam} A \leq r(\alpha^-), \overline{U_\alpha} \cap A \neq \varnothing \right\}$$

has at most C elements.

Proof. Let $w = \operatorname{diam} U$, $p = \operatorname{volume}(U) = \mathcal{L}^d(U)$, $t = \mathcal{L}^d(B_1(0))$. Then I claim that the constant

$$C = \frac{t(1+w)^d}{p r_{\min}^d}$$

has the property specified. Let A be a nonempty bounded set. Write

$$H = \left\{ \alpha \in E^{(*)} : r(\alpha) < \operatorname{diam} A \leq r(\alpha^-), \overline{U_\alpha} \cap A \neq \varnothing \right\}.$$

Let m be the number of elements in H. We must show that $m \leq C$.

As α ranges through H, the sets U_α are disjoint by the open set condition. Now, each such U_α has diameter $wr(\alpha) < w \operatorname{diam} A$, and its closure meets A. So all of these sets U_α are contained in a ball with center in A and radius $(1 + w) \operatorname{diam} A$. Now, each such U_α has volume $pr(\alpha)^d \geq p(\operatorname{diam} A)^d r_{\min}^d$. So the sum of the volumes of the U_α must be $\leq$ the volume of the large ball:

$$mp(\operatorname{diam} A)^d r_{\min}^d \leq t((1 + w) \operatorname{diam} A)^d,$$

or

$$m \leq \frac{t(1+w)^d}{pr_{\min}^d} = C$$

as claimed.   ☺

(5.2.5) Theorem. *Let (f_e) realize the ratio list (r_e) in Euclidean space $\mathbb{R}^d$. Suppose the open set condition is satisfied. Let $\mathcal{T}$ be the self-similar measure corresponding to weights (p_e). Let s_0 be the similarity dimension. Then $\operatorname{Dim}^* \mathcal{T} = \operatorname{Dim}_* \mathcal{T} = \operatorname{dim}^* \mathcal{T} = \operatorname{dim}_* \mathcal{T} = s_0$.*

Proof. It suffices to prove $\operatorname{dim}_* \mathcal{T} \geq s_0$, since the other three dimensions are all $\geq \operatorname{dim}_* \mathcal{T}$, and by Corollary 5.2.3, they are all $\leq s_0$.

Let F be any Borel set in $\mathbb{R}^d$ with $\mathcal{T}(F) = a > 0$. I must show that $\operatorname{dim} F \geq s_0$. Let $\delta > 0$ be given. We will investigate the $(s_0 - \delta)$-dimensional Hausdorff measure of F. Let $\widehat{F} = h^{-1}[F]$. Then let

$$S_k = \left\{ \sigma \in \widehat{F} : \frac{\log p(\sigma{\restriction}k)}{\log r(\sigma{\restriction}k)} \geq s_0 - \delta \right\},$$

$$T_N = \bigcap_{k=N}^{\infty} S_k.$$

Now, $\widehat{F}$ is contained (except for a set of measure zero) in the increasing union $\bigcup_{N=1}^{\infty} T_N$ and $\mathcal{M}(\widehat{F}) = a > 0$, so by the countable additivity of the measure,

$$\lim_{N \to \infty} \mathcal{M}(T_N) = a.$$

Choose N so large that $\mathcal{M}(T_N) > a/2$; then choose $\varepsilon > 0$ so small that $\varepsilon < r(\alpha)$ for all α of length N.

Now suppose that $\{A_i\}$ is a countable cover of F by sets with diam $A_i < \varepsilon$. For each i, let

$$H_i = \left\{ \alpha \in E^{(*)} : r(\alpha) < \operatorname{diam} A_i \leq r(\alpha^-), \overline{U_\alpha} \cap A_i \cap F \neq \varnothing \right\}.$$

These sets are contained in sets of the form given in the lemma, so each H_i contains at most C members. If we write $H = \bigcup_{i=1}^{\infty} H_i$, then

$$\sum_{\alpha \in H} r(\alpha)^{s_0 - \delta} \leq C \sum_i (\operatorname{diam} A_i)^{s_0 - \delta}.$$

Now, $\{ [\alpha] : \alpha \in H \}$ covers $\widehat{F} \supseteq T_N$. We need to construct a cover more efficient than H. First, there is no need for the sets that do not meet T_N: let $H' = \{ \alpha \in H : [\alpha] \cap T_n \neq \varnothing \}$. Also, we need to cover the set only once: if two cylinders $[\alpha]$ are not disjoint, then one of them is contained in the other, so we may discard the smaller one. So there is a set $H'' \subseteq H'$ such that $\{ [\alpha] : \alpha \in H'' \}$ is a disjoint cover of T_N.

Now, for each $\alpha \in H''$ there exist $\sigma \in T_N$ and $k \geq N$ such that $\sigma{\restriction}k = \alpha$. Then $\sigma \in S_k$, so

$$\begin{aligned} \frac{\log p(\alpha)}{\log r(\alpha)} &\geq s_0 - \delta, \\ \log p(\alpha) &\leq (s_0 - \delta) \log r(\alpha), \\ p(\alpha) &\leq r(\alpha)^{s_0 - \delta}. \end{aligned}$$

Thus, if $\alpha \in H''$, then $\mathcal{M}([\alpha]) = p(\alpha) \leq r(\alpha)^{s_0-\delta}$. Now,

$$\begin{aligned} \frac{a}{2} \leq \mathcal{M}(T_N) \leq \mathcal{M}\left(\bigcup_{\alpha \in H''} [\alpha] \right) &= \sum_{\alpha \in H''} \mathcal{M}([\alpha]) \\ &\leq \sum_{\alpha \in H''} r(\alpha)^{s_0-\delta} \leq \sum_{\alpha \in H} r(\alpha)^{s_0-\delta} \\ &\leq C \sum_i (\operatorname{diam} A_i)^{s_0-\delta}. \end{aligned}$$

Thus we have $\mathcal{H}^{s_0-\delta}(F) \geq a/2C > 0$. So $\dim F \geq s_0 - \delta$. This is true for all $\delta > 0$, so $\dim F \geq s_0$. Now F is any set with positive $\mathcal{T}$-measure, so $\dim_* \mathcal{T} \geq s_0$. ☺

(5.2.6) Exercise. Show that Lemma 5.2.4 remains true in a metric space S if there is a finite Federer measure $\mathcal{M}$ whose support includes the attractor K.

5.3 Random Cantor Sets

A good example of a "statistically self-similar set" is the random Cantor set in the line.

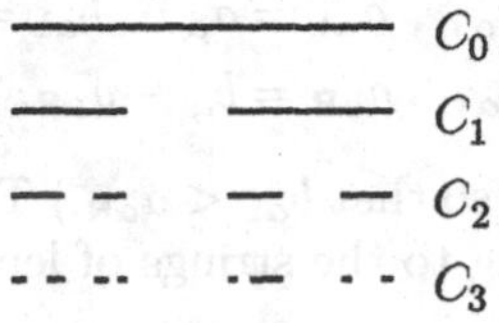

Figure 5.3.1. Construction of a random Cantor set.

Begin with the unit interval $C_0 = [0,1]$. Choose "at random" two numbers x, y with $x > 0$, $y > 0$, $x + y < 1$. (The meaning of "at random" is considered further below.) The next stage consists of two closed intervals with lengths x and y, like this:

$$C_1 = [0, x] \cup [1 - y, 1].$$

Now repeat the step using each of these two intervals: for the left interval, choose at random x', y' with $x' > 0$, $y' > 0$, $x' + y' < 1$, and for the right interval choose at random x'', y'' with $x'' > 0$, $y'' > 0$, $x'' + y'' < 1$. Inside the left interval (with length x) we use two intervals with length xx' and xy'; inside the right interval (with length y) we use two intervals with length yx'' and yy'', like this:

$$C_2 = [0, xx'] \cup [x - xy', x] \cup [1 - y, 1 - y + yx''] \cup [1 - yy'', 1].$$

Now we continue in this way. The set C_k will consist of 2^k intervals with random lengths. The next set C_{k+1} is constructed from C_k by replacing each of these intervals by two smaller intervals, one at each end, with random lengths chosen to total less than the length of the interval they replace. The "random Cantor set" constructed is the intersection $C = \bigcap_{k=0}^{\infty} C_k$.

The Construction. In order to avoid runaway notation, let us use the two-letter alphabet $E = \{\mathbf{L}, \mathbf{R}\}$ (for "left" and "right"). Each of the intervals in the construction corresponds to a finite string from this alphabet, and we will label the random variables accordingly. (So $x = x_\Lambda$, $x' = x_\mathbf{L}$, $x'' = x_\mathbf{R}$, and similarly for y.)

The formal construction goes like this. Let $\mathcal{M}$ be a probability measure on the open triangle

$$T = \{ (x, y) : x > 0, y > 0, x + y < 1 \}.$$

Now, the set $E^{(*)}$ of finite strings is a countable set. For each $\alpha \in E^{(*)}$ let (x_α, y_α) be a random element of T with distribution $\mathcal{M}$. These random variables (one for each α) should be independent. Then define our sets recursively. Begin with the empty string Λ:

$$a_\Lambda = 0, \quad b_\Lambda = 1, \quad u_\Lambda = 1.$$

Now, $C_0 = [a_\Lambda, b_\Lambda] = [0,1]$ consists of a single interval of length $u_\Lambda = 1$. Suppose that $\alpha \in E^{(*)}$ is a finite string, and that a_α, b_α, and $u_\alpha = b_\alpha - a_\alpha$ have been defined. Let

$$u_{\alpha\mathbf{L}} = u_\alpha x_\alpha, \quad a_{\alpha\mathbf{L}} = a_\alpha, \quad b_{\alpha\mathbf{L}} = a_\alpha + u_{\alpha\mathbf{L}},$$

$$u_{\alpha\mathbf{R}} = u_\alpha y_\alpha, \quad a_{\alpha\mathbf{R}} = b_\alpha - u_{\alpha\mathbf{R}}, \quad b_{\alpha\mathbf{R}} = b_\alpha.$$

(Note that $u_{\alpha\mathbf{L}} + u_{\alpha\mathbf{R}} < u_\alpha$, so that $b_{\alpha\mathbf{L}} < a_{\alpha\mathbf{R}}$.) Then for $k \in \mathbb{N}$ let C_k consist of 2^k intervals corresponding to the strings of length k:

$$C_k = \bigcup_{|\alpha|=k} [a_\alpha, b_\alpha].$$

Note that $C_0 \supseteq C_1 \supseteq C_2 \supseteq \cdots$. Finally,

$$C = \bigcap_{k=0}^{\infty} C_k$$

is the "random Cantor set" corresponding to the measure $\mathcal{M}$.

(5.3.2) Exercise. Show that the set C is indeed properly measurable to qualify as a random element of $\mathfrak{K}(\mathbb{R})$.

This random set C has a random version of the usual self-similarity property. There are two random similarities of $\mathbb{R}$ given by

$$f_{\mathsf{L}}(t) = x_\Lambda t; \qquad f_{\mathsf{R}}(t) = 1 - y_\Lambda(1 - t).$$

But the i.i.d. family

$$\left\{ (x_\alpha, y_\alpha) \right\}_{\alpha \in E(*)}$$

has the same joint distribution as the family

$$\left\{ (x_{\mathsf{L}\alpha}, y_{\mathsf{L}\alpha}) \right\}_{\alpha \in E(*)}$$

and as the family

$$\left\{ (x_{\mathsf{R}\alpha}, y_{\mathsf{R}\alpha}) \right\}_{\alpha \in E(*)}.$$

So in fact, C is made up of two parts, $C_{\mathsf{L}} = [a_{\mathsf{L}}, b_{\mathsf{L}}] \cap C$ and $C_{\mathsf{R}} = [a_{\mathsf{R}}, b_{\mathsf{R}}] \cap C$; each of these two parts is similar to an instance of the random Cantor set C. (Although C_{L} is not similar to C, it is similar to what C might look like in another universe, where the random choices came out differently.) More technically, the distribution of C (a measure on $\mathfrak{K}(\mathbb{R})$) is equal to the distribution of $f_{\mathsf{L}}^{-1}[C_{\mathsf{L}}]$, and to the distribution of $f_{\mathsf{R}}^{-1}[C_{\mathsf{R}}]$.

Examples. Let $\mathcal{M}$ be the point mass concentrated at the point $(1/3, 1/3)$. Then the construction is deterministic (not really random), and the result is the ordinary ternary Cantor set.

Now a simple (but genuinely random) example: Let $\mathcal{M}$ be the measure with mass $1/2$ at the point $(1/4, 1/4)$ and mass $1/2$ at the point $(1/8, 1/8)$. So at each stage, we flip a fair coin; if it comes up heads, we replace an interval by two intervals of $1/4$ the length; if it comes up tails, we replace it by two intervals of $1/8$ the length.

(5.3.3) Exercise. Let $\mathcal{M}$ be the two-point measure just described. (a) What is the expected total length of the level k set C_k? (b) For a fixed exponent s, compute the expected value of the sum of the sth powers of the lengths of the intervals in C_k,

$$\mathbb{E}\left[\sum_{|\alpha|=k} u_\alpha^s \right].$$

(c) What happens to this as $k \to \infty$?

For a third example, let $\mathcal{M}$ be uniformly distributed on the triangle T. That is, $\mathcal{M}$ is 2 times area.

(5.3.4) Exercise. Find the critical value of the exponent s so that

$$2 \int_T (x^s + y^s)\, dx\, dy = 1.$$

Convergence. In general (as in the special cases of the exercises),

$$\mathbb{E}\left[\sum_{|\alpha|=k} u_\alpha^s \right] = \Phi(s)^k,$$

where $\Phi(s) = \int_T (x^s + y^s)\, \mathcal{M}(d(x,y))$. This is proved by induction on k.

Now, Φ is continuous, $\Phi(0) = 2$, $\Phi(1) = \int (x+y)\, \mathcal{M}(d(x,y)) < 1$ (since $x + y < 1$), and Φ is strictly decreasing. So there is a unique s_0 such that $\Phi(s_0) = 1$. For this value $s = s_0$, we have

$$\mathbb{E}\left[\sum_{|\alpha|=k} u_\alpha^s \right] = 1$$

for all k. This critical value should be the dimension of the attractor C (with probability one).

But to evaluate the fractal dimension of the set C we need to know about the convergence of the random variable

$$Y(k) = \sum_{|\alpha|=k} u_\alpha^s$$

itself, not just its expectation.

(5.3.5) Theorem. *Let s be the solution of $\Phi(s) = 1$, and let*

$$Y(k) = \sum_{|\alpha|=k} u_\alpha^s.$$

This sequence is a nonnegative martingale. Thus, it converges with probability one.

Proof. For a stochastic basis, let $\mathcal{F}(k)$ be the smallest σ-algebra making x_α, y_α measurable for all α with $|\alpha| < k$. So $Y(k)$ is $\mathcal{F}(k)$-measurable. Now let α be a string with $|\alpha| = k$. Its two successors are $\alpha \mathbf{L}$ and $\alpha \mathbf{R}$. By definition $u_{\alpha\mathbf{L}} = u_\alpha x_\alpha$, and $u_{\alpha\mathbf{R}} = u_\alpha y_\alpha$, but u_α is $\mathcal{F}(k)$-measurable, while (x_α, y_α) is independent of $\mathcal{F}(k)$. Thus

$$\mathbb{E}\left[u_{\alpha\mathbf{L}}^s + u_{\alpha\mathbf{R}}^s \mid \mathcal{F}(k) \right] = \mathbb{E}\left[u_\alpha^s (x_\alpha^s + y_\alpha^s) \mid \mathcal{F}(k) \right] = u_\alpha^s \mathbb{E}\left[x_\alpha^s + y_\alpha^s \right] = u_\alpha^s.$$

Summing over all α of length k, we get

$$\mathbb{E}\left[Y(k+1)\,|\,\mathcal{F}(k)\right] = Y(k).$$

That is, the sequence $Y(k)$ is a martingale.

Nonnegative martingales converge by Theorem 4.4.1.   ☺

(5.3.6) Exercise. The martingale $Y(k)$ is mean-square bounded. Write $B = \mathbb{E}\left[\left(x_\alpha^s + y_\alpha^s\right)^2\right]$, so that $B \geq 1$. Then show that

$$\mathbb{E}\left[Y(k)^2\right] = 1 + (B - 1)\sum_{j=0}^{k-1}\Phi(2s)^j.$$

Finally, $\Phi(2s) < 1$, so the geometric series converges to a finite bound.

Now the upper bound estimate can be done:

(5.3.7) Theorem. *Let C be the random Cantor set generated from a measure $\mathcal{M}$, and let s be the value such that $\int_T (x^s + y^s)\,\mathcal{M}(d(x,y)) = 1$. Then $\dim C \leq s$ with probability one.*

Proof. Using the notation of the previous proof, write $Z = \lim_k Y(k)$. Then with probability one, Z exists and is finite. Now, $\sum_{|\alpha|=k} u_\alpha$ decreases as k increases, and $\mathbb{E}\left[\sum_{|\alpha|=k} u_\alpha\right] = \Phi(1)^k \to 0$. Therefore, by the Monotone Convergence Theorem 2.2.5, $\mathbb{E}\left[\lim_k \sum_{|\alpha|=k} u_\alpha\right] = 0$, so $\lim_k \sum_{|\alpha|=k} u_\alpha = 0$ with probability one. Thus with probability one, $\varepsilon_k \to 0$, where $\varepsilon_k = \max\left\{u_\alpha : |\alpha| = k\right\}$.

Now, for a given k, the set C is covered by the intervals $[a_\alpha, b_\alpha]$ with $|\alpha| = k$. So

$$\mathcal{H}^s_{\varepsilon_k}(C) \leq \sum_{|\alpha|=k} u_\alpha^s = Y(k).$$

Now let $k \to \infty$ to obtain

$$\mathcal{H}^s(C) \leq \lim_k Y(k) = Z < \infty$$

with probability one. Thus $\dim C \leq s$.   ☺

Lower Bound. Let us begin by computing the Hausdorff dimension for a string space. The alphabet is $E = \{\mathsf{L}, \mathsf{R}\}$, and the string space $E^{(\omega)}$ consists of infinite strings from this alphabet.

A *random* metric ρ may be defined on this string space in the usual way. Let the diameter of the cylinder $[\alpha]$ be u_α. More precisely, if $\sigma, \tau \in E^{(\omega)}$, then $\rho(\sigma, \tau) = 0$ if $\sigma = \tau$, and otherwise $\rho(\sigma, \tau) = u_\alpha$, where α is the longest common prefix of the two strings σ and τ. The conditions for a metric may then be checked. They are satisfied with probability one.

(5.3.8) Exercise. Let s be the value such that $\int_T (x^s + y^s)\, \mathcal{M}(d(x,y)) = 1$. Verify that with probability one, the upper bound estimate $\dim E^{(\omega)} \leq s$ holds for the metric ρ defined above.

For the lower bound, we will use the "potential" method from §3.2. First we will define a random measure $\mathcal{N}$ on $E^{(\omega)}$. For a finite string α, let

$$\mathcal{N}([\alpha]) = \lim_k \sum_{\substack{\beta \geq \alpha \\ |\beta| = k}} u_\beta^s.$$

This converges (it is a martingale) for the same reasons as in (5.3.5). In fact, the limit is u_α^s times a random variable Z_α with the same distribution as Z, but independent of u_α. Note that the martingale $Y(k)$ is mean-square bounded (by Exercise 5.3.6), so $\mathbb{E}[Z] = 1$ by Proposition 2.2.12. Similarly, $\mathbb{E}[\mathcal{N}([\alpha]) \mid \mathcal{F}(k)] = u_\alpha^s$.

(5.3.9) Exercise. Show that $Z_\alpha = x_\alpha^s Z_{\alpha L} + y_\alpha^s Z_{\alpha R}$. Verify that $\mathcal{N}$ has the properties to ensure that it can be extended to a measure.

(5.3.10) Exercise. Show that $Z_{\alpha R}$ and $Z_{\alpha L}$ are independent of each other and of $u_\alpha, x_\alpha, y_\alpha$.

(5.3.11) Proposition. *Let s be the value such that $\int_T (x^s + y^s)\, \mathcal{M}(d(x,y)) = 1$. Then with probability one, the lower bound estimate $\dim E^{(\omega)} \geq s$ holds for the metric ρ defined above.*

Proof. For $t > 0$, write $\Phi(t) = \int_T (x^t + y^t)\, \mathcal{M}(d(x,y))$ as before. Thus $\Phi(s) = 1$, $\Phi(t) > 1$ if $t < s$, and $\Phi(t) < 1$ if $t > s$. Now we will consider the t-energy integral

$$I^t(\mathcal{N}) = \int \int \frac{\mathcal{N}(d\sigma)\mathcal{N}(d\tau)}{\rho(\sigma,\tau)^t}.$$

To do this, consider a finite string $\alpha \in E^{(*)}$. Now, in the Cartesian product space $E^{(\omega)} \times E^{(\omega)}$, let

$$A_\alpha = \{ (\sigma, \tau) : \alpha \text{ is the longest common prefix of } \sigma \text{ and } \tau \}$$
$$= ([\alpha L] \times [\alpha R]) \cup ([\alpha R] \times [\alpha L]),$$

so its $\mathcal{N} \otimes \mathcal{N}$-measure is

$$u_\alpha^s x_\alpha^s Z_{\alpha L} u_\alpha^s y_\alpha^s Z_{\alpha R} + u_\alpha^s y_\alpha^s Z_{\alpha R} u_\alpha^s x_\alpha^s Z_{\alpha L} = 2 u_\alpha^{2s} x_\alpha^s y_\alpha^s Z_{\alpha L} Z_{\alpha R}.$$

This is, of course, a random real number. Now, if $(\sigma, \tau) \in A_\alpha$, then $\rho(\sigma, \tau) = u_\alpha$.

Now compute the t-energy:

$$I^t(\mathcal{N}) = \int \int \frac{\mathcal{N}(d\sigma)\mathcal{N}(d\tau)}{\rho(\sigma,\tau)^t} = \sum_{\alpha \in E^{(*)}} \frac{2 u_\alpha^{2s} x_\alpha^s y_\alpha^s Z_{\alpha L} Z_{\alpha R}}{u_\alpha^t}.$$

This, too, is a random real number. But by the independence properties (Exercise 5.3.10), its expectation is

$$\mathbb{E}\left[I^t(\mathcal{N})\right] = \sum_\alpha \mathbb{E}\left[2u_\alpha^{2s-t}x_\alpha^s y_\alpha^s\right] = \sum_\alpha \mathbb{E}\left[u_\alpha^{2s-t}\right]\mathbb{E}\left[2x_\alpha^s y_\alpha^s\right].$$

Now let us write $A = \mathbb{E}\left[2x_\alpha^s y_\alpha^s\right]$; it does not depend on α, by the identical distribution assumption. So we get

$$\mathbb{E}\left[I^t(\mathcal{N})\right] = A\sum_\alpha \mathbb{E}\left[u_\alpha^{2s-t}\right]$$

$$= A\sum_{k=0}^\infty \sum_{|\alpha|=k} \mathbb{E}\left[u_\alpha^{2s-t}\right]$$

$$= A\sum_{k=0}^\infty \Phi(2s-t)^k.$$

Now, if $t < s$, then this geometric series converges, so the expected t-energy is finite, $\mathbb{E}\left[I^t(\mathcal{N})\right] < \infty$. And therefore with probability one, $I^t(\mathcal{N}) < \infty$, so by Corollary 3.2.5, $\dim E^{(\omega)} \geq t$. This is true for any $t < s$, so (using a sequence of values that increases to s) we may conclude that $\dim E^{(\omega)} \geq s$ with probability one.   ☺

Dimension of C Itself. Now what about the Hausdorff dimension of C? We will put a mild assumption on the measure $\mathcal{M}$ for this calculation. The computation follows the preceding one in Proposition 5.3.11.

Let $\mathcal{N}$ be the random measure in $E^{(\omega)}$ just constructed. Let $h: E^{(\omega)} \to C$ be the (random) model map. Now, if $\sigma, \tau \in E^{(\omega)}$ have longest common prefix α, then $\rho(\sigma, \tau) = u_\alpha$. But one of the two images $h(\sigma), h(\tau)$ is in the interval $[a_{\alpha\mathsf{L}}, b_{\alpha\mathsf{L}}]$ and the other is in $[a_{\alpha\mathsf{R}}, b_{\alpha\mathsf{R}}]$, so

$$|h(\sigma) - h(\tau)| \geq a_{\alpha\mathsf{R}} - b_{\alpha\mathsf{L}} = u_\alpha(1 - x_\alpha - y_\alpha).$$

Using this lower bound, we may estimate the t-energy for the image measure $h_*(\mathcal{N})$ as

$$I^t\big(h_*(\mathcal{N})\big) = \int\int \frac{\mathcal{N}(d\sigma)\,\mathcal{N}(d\tau)}{|h(\sigma)-h(\tau)|^t} \leq \sum_\alpha \frac{2u_\alpha^{2s}x_\alpha^s y_\alpha^s Z_{\alpha\mathsf{L}}Z_{\alpha\mathsf{R}}}{u_\alpha^t(1-x_\alpha-y_\alpha)^t}.$$

The expectation is then estimated as

$$\mathbb{E}\left[I^t\big(h_*(\mathcal{N})\big)\right] \leq B\sum_k \Phi(2s-t)^k,$$

where

$$B = \mathbb{E}\left[\frac{2x_\alpha^s y_\alpha^s}{(1-x_\alpha-y_\alpha)^t}\right]$$

is independent of α. So we get this result:

(5.3.12) Theorem. *Let* $\mathcal{M}$ *be a measure on the triangle* T. *Let* s *be the solution of* $\int_T (x^s + y^s)\, \mathcal{M}(d(x,y)) = 1$. *Assume that*

$$\int_T \frac{x^s y^s}{(1 - x - y)^t}\, \mathcal{M}(d(x,y)) < \infty$$

for all $t < s$. *Then with probability one,* $\dim C = s$.

(5.3.13) Corollary. *If there is a constant* $c < 1$ *such that* $\mathcal{M}\{x + y < c\} = 1$, *then* $\dim C = s$.

(5.3.14) Exercise. Show that

$$\int_0^1 \int_0^{1-y} \frac{dx\, dy}{(1 - x - y)^t} < \infty$$

for all $t < 1$. Conclude that $\dim C = (-3 + \sqrt{17})/2$ for the random Cantor set C based on the uniform distribution on T.

(5.3.15) Exercise. Investigate the packing dimension of the random Cantor set.

5.4 Statistical Self-Similarity

Now we will discuss the idea of statistical self-similarity. The random Cantor set in the preceding section should be a good example.

Diameter System. Before we define a concrete statistically self-similar set, we define the system of diameters to be used. This corresponds to the presentation used in [*MTFG*] where we begin with a "ratio list."

There is a set E of labels or letters; the set E could be countably infinite, although we will assume that only finitely many letters are used at a time. A **ratio vector** is a list $(a(e))_{e \in E}$ where $a(e) \in [0, 1)$ and all but finitely many are 0. (We write by convention $a(e) = 0$ when the letter e is not used.) Let V be the set of all ratio vectors. We may consider the set V to be a Borel space by considering it a subset of the countable product $[0, 1)^{\mathbb{N}}$.

Next, we postulate a probability measure $\mathcal{M}$ on the set V. Now we want a family a_α of random ratio vectors, all with the distribution $\mathcal{M}$ and independent of each other. There is potentially one random ratio vector for each finite string α, although many of them will not be used. Write $a_\alpha(e)$ for the components of a_α. The **random diameter system** (u_α) corresponding to these data is constructed recursively as follows. Begin with the empty string Λ, and define $u_\Lambda = 1$. If u_α has been defined, its children are $u_{\alpha e} = u_\alpha a_\alpha(e)$ for $e \in E$. (Of course, if $u_\alpha = 0$, then its children are all 0, so the random ratio vector a_α is not needed.)

A **diameter system** (whether constructed randomly or not) involves numbers $u_\alpha \geq 0$ such that $u_\Lambda = 1$; a child $u_{\alpha e}$ is smaller than its parent u_α; and for each given α, only finitely many of the children $u_{\alpha e}$ are nonzero.

Now, given a diameter system u_α, we may define a string space. Given $k \in \mathbb{N}$, let $E^{(k)}$ consist of all those strings α of length k such that $u_\alpha > 0$. Let $E^{(\omega)}$ consist of all the infinite strings σ such that $u_{\sigma \restriction k} > 0$ for all k. In order to get an interesting case, we will want $E^{(\omega)} \neq \emptyset$, but the definition allows the possibility of the empty set.

For a finite string α of length k, the corresponding cylinder is $[\alpha] = \{\sigma \in E^{(\omega)} : \sigma \restriction k = \alpha\}$. The set of all cylinders constitutes a base for a topology for this string space. It is compact. The normal way to put a metric on the string space $E^{(\omega)}$ is to postulate that the cylinder α has diameter u_α. More correctly, if $\sigma, \tau \in E^{(\omega)}$, define $\rho(\sigma, \tau) = 0$ if $\sigma = \tau$, and $\rho(\sigma, \tau) = u_\alpha$ if α is the longest common prefix of σ and τ. The closed balls for this metric should be exactly the cylinders $[\alpha]$, so the metric is compatible with the topology specified before. We will have diam $[\alpha] = u_\alpha$ for all α provided that each nonzero u_α has at least two nonzero children $u_{\alpha e}$. But the possibility of only one child is permitted in the definition (or indeed, no children at all).

(5.4.1) Exercise. Show that $E^{(\omega)}$ is compact.

In our random Cantor set example, the random diameter system u_α is constructed from a random ratio vector as described above: Two of the letters are **L, R**; any string involving another letter is assigned ratio 0. Then our measure $\mathcal{M}$ is concentrated on the subset T of the full set V.

Realization of a Diameter System. Suppose a diameter system (u_α) is given (constructed randomly or not). A **realization** of it in a metric space S is constructed as follows. Begin with a bounded nonempty closed set J. For convenience, we suppose diam $J = 1$. Write $J_\Lambda = J$. Then, for each letter e such that $u_e \neq 0$, we want to place a set J_e inside J, similar to J, but shrunk by the factor u_e.[1] Continuing recursively, suppose that the set J_α has been placed, similar to J, but with diameter $u_\alpha > 0$. Then for each letter e with $u_{\alpha e} > 0$ we want to place a set $J_{\alpha e}$ inside J_α; the sets $J_{\alpha e}$ are all similar to J, but shrunk to have diameter $u_{\alpha e}$. Perhaps we will want the sets $J_{\alpha e}$ to be disjoint, or at least nonoverlapping. But beyond that, how the subsets are placed is not specified in this definition.

The construction object of this realization is

$$K = \bigcap_{k=0}^{\infty} \bigcup_{|\alpha|=k} J_\alpha.$$

(If $\lim_k u_{\sigma \restriction k} = 0$ for all $\sigma \in E^{(\omega)}$ and the metric space S is complete, then K will be compact. We will see below (in the proof of (5.4.11)) that $\lim_k u_{\sigma \restriction k} = 0$

---

[1] When $u_e = 0$, we may sometimes say by convention that $J_e = \emptyset$.

holds with probability one when (u_α) is constructed in the random manner described above.) If β is any finite string (with $u_\beta > 0$), write

$$K_\beta = \bigcap_{k=0}^\infty \bigcup_{|\alpha|=k} J_{\beta\alpha}.$$

Now we can see the decomposition of K into its parts:

$$K = \bigcup_{\substack{e \in E \\ u_e > 0}} K_e.$$

Each of the parts K_e is the result of a construction like the construction of K, but according to a slightly different diameter system. (The diameter system that describes K_e is (v_α), where $v_\alpha = u_{e\alpha}/u_e$.) If the diameter system (u_α) is constructed from a random ratio vector in the random manner described above, then the diameter systems $(u_{e\alpha}/u_e)$ are all independent and distributed identically to the original diameter system (u_α).

Statistically Self-Similar Set. Suppose a diameter system is constructed from a random ratio vector using the method above. Using these ratios, we may somehow specify a method for placing the subsets $J_{\alpha e}$ inside of J_α. This method may be random, but if it is, it should be independent of all choices except the set J_α and the ratios $u_{\alpha e}$. What exactly is done here may depend on the example at hand; I think a technical formulation of the restrictions will have to wait for more experience.

   After all of these dry definitions, let us try out a few examples of statistically self-similar sets. The first example is of course the random Cantor set from the preceding section. Another class of examples will be the deterministic self-similar sets—the measure $\mathcal{M}$ is concentrated on a single element of V, and that is the ratio list of the construction.

Keep Squares at Random (3 of 4). In this example we construct a set in the plane. Begin with the unit square. Subdivide it into four smaller squares by bisecting the sides. Then select one of the four squares at random to delete. Keep the other three squares. Continue. In Figure 5.4.2 we show the first two stages and the limit.

Figure 5.4.2. Keep 3 of 4 squares.

What is the ratio vector used in this example? Let us use letters $\mathbf{A}, \mathbf{B}, \mathbf{C}, \mathbf{D}$. Four ratio vectors are used:

$$b_1(\mathbf{A}) = b_1(\mathbf{B}) = b_2(\mathbf{C}) = 1/2, \ b_1(e) = 0 \text{ for all other letters;}$$
$$b_2(\mathbf{A}) = b_2(\mathbf{B}) = b_2(\mathbf{D}) = 1/2, \ b_2(e) = 0 \text{ for all other letters;}$$
$$b_3(\mathbf{A}) = b_3(\mathbf{C}) = b_3(\mathbf{D}) = 1/2, \ b_3(e) = 0 \text{ for all other letters;}$$
$$b_4(\mathbf{B}) = b_4(\mathbf{C}) = b_4(\mathbf{D}) = 1/2, \ b_4(e) = 0 \text{ for all other letters.}$$

The measure $\mathcal{M}$ gives each of these measure $1/4$, and all other ratio vectors measure 0.

Keep Squares at Random (independent). Here is another way to keep squares. There is a parameter $p \in (0, 1)$. Subdivide the square into four parts as usual. But now, for each of the four parts, decide whether to keep it independently of the other parts: keep it with probability p. Now the measure $\mathcal{M}$ on the set V of ratio vectors will be concentrated on 16 particular ratio vectors. Ignoring all letters except $\mathbf{A}, \mathbf{B}, \mathbf{C}, \mathbf{D}$, the measure $\mathcal{M}$ assigns

measure p^4 to vector $(1/2, 1/2, 1/2, 1/2)$,

measure $p^3(1 - p)$ to vector $(1/2, 1/2, 1/2, 0)$,

measure $p^3(1 - p)$ to vector $(1/2, 1/2, 0, 1/2)$,

. . .

measure $(1 - p)^4$ to vector $(0, 0, 0, 0)$.

Figure 5.4.3 illustrates the result. For the pictures, we have used $p = 0.9, 0.7$, and 0.5.

Figure 5.4.3. Keeping squares at random, $p = 0.9, 0.7, 0.5$.

(5.4.4) Exercise. In the statistically self-similar set just described, find the probability that the entire set K is empty (the probability depends on p, of course). [As a hint, I provide the graph in Figure 5.4.5.]

Uniform Distribution in an Interval. In this example, we construct a random set $K \in \mathfrak{P}([0, 1])$ according to the usual method. But it turns out that with probability one, that set contains only a single element: $K = \{X\}$; so we have actually specified a real random variable X.

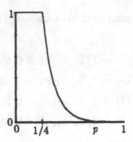

Figure 5.4.5. Probability that K is empty.

Let $\mathbf{0}, \mathbf{1}$ be two of the letters. Define two ratio vectors b_0, b_1 by

$$b_0(\mathbf{0}) = 1/2, \quad b(e) = 0 \text{ for all other letters;}$$
$$b_1(\mathbf{1}) = 1/2, \quad b(e) = 0 \text{ for all other letters.}$$

The measure $\mathcal{M}$ is concentrated on these two points, assigning each one probability $1/2$. We may realize the construction in $[0, 1]$ in such a way that the string σ corresponding to a point x is the binary expansion of the point.

Here is a more informal description of our random construction. Begin with the unit interval $[0, 1]$. Then divide it into its two halves, $[0, 1/2]$ and $[1/2, 1]$. Flip a fair coin, and use the result to choose one of the halves. Throw out the other half. In the half we have chosen, repeat the construction.

(5.4.6) Exercise. Verify that we have a construction of a statistically self-similar set K and that with probability one, $K = \{X\}$, where X is uniformly distributed on $[0, 1]$.

Circle of Circles. This will be a random set in the plane. There is a parameter $r \in (0, 1)$. We begin at the origin with radius 1 in mind. The set $S_0 = \{0\}$ is our beginning approximation. Think of a circle, centered at our point with radius 1. Then we choose an integer $n \geq 2$ at random. (The probability distribution used here is another parameter.) We choose a point on our circle at random, and starting at that point, select the n equally spaced points around the circle. The next approximation S_1 is the set of these n points. Those points are our next centers, and the new radius at each of those centers is[2] $r \sin(\pi/n)$. Repeat at each of those centers. This sequence of sets S_k converges to a random set S.

In Figure 5.4.7, three different ways to visualize this process have been tried. The first one is supposed to show the random set K itself. In the second one, we included not only the limit points, but also some line segments: from the center of each circle to the n points on its circumference. In the third one, I went only up to stage 4, and then drew the circles. Random selections were

---

[2] The value $\sin(\pi/n)$ would mean that the circles centered at the n points are mutually tangent.

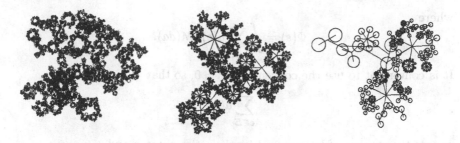

Figure 5.4.7. Circle of circles.

different in these three cases, so they show different instances of the random construction.

Similarity Dimension. Associated with a (nonrandom) ratio list $(r_e)_{e \in E}$ is a number called its similarity dimension, the solution s of the equation $\sum_{e \in E} r_e^s = 1$. In like manner we associate a similarity dimension with the diameter scheme used in the construction of a statistically self-similar set. The similarity dimension is a deterministic number, not a random number.

Suppose the diameter system (u_α) is constructed based on a probability measure $\mathcal{M}$ on the set V of ratio vectors. For a number $s > 0$ and a natural number k, consider the expected value

$$\mathbb{E}\left[\sum_{|\alpha|=k} u_\alpha^s\right].$$

(This is a useful sum to consider, since our random set K is covered by the sets J_α with $|\alpha| = k$, and u_α is the diameter of J_α.) Now, the random diameter system (u_α) is constructed recursively using $u_{\alpha e} = u_\alpha a_\alpha(e)$, where the random ratio vectors (a_α) are independent of the ratios u_α. Therefore,

$$\mathbb{E}\left[\sum_{|\alpha|=k+1} u_\alpha^s\right] = \mathbb{E}\left[\sum_{|\alpha|=k} \sum_{e \in E} u_{\alpha e}^s\right]$$

$$= \mathbb{E}\left[\sum_{|\alpha|=k} \sum_{e \in E} u_\alpha^s a_\alpha(e)^s\right]$$

$$= \mathbb{E}\left[\sum_{|\alpha|=k} u_\alpha^s\right] \mathbb{E}\left[\sum_{e \in E} a_\alpha(e)^s\right].$$

The ratio vectors (a_α) are all supposed to have the distribution $\mathcal{M}$. So we have (by induction, starting with $u_\Lambda = 1$)

$$\mathbb{E}\left[\sum_{|\alpha|=k} u_\alpha^s\right] = \Phi(s)^k,$$

where

$$\Phi(s) = \int \sum_{e \in E} a(e)^s \, \mathcal{M}(da).$$

It is convenient to use the convention $0^0 = 0$, so that

$$\sum_{e \in E} a(e)^0$$

counts the number of letters used by the ratio vector a and

$$\Phi(0) = \int \sum_{e \in E} a(e)^0 \, \mathcal{M}(da)$$

is the expected number of letters used.

(5.4.8) Exercise. Let $\mathcal{M}$ be a probability measure on V. Define

$$\Phi(s) = \int \sum_{e \in E} a(e)^s \, \mathcal{M}(da)$$

for $s \in [0, \infty)$. Then:

(a) Φ is nonincreasing: if $s_1 < s_2$, then $\Phi(s_1) \geq \Phi(s_2)$.
(b) Φ is right continuous: $\lim_{s \searrow s_0} \Phi(s) = \Phi(s_0)$.
(c) If $\Phi(s) < \infty$ for some $s < s_0$, then Φ is continuous at s_0.
(d) If $\Phi(s) < \infty$ for some s, then Φ is strictly decreasing where it is finite, and $\lim_{s \to \infty} \Phi(s) = 0$.

(5.4.9) Definition. The **similarity dimension** associated with the random diameter system built using the measure $\mathcal{M}$ is

$$s_0 = \inf \{ s : \Phi(s) \leq 1 \},$$

where

$$\Phi(s) = \int \sum_{e \in E} a(e)^s \, \mathcal{M}(da).$$

Two unusual cases are possible: (a) $\Phi(s) = \infty$ for all s, so that $s_0 = \infty$; (b) $\Phi(0) \leq 1$, so that $s_0 = 0$. The set $\{ s : \Phi(s) \leq 1 \}$ is closed, so $\Phi(s_0) \leq 1$ except possibly in these unusual cases.

The following example shows that it is possible for $\Phi(s_0) < 1$; this is consistent with the stipulations in Exercise 5.4.8 only if $\Phi(s) = \infty$ for all $s < s_0$, so that Φ need not be left-continuous at s_0.

Here is the example in $[0, 1]$, taken from [188]. Choose a natural number N at random with $\mathbb{P}\{N = n\} = 2^{-n}$ for $n = 1, 2, \cdots$ (a geometric distribution). Partition the interval $J_\Lambda = [0, 1]$ into $2^{N^2 + N}$ intervals of equal length, and keep every other one: the subintervals are $J_i = [(2i - 1)/2^{N^2+N}, 2i/2^{N^2+N}]$

for $i = 1, 2, \cdots, 2^{N^2+N-1}$. Repeat on each of these intervals J_i as usual. Then we have

$$\Phi(s) = \sum_{n=1}^{\infty} 2^{-n} \left(\frac{1}{2^{n^2+n}} \right)^s 2^{n^2+n-1} = \sum_{n=1}^{\infty} 2^{n^2(1-s)-ns-1}.$$

This gives us $\Phi(s) = \infty$ for all $s < 1$ and $\Phi(1) = 1/2$.

This example is also of interest since it yields a random set $K \subseteq [0, 1]$, topologically a Cantor set, with Lebesgue measure 0 but similarity dimension (and Hausdorff dimension) 1. In the nonrandom self-similar case, when K has Hausdorff dimension s the Hausdorff measure in that dimension is positive and finite: $0 < \mathcal{H}^s(K) < \infty$. But in the statistically self-similar case, when K has Hausdorff dimension s it is often the case that the Hausdorff measure in that dimension is zero: $\mathcal{H}^s(K) = 0$.

(5.4.10) Exercise. What is the similarity dimension associated with each of the following? (1) "Keep squares at random (3 of 4)"; (2) "Keep squares at random (independent)"; (3) "Uniform distribution on an interval."

(5.4.11) Theorem. *Suppose K is a random self-similar set realizing the diameter system built from a probability measure $\mathcal{M}$ on V. Then with probability one, the Hausdorff dimension satisfies $\dim K \leq s_0$, where s_0 is the similarity dimension $s_0 = \inf \{ s : \Phi(s) \leq 1 \}$, with $\Phi(s) = \int \sum_{e \in E} a(e)^s \, \mathcal{M}(da)$.*

Proof. Let (u_α) be the diameter system. The set K is covered by $\bigcup_{|\alpha|=k} J_\alpha$ for any $k \in \mathbb{N}$. Let $s > s_0$, so that $\Phi(s) < 1$. Then

$$\mathbb{E}\left[\sum_{|\alpha|=k} u_\alpha^s \right] = \Phi(s)^k \to 0,$$

so $\mathbb{E}[\max \{ u_\alpha^s : |\alpha| = k \}] \to 0$. Now, $\varepsilon_k^s = \max \{ u_\alpha^s : |\alpha| = k \}$ decreases as k increases, so by the Monotone Convergence Theorem $\mathbb{E}[\lim_k \varepsilon_k^s] = 0$, so that $\lim_k \varepsilon_k = 0$ with probability one. The cover $\bigcup_{|\alpha|=k} J_\alpha$ shows us that

$$\mathcal{H}_{\varepsilon_k}^s(K) \leq \sum_{|\alpha|=k} u_\alpha^s,$$

so that $\mathbb{E}\left[\mathcal{H}_{\varepsilon_k}^s(K) \right] \leq \Phi(s)^k$. Now, $\mathcal{H}_{\varepsilon_k}^s(K)$ increases when k increases, so by the Monotone Convergence Theorem 2.2.5 we may take the limit: $\mathbb{E}[\mathcal{H}^s(K)] = 0$. Therefore, $\mathcal{H}^s(K) = 0$ with probability one. This is true for all $s > s_0$ (and it is enough to use countably many such s), so $\dim K \leq s_0$ with probability one. ☺

(5.4.12) Exercise. Suppose $\Phi(0) < 1$. Show that $K = \varnothing$ with probability one. What can you say if $\Phi(0) = 1$?[3]

---

[3] Each node has one child on average. This could be called ZPG (zero population growth).

There is a corresponding lower bound theorem, which requires some appropriate nonoverlap condition [188]. That proof will not be done here. If $\Phi(s_0) = 1$, where s_0 is the similarity dimension, then the computation in the string space should be possible by following the martingale method from §5.3.

(5.4.13) Exercise. Suppose $\Phi(s_0) = 1$. Then with probability one, the string space $E^{(\omega)}$ has Hausdorff dimension s_0.

(5.4.14) Exercise. Investigate an upper bound for the packing dimension like the upper bound in Theorem 5.4.11.

The Poisson Process. Here is a construction for a random subset of $[0, 1]$. However, it does not exactly fit the construction method described in this section.

We begin with a parameter $n \in \mathbb{N}$. It should be chosen at random, using a Poisson distribution of mean 1. The value n is the number of points that will be in our final set K. Once n is known, divide the interval $[0, 1]$ into its two halves, $[0, 1/2]$ and $[1/2, 1]$. Now use a binomial distribution with parameters $1/2, n$ to choose k. This means we will put k of our points in the left half $[0, 1/2]$ and the remaining $n - k$ in the right half $[1/2, 1]$. Continue in this way. If an interval is supposed to have 0 points in it, throw it away. If an interval is supposed to have m points in it, divide them between the two halves using a binomial distribution with parameters $1/2, m$. At each stage, the sum of the integers associated with all of the intervals remains unchanged, but the locations of the points become more precisely known. The limiting random set K that is constructed in this way is known as the **Poisson process**.

(5.4.15) Proposition. *Write $n(i, k)$ for the number of points of K in the interval $[(i - 1)/2^k, i/2^k]$. Then for each $k \in \mathbb{N}$, the random integers $(n(i, k))_{1 \le i \le 2^k}$ are independent and all have Poisson distribution with mean 2^{-k}.*

Proof. This is proved by induction on k. For $k = 0$, it is the original construction of the number of points. Assume that the assertion is true for a given k. This means that for natural numbers $j_1, j_2, \cdots, j_{2^k}$, we have

$$\mathbb{P}\{n(1, k) = j_1, n(2, k) = j_2, \cdots, n(2^k, k) = j_{2^k}\} = \prod_{i=1}^{2^k} \frac{\exp(-2^{-k})2^{-kj_i}}{j_i!}.$$

Consider the next value, $k + 1$. Conditioned on the value of $n(i, k)$, the two numbers $n(2i - 1, k + 1)$ and $n(2i, k + 1)$ are distributed binomially with parameters $1/2, n(i, k)$, so that their sum is $n(i, k)$. That is,

$$\mathbb{P}\left[n(2i - 1, k + 1) = l_i, n(2i) = m_i \mid n(i, k) = j_i\right] = 0$$

if $l_i + m_i \ne j_i$, and

$$\mathbb{P}\left[n(2i-1,k+1)=l_i, n(2i)=m_i \mid n(i,k)=j_i\right]=\binom{j_i}{l_i}2^{-j_i}$$

if $l_i + m_i = j_i$. Thus we get for any natural numbers $l_1, \cdots, l_{2^k}, m_1, \cdots, m_{2^k}$,

$$\mathbb{P}\left(\bigcap_{i=1}^{2^k}\{n(2i-1,k+1)=l_i, n(2i,k+1)=m_i\}\right)$$

$$=\prod_{i=1}^{2^k}\frac{\exp(-2^{-k})2^{-k(l_i+m_i)}}{(l_i+m_i)!}\binom{l_i+m_i}{l_i}2^{-l_i-m_i}$$

$$=\prod_{i=1}^{2^k}\frac{\exp(-2^{-k-1})2^{-(k+1)l_i}}{l_i!}\frac{\exp(-2^{-k-1})2^{-(k+1)m_i}}{m_i!}.$$

This shows that the numbers $n(i, k+1)$ satisfy the conditions as required. ☺

(5.4.16) Proposition. *For any interval $I \subseteq [0,1]$, write $n(I)$ for the number of points of the Poisson process K that are in I. Then $n(I)$ itself has Poisson distribution with mean $\mathcal{L}(I)$. If $I_1, I_2, \cdots, I_m$ are nonoverlapping intervals, then the random variables $n(I_1), n(I_2), \cdots, n(I_m)$ are independent.*

Proof. First, suppose the intervals I_j all have dyadic rationals as endpoints. Let 2^k be the largest denominator involved in any of the endpoints. Then each $n(I_j)$ is a sum of some of the numbers $n(i,k)$. The number of terms in the sum is $\mathcal{L}(I_j)/2^{-k}$. Because the sum of independent Poisson random variables is a Poisson random variable and the means add, we conclude that $n(I_j)$ has Poisson distribution with mean $\mathcal{L}(I_j)$. The intervals $I_1, I_2, \cdots, I_m$ are nonoverlapping, so the sums of the $n(i,k)$ that contribute to each $n(I_j)$ consist of different summands. So the $n(I_j)$ are independent.

For the case where the intervals I_j need not have dyadic endpoints, approximate by intervals with dyadic endpoints and use the Dominated Convergence Theorem to pass to the limit. ☺

(5.4.17) Exercise. There is only one probability distribution on $\mathfrak{K}([0,1])$ satisfying the conditions of Proposition 5.4.16.

The full Poisson process is a random subset of the entire line $\mathbb{R}$ with the property described in Proposition 5.4.16. To construct it, let K_n be i.i.d. copies of the Poisson process in $[0,1]$ as just constructed (one for each integer $n \in \mathbb{Z}$). The set F is then obtained by using these sets K_n in each of the unit-length intervals determined by the integers:

$$F=\bigcup_{n\in\mathbb{Z}}(n+K_n).$$

Now, F is a random closed set but probably not a compact set. With probability one it has only finitely many points in any bounded interval of $\mathbb{R}$.

(5.4.18) Exercise. Let F be the Poisson process in $\mathbb{R}$. For each interval $I \subseteq \mathbb{R}$, write $n(I)$ for the number of points of F that fall in I. Then $n(I)$ has Poisson distribution with mean $\mathcal{L}(I)$. If $I_i, I_2, \cdots, I_m$ are nonoverlapping intervals, then the random variables $n(I_1), n(I_2), \cdots, n(I_m)$ are independent.

Now, how about a Poisson process in the plane $\mathbb{R}^2$? I will leave it to you to figure out a way to do it.

(5.4.19) Exercise. Construct a random closed set F in $\mathbb{R}^2$ with this property: For each bounded open set $U \subseteq \mathbb{R}^2$, write $n(U)$ for the number of points in $F \cap U$. Then $n(U)$ has Poisson distribution with mean $\mathcal{L}^2(U)$. If $U_1, U_2, \cdots, U_m$ are nonoverlapping bounded open sets, then the random variables $n(U_1), n(U_2), \cdots, n(U_m)$ are independent.

Statistically Self-Similar Measures. When a ratio vector $(a(e))_{e \in E}$ is selected, suppose we also pick a set of weights (probabilities) $(p(e))_{e \in E}$ such that $p(e) = 0$ whenever $a(e) = 0$ and $\sum_e p(e) = 1$. Let V' be the set of all such pairs (a, p) of ratio vectors with weights. Suppose $\mathcal{M}$ is a measure on V'. Select a family (a_α, p_α) of independent ratio vectors with weights using the distribution $\mathcal{M}$. Constructing the string space $E^{(\omega)}$ as before, we may define a measure $\mathcal{N}$ on it by specifying recursively $\mathcal{N}([\Lambda]) = 1$ and $\mathcal{N}([\alpha e]) = \mathcal{N}([\alpha])p_\alpha(e)$. The corresponding measure on the construction object K is what might be called a **statistically self-similar measure**.

However, we do not investigate further here.

5.5 Statistically Self-Affine Graphs

Suppose $f \colon [\alpha, \beta] \to \mathbb{R}$ is a random function, say a random element of $\mathfrak{C}([\alpha, \beta])$ or perhaps $\mathrm{Lip}([\alpha, \beta])$. Then the graph of that random function will be a random set in $\mathbb{R}^2$ (perhaps a random element of $\mathfrak{K}(\mathbb{R}^2)$). A statistically self-similar set (like a self-similar set) is unlikely to be the graph of a continuous function. But (as with the nonrandom case) there are many interesting examples where the graph is statistically self-affine.

The definition of a statistically self-affine graph will follow the definition of self-affine graph, except that some (or all) of the elements involved may be random.

First, an interval $[\alpha, \beta]$ will be the domain of our function. The endpoints α, β may be random. Next, we will need knots that partition it: $\alpha = x_0 < x_1 < \cdots < x_n = \beta$. They may be random; even their number n may be random. (In most of the examples considered here, the above will all be nonrandom.) Next, we will need vertical coordinates, $y_0, y_1, \cdots, y_n$. Here is where the random choice is most commonly seen. Finally, there will be vertical scaling coefficients d_i, possibly random. All of these data will allow the construction of an iterated function system consisting of n affine transformations $F_1, F_2, \cdots, F_n$; say F_i

has Barnsley coefficients $(a_i, 0, c_i, d_i, e_i, f_i)$. We will want $a_i x + e_i$ to map the interval $[\alpha, \beta]$ onto $[x_{i-1}, x_i]$. The coefficients c_i, f_i must be chosen such that $F_i(x_0, y_0)$ and $F_i(x_n, y_n)$ are the points (x_{i-1}, y_{i-1}) and (x_i, y_i) in the appropriate order.

The entire graph (with domain $[\alpha, \beta]$) is made up of the n parts (with domains $[x_{i-1}, x_i]$). The entire procedure is repeated on each of the parts, starting with the interval $[x_{i-1}, x_i]$ rather than the interval $[\alpha, \beta]$. Normally, the choices made for each of the parts should be independent of the others (given the items already chosen).

The exact details of how this is done will vary, depending on who is writing and what the purpose is. Let us look at a few examples.

Random Midpoint Displacement. A simple construction was proposed by Mandelbrot to model rugged terrain. (Chapter 26 of [176] is called "Random Midpoint Displacement Curves.") A one-dimensional version of that model is described here.

Begin with the domain $[0, 1]$. We intend to construct a random function $V : [0, 1] \to \mathbb{R}$. There is one parameter r for the example, $0 < r < 1$. The starting endpoints are $(0, 0), (1, 0)$; the first approximating polygon is the constant function $V_0(x) = 0$. The main step is the displacement of the midpoint. The midpoint here is at $(1/2, 0)$; we will displace it vertically by a random amount. To be explicit, we choose the vertical displacement uniformly in the interval $[-1, 1]$. The vertical scaling coefficients will both be r, but that does not appear directly in the new approximation. So our knots are at $0, 1/2, 1$, and the vertical coordinates are $y_0 = 0, y_1, y_2 = 0$, where y_1 is chosen uniformly in $[-1, 1]$. So our affine transformations are

$$F_0 \text{ with Barnsley coefficients } (1/2, 0, y_1, r, 0, 0);$$
$$F_1 \text{ with Barnsley coefficients } (1/2, 0, -y_1, r, 1/2, y_1).$$

The approximation V_1 is piecewise linear, with vertices $(0, 0), (1/2, y_1), (1, 0)$.

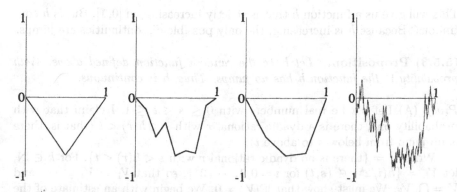

Figure 5.5.1. Random midpoint displacement.

Now we want to repeat with each of the small intervals $[0, 1/2]$, $[1/2, 1]$. The vertical scaling coefficients r now come into play: at the new midpoints $1/4, 3/4$ we choose new displacements uniformly in the interval $[-r, r]$. These displacements are chosen independent of each other (and of the previous choice y_1). The value of $V_2(1/4)$ is thus uniformly distributed in the interval $[V_1(1/4) - r, V_1(1/4) + r]$. The vertical scaling coefficients are again r. At the next stage, then, the displacements will be distributed uniformly in $[-r^2, r^2]$.

Our final result is the limit $V(x) = \lim_{k \to \infty} V_k(x)$. Of course this limit exists; in fact, the sequence V_k converges uniformly. This may be seen because (by induction) $|V_{k+1}(x) - V_k(x)| \le r^k$ and the series $\sum r^k$ converges. Figure 5.5.1 illustrates this process; here we have used $r = 3/4$.

(5.5.2) Exercise. Let $x_0 \in (0, 1)$ be given. Compute the mean $\mathbb{E}[V(x_0)]$. Compute the variance $\mathrm{Var}[V(x_0)]$. The expansion of x_0 in base 2 may be a useful ingredient for the computations.

Random Homeomorphism. A **homeomorphism** from one metric space S to another T is a bijective function $h: S \to T$ such that both h and h^{-1} are continuous. Thus, a homeomorphism of $[0, 1]$ to itself is a continuous strictly increasing (or strictly decreasing) function of $[0, 1]$ onto itself (since the inverse will necessarily also be continuous).

A "random homeomorphism" h of $[0, 1]$ was constructed in [117] as follows. Begin with $h(0) = 0$ and $h(1) = 1$. Then choose $h(1/2)$ at random, uniformly distributed in the interval $(0, 1)$. Continue as usual: If $h((i-1)/2^k)$ and $h(i/2^k)$ have been chosen, then the midpoint value $h((2i - 1)/2^{k+1})$ should be chosen at random, using uniform distribution on the interval $\big(h((i-1)/2^k), h(i/2^k)\big)$, independent of all other previous choices.

These choices lead to a function h on the set of dyadic rationals. Certainly, h is strictly increasing there. For $x \in [0, 1]$ not a dyadic rational, let

$$h(x) = \sup \{ h(r) : r < x, \ r \text{ dyadic rational} \}.$$

This will give us a function h that is strictly increasing on $[0, 1]$. But is h continuous? Because h is increasing, the only possible discontinuities are jumps.

(5.5.3) Proposition. *Let h be the random function defined above. With probability 1, the function h has no jumps. Thus, h is continuous.*

Proof. (A) Let s, t be real numbers with $0 \le s < t \le 1$. I claim that with probability one, there is a dyadic rational r with $s < h(r) < t$ (that is, there is no jump from below s to above t).

Write $V = \{$there is no dyadic rational r with $s < h(r) < t\}$. For $k \in \mathbb{N}$, let $V_k = \{h(i/2^k) \notin (s, t)$ for $i = 0, 1, \cdots, 2^k\}$, so that $V_1 \supseteq V_2 \supseteq \cdots$ and $V = \bigcap_k V_k$. We must show that $\mathbb{P}(V) = 0$. We begin with an estimate of the conditional probability $\mathbb{P}[V_{k+1} \mid V_k]$. In the event V_k, there is i with

$$h\left(\frac{i-1}{2^k}\right) \leq s < t \leq h\left(\frac{i}{2^k}\right).$$

Conditioned on this, the probability that the midpoint value $h((2i-1)/2^{k+1})$ is not in (s,t) is

$$1 - \frac{t-s}{h(i/2^k) - h((i-1)/2^k)} < 1 - \frac{t-s}{1}.$$

Thus, $\mathbb{P}\left[V_{k+1} \mid V_k\right] \leq 1 - (t-s)$. But now

$$\mathbb{P}(V_k) = \mathbb{P}(V_0)\mathbb{P}\left[V_1 \mid V_0\right]\mathbb{P}\left[V_2 \mid V_1\right] \cdots \mathbb{P}\left[V_k \mid V_{k-1}\right] \leq \left(1 - (t-s)\right)^k.$$

This converges to 0 as $k \to \infty$, so $\mathbb{P}(V) = 0$. This proves the claim.

(B) If (s,t) is a pair of rational numbers with $0 \leq s < t \leq 1$, let

$$W(s,t) = \{\text{the function } h \text{ has a jump from below } s \text{ to above } t\}.$$

As we have just seen, $\mathbb{P}(W(s,t)) = 0$. But there are countably many such pairs (s,t), and the event $W = \{\text{the function } h \text{ has a jump}\}$ is their union, so $\mathbb{P}(W) = 0$.   ☺

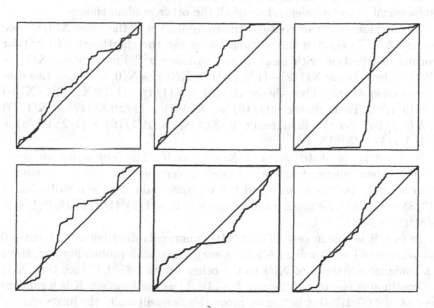

Figure 5.5.4. Random homeomorphism (6 instances).

For Figure 5.5.4 I programmed this algorithm and ran it several times. Do you think I programmed it correctly? What would you look for in the pictures as evidence that the program is correct (or incorrect)? Try programming it yourself, and see whether your pictures look similar to mine.

(5.5.5) Exercise. The random homeomorphism constructed above has a statistically self-affine graph. Is it the c.d.f. of a statistically self-similar measure in $[0, 1]$?

(5.5.6) Exercise. If h is the random homeomorphism defined here, then its inverse h^{-1} is another random homeomorphism of $[0, 1]$. Describe the random affine iterated function system that describes the graph of h^{-1}. Does h^{-1} have the same distribution as h?

Brownian Motion. The "Brownian motion" function will be discussed more fully in Section 5.6. But here we construct the basic version of it as a random function with statistically self-similar graph. We will use a "midpoint displacement" method like the one discussed above.

We begin by constructing Brownian motion in $\mathbb{R}$ with parameter set $[0, 1]$. It will be a random function $\mathbf{X} \colon [0, 1] \to \mathbb{R}$. Begin by specifying $\mathbf{X}(0) = 0$. Then choose $\mathbf{X}(1)$ at random using a normal distribution with mean 0 and variance 1. After $\mathbf{X}((i-1)/2^k)$ and $\mathbf{X}(i/2^k)$ have been chosen, we choose the midpoint value $X((2i-1)/2^{k+1})$ at random using a normal distribution with mean $(1/2)(\mathbf{X}((i-1)/2^k) + \mathbf{X}(i/2^k))$ and variance $1/2^{k+2}$. Or we may think of this as a midpoint displacement with mean 0 and variance $1/2^{k+2}$. This displacement is to be independent of all the other random choices.

As an example, let us consider the computation of the value $\mathbf{X}(3/8)$. See Figure 5.5.7. To shorten the writing, let us use the shorthand $N(\mu, \sigma^2)$ for "normal distribution with mean μ and variance σ^2." First, choose $\mathbf{X}(1)$ as $N(0, 1)$. Then choose $\mathbf{X}(1/2) - (1/2)(\mathbf{X}(1) + \mathbf{X}(0))$ as $N(0, 1/4)$. Call this random variable $A(1/4)$. Then choose $A(1/8) = \mathbf{X}(1/4) - (1/2)(\mathbf{X}(1/2) + \mathbf{X}(0))$ as $N(0, 1/8)$. Then choose $A(1/16) = \mathbf{X}(3/8) - (1/2)(\mathbf{X}(1/2) + \mathbf{X}(1/4))$ as $N(0, 1/16)$. So the final result is $\mathbf{X}(3/8) = A(1/16) + (1/2)A(1/8) + (3/4)A(1/4) + (3/8)\mathbf{X}(1)$.

So what is the distribution of $\mathbf{X}(3/8)$ itself? The four terms are independent; when independent normal random variables are added, the sum is again normal, the means add, and the variances add. So the distribution of $\mathbf{X}(3/8)$ is normal with mean 0 and variance $1/16 + (1/2)^2(1/8) + (3/4)^2(1/4) + (3/8)^2(1) = 3/8$.

In fact, it is true in general that $\mathbf{X}(t)$ is normally distributed with mean 0 and variance t (see Corollary 5.5.15 below). Also, with probability one, there is a (unique) extension of $\mathbf{X}(t)$ to the other values $t \in [0, 1]$ such that $\mathbf{X}(t)$ is a continuous function (Theorem 5.5.13). Thus the function $\mathbf{X}$ is a random element of $\mathfrak{C}([0, 1])$. But before we prove this, we will study the increments of the function defined so far.

A **dyadic rational** is a number of the form $i/2^k$, where $i \in \mathbb{Z}, k \in \mathbb{N}$. At this point, we are interested in dyadic rationals in $[0, 1]$, so the restrictions are $k \geq 0, 0 \leq i \leq 2^k$. The **increment** of $\mathbf{X}$ on the interval $[s, t]$ is the difference $\mathbf{X}(t) - \mathbf{X}(s)$. Two intervals are **nonoverlapping** iff their intersection is at most one point.

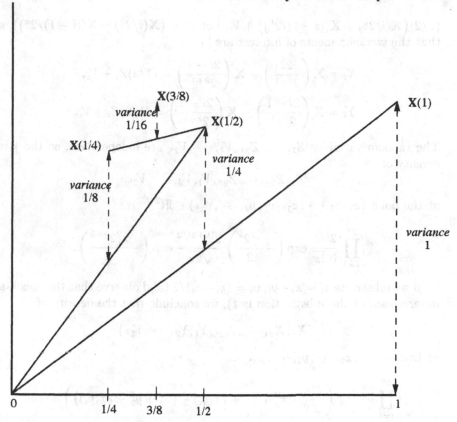

Figure 5.5.7. Computation of $\mathbf{X}(3/8)$.

(5.5.8) Theorem. *Let $\mathbf{X}(t)$ be defined for dyadic rationals $t \in [0,1]$ as above. Then for every $k \in \mathbb{N}$, the increments*

$$\mathbf{X}\left(\frac{1}{2^k}\right) - \mathbf{X}\left(\frac{0}{2^k}\right), \mathbf{X}\left(\frac{2}{2^k}\right) - \mathbf{X}\left(\frac{1}{2^k}\right), \cdots ,$$

$$\mathbf{X}\left(\frac{i}{2^k}\right) - \mathbf{X}\left(\frac{i-1}{2^k}\right), \cdots , \mathbf{X}\left(\frac{2^k}{2^k}\right) - \mathbf{X}\left(\frac{2^k-1}{2^k}\right)$$

are independent, and each of these random variables has normal distribution with mean 0 and variance $1/2^k$.

Proof. The proof proceeds by induction on k. The case $k = 0$ states merely that $\mathbf{X}(1) - \mathbf{X}(0)$ is normally distributed with mean 0 and variance 1; this is the definition of $\mathbf{X}(1)$.

Now suppose the statement is true for a certain value of k. Consider dyadic rationals with denominator $k + 1$. The interval $[(i - 1)/2^k, i/2^k]$ is subdivided into two parts. Write V_i for the new random variable with $N(0, 1/2^{k+2})$ distribution, so that the midpoint value is $\mathbf{X}((2i - 1)/2^{k+1}) =$

$(1/2)\big(\mathbf{X}(i/2^k) + \mathbf{X}((i-1)/2^k)\big) + V_i$. Let $Z_i = \big(\mathbf{X}(i/2^k) - \mathbf{X}((i-1)/2^k)\big)$, so that the two increments of interest are

$$X_i = \mathbf{X}\left(\frac{2i}{2^{k+1}}\right) - \mathbf{X}\left(\frac{2i-1}{2^{k+1}}\right) = (1/2)Z_i - V_i,$$

$$Y_i = \mathbf{X}\left(\frac{2i-1}{2^{k+1}}\right) - \mathbf{X}\left(\frac{2i-2}{2^{k+1}}\right) = (1/2)Z_i + V_i.$$

The random variables $Z_1, \cdots, Z_{2^k}, V_1, \cdots, V_{2^k}$ are independent, so the joint density of

$$Z_1, Z_2, \cdots, Z_{2^k}, V_1, V_2, \cdots, V_{2^k}$$

at the point $(z_1, z_2, \cdots, z_{2^k}, v_1, v_2, \cdots, v_{2^k}) \in \mathbb{R}^{2 \cdot 2^k}$ is

$$\prod_{i=1}^{2^k} \frac{2^{k/2}}{\sqrt{2\pi}} \exp\left(-\frac{z_i^2 2^k}{2}\right) \frac{2^{(k+2)/2}}{\sqrt{2\pi}} \exp\left(-\frac{v_i^2 2^{k+2}}{2}\right).$$

So if we substitute $z_i = x_i + y_i$, $v_i = (y_i - x_i)/2$ (and observe that the Jacobian determinant of the substitution is 1), we conclude that the density of

$$(X_1, X_2, \cdots, X_{2^k}, Y_1, Y_2, \cdots, Y_{2^k})$$

at $(x_1, x_2, \cdots, x_{2^k}, y_1, y_2, \cdots, y_{2^k})$ is

$$\prod_{i=1}^{2^k} \frac{2^k}{\pi} \exp\left(\frac{2^k}{2}\big(-(x_i^2 + y_i^2 + 2x_iy_i) - (x_i^2 + y_i^2 - 2x_iy_i)\big)\right)$$

$$= \prod_{i=1}^{2^k} \frac{2^{(k+1)/2}}{\sqrt{2\pi}} \exp\left(-\frac{x_i^2 2^{k+1}}{2}\right) \frac{2^{(k+1)/2}}{\sqrt{2\pi}} \exp\left(-\frac{y_i^2 2^{k+1}}{2}\right).$$

Thus these $2 \cdot 2^k$ random variables are independent with normal distribution of mean 0 and variance $1/2^{k+1}$. This completes the inductive proof.   ☺

(5.5.9) Corollary. *The increments of the function $\mathbf{X}$ on nonoverlapping intervals (with dyadic endpoints) are independent. An increment $\mathbf{X}(t) - \mathbf{X}(s)$ is normally distributed with mean 0 and variance $t - s$.*

Proof. Let $I_1, I_2, \cdots, I_n$ be nonoverlapping intervals with dyadic endpoints. Let 2^k be the largest denominator involved in any of the endpoints. Then each increment on one of the intervals I_j is the sum of increments $Z_i = \mathbf{X}(i/2^k) - \mathbf{X}((i-1)/2^k)$. Since the I_j are nonoverlapping, the corresponding increments involve disjoint sets of the Z_i's. But we know that the Z_i's are independent, so these sums are also independent.

The increment on an interval $[s, t]$ is the sum of independent increments Z_i with mean zero and variances that add to $t - s$. So the increment on $[s, t]$ has mean zero and variance $t - s$.   ☺

Again write D for the set of dyadic rationals in $[0, 1]$.

(5.5.10) Corollary. (a) *For any* $\lambda > 0$,

$$\mathbb{P}\left\{\sup_{\substack{0 \leq t \leq 1 \\ t \in D}} |\mathbf{X}(t)| > \lambda\right\} \leq \frac{1}{\lambda^2}.$$

(b) $\sup\{\,|\mathbf{X}(t)| : 0 \leq t \leq 1, t \in D\,\}$ *is an integrable random variable.*

Proof. (a) Write

$$A = \left\{\sup_{\substack{0 \leq t \leq 1 \\ t \in D}} |\mathbf{X}(t)| > \lambda\right\}$$

and for $k \in \mathbb{N}$ write

$$A_k = \left\{\sup_{0 \leq i \leq 2^k} \left|\mathbf{X}\left(\frac{i}{2^k}\right)\right| > \lambda\right\}.$$

Then $A_k \nearrow A$ as $k \to \infty$. Now, the values $\mathbf{X}(i/2^k)$ are the sums of a family of increments with mean 0 and variance $1/2^k$. So by the maximal inequality (4.3.11), we have

$$\mathbb{P}(A_k) \leq \frac{1}{\lambda^2} \sum_{i=1}^{2^k} \frac{1}{2^k} = \frac{1}{\lambda^2}.$$

Therefore, $\mathbb{P}(A) = \lim_k \mathbb{P}(A_k) \leq 1/\lambda^2$.

(b) Let $T = \sup\{\,|\mathbf{X}(t)| : 0 \leq t \leq 1, t \in D\,\}$. Then

$$\mathbb{E}[T] = \int_0^\infty \mathbb{P}\{T > \lambda\}\, d\lambda \leq \int_0^1 1\, d\lambda + \int_1^\infty \frac{1}{\lambda^2}\, d\lambda < \infty. \qquad \text{☺}$$

We next consider the reason that the definition of $\mathbf{X}(t)$ may be extended continuously to the rest of the values $t \in [0, 1]$. Our first attempt, although very natural, has a hole in the reasoning.

(5.5.11) Proposition. *Let* $x \in (0, 1)$. *Write* D *for the set of dyadic rationals in* $[0, 1]$. *Then with probability one, the limits*

$$\lim_{t \nearrow x, t \in D} \mathbf{X}(t), \qquad \lim_{t \searrow x, t \in D} \mathbf{X}(t)$$

exist and are equal. If x *is a dyadic rational, these limits are both* $\mathbf{X}(x)$.

Proof (?). If t_n is a sequence of dyadic rationals with $t_n \nearrow x$, then (write $t_0 = 0$ for convenience) the increments $\mathbf{X}(t_n) - \mathbf{X}(t_{n-1})$ are independent, with mean zero, and with summable variances $\sum_n (t_n - t_{n-1}) < \infty$. So by Corollary 4.3.12, $\mathbf{X}(t_n)$ converges. This is true for any sequence increasing to x, so $\lim_{t \nearrow x, t \in D} \mathbf{X}(t)$ exists. Similar reasoning applies to a sequence t'_n decreasing to x. If $t_n \nearrow x$ and $t'_n \searrow x$, then $\mathbf{X}(t'_n) - \mathbf{X}(t_n) \to 0$ by Proposition

4.2.22. This shows that the two limits agree. If x is a dyadic rational, repeat this with $\mathbf{X}(t_n') - \mathbf{X}(x)$ and $\mathbf{X}(x) - \mathbf{X}(t_n)$ to show that the limits are $\mathbf{X}(x)$.

☺

(5.5.12) Exercise. There are questionable elements of the preceding proof. The exceptional events of measure zero cause problems in the reasoning. (a) Give an example of a random function $h \colon [0,1] \to \mathbb{R}$ such that

$$\text{for every } x \in (0,1),\ \mathbb{P}\{h \text{ is continuous at } x\} = 1$$

but

$$\mathbb{P}\{\text{for every } x \in (0,1),\ h \text{ is continuous at } x\} = 0.$$

(b) Give an example of a random function $h \colon [0,1] \to \mathbb{R}$ such that

$$\text{for every sequence } x_n \nearrow 1,\ \mathbb{P}\{\lim_n h(x_n) = 0\} = 1$$

but

$$\mathbb{P}\{\lim_{x \nearrow 1} h(x) = 0\} = 0.$$

Because of this difficulty with the events of probability 0, we will need to work a little harder to get continuity. In fact, if we do it right, we can prove a Hölder condition at the same time.

(5.5.13) Theorem. *Let* $\big(\mathbf{X}(t)\big)_{t \in D}$ *be the random function defined above. Define* $\mathbf{X}(t) = \limsup_{t \nearrow x} \mathbf{X}(t)$ *for* $x \in [0,1] \setminus D$. *With probability one, this extension satisfies Hölder conditions of all orders* $< 1/2$: *that is, for any* $\gamma < 1/2$, *there are constants* C *and* $\varepsilon > 0$ *such that* $|\mathbf{X}(t) - \mathbf{X}(s)| \le C|t - s|^\gamma$ *for all* s *and* t *with* $0 \le s \le t \le 1$, $t - s < \varepsilon$.

Proof. (a) For $m \in \mathbb{N}$, we will prove a Hölder condition of order $\gamma = 1/2 - 1/m$. (If such a Hölder condition holds for dyadic rationals s, t, then the same condition holds for the extension to the whole interval $[0,1]$, and that extension is therefore continuous.) For the standard normal distribution $\mathbf{X}(1)$, the moment $M = \mathbb{E}\big[|\mathbf{X}(1)|^{2m}\big]$ is finite; if $s < t$ are dyadic rationals, then the increment $\mathbf{X}(t) - \mathbf{X}(s)$ has the same distribution as $\sqrt{t - s}\,\mathbf{X}(1)$, so

$$\mathbb{E}\big[|\mathbf{X}(t) - \mathbf{X}(s)|^{2m}\big] = M|t - s|^m.$$

Write $\alpha = m - 1$ and $\beta = 2m$, so that

$$\mathbb{E}\big[|\mathbf{X}(t) - \mathbf{X}(s)|^\beta\big] \le M|t - s|^{1+\alpha}$$

and $\gamma < \alpha/\beta$. This is what we will use in the remainder of the proof. (It is known as **Kolmogorov's continuity criterion.**)

(b) Now, $\mathbb{P}\{|Y| \ge \lambda\} \le (1/\lambda^\beta)\mathbb{E}\big[|Y|^\beta\big]$ for any Y. Let $\delta > 0$. Fix $k \in \mathbb{N}$, and let $R = \big\{(i,j) \in \mathbb{N} \times \mathbb{N} : 0 \le i, j \le 2^k, 0 < j - i \le 2^{k\delta}\big\}$. The number of elements in R is at most $2^k 2^{k\delta}$. Now,

$$\mathbb{P}\left\{\left|\mathbf{X}\left(\frac{j}{2^k}\right) - \mathbf{X}\left(\frac{i}{2^k}\right)\right| > \left(\frac{j-i}{2^k}\right)^{\gamma} \text{ for some } (i,j) \in R\right\}$$

$$\leq \sum_{(i,j)\in R} \left(\frac{j-i}{2^k}\right)^{-\beta\gamma} \mathbb{E}\left[\left|\mathbf{X}\left(\frac{j}{2^k}\right) - \mathbf{X}\left(\frac{i}{2^k}\right)\right|^{\beta}\right]$$

$$\leq \sum_{(i,j)\in R} M\left(\frac{j-i}{2^k}\right)^{-\beta\gamma+1+\alpha}$$

$$\leq M2^k 2^{k\delta}\left(2^{k\delta}2^{-k}\right)^{-\beta\gamma+1+\alpha}$$

$$= M2^{-k\tau},$$

where $\tau = (1-\delta)(-\beta\gamma+1+\alpha) - (1+\delta)$. Now, since $\gamma < \alpha/\beta$, we have $-\beta\gamma+1+\alpha > 1$, so for δ small enough, $\tau > 0$. Then the series $\sum_k M2^{-k\tau}$ converges, so by the Borel–Cantelli Lemma, with probability one, there is $K \in \mathbb{N}$ such that for all $k > K$ and all i, j with $0 < j - i \leq 2^{k\delta}$,

$$\left|\mathbf{X}\left(\frac{j}{2^k}\right) - \mathbf{X}\left(\frac{i}{2^k}\right)\right| \leq \left(\frac{j-i}{2^k}\right)^{\gamma}.$$

(c) Now we would be finished, except that the value of ε we have obtained, $2^{k\delta}/2^k$, depends on k. So lastly, we must find an ε independent of k. To do this, we will need to compensate with a constant C.

Let $\varepsilon = 2^{-(1-\delta)K}$. Let s, t be dyadic rationals with $0 < t - s < \varepsilon$. Pick $k \in \mathbb{N}$ such that

$$2^{-(k+1)(1-\delta)} \leq t - s < 2^{-k(1-\delta)}.$$

Note that $k \geq K$. Now choose i and j such that

$$\frac{i-1}{2^k} < s \leq \frac{i}{2^k} \leq \frac{j}{2^k} \leq t < \frac{j+1}{2^k}.$$

Now, $j/2^k - i/2^k \leq t - s$, so

$$\left|\mathbf{X}\left(\frac{j}{2^k}\right) - \mathbf{X}\left(\frac{i}{2^k}\right)\right| \leq \left(\frac{j-i}{2^k}\right)^{\gamma} \leq (t-s)^{\gamma}.$$

Next,

$$t = \frac{j}{2^k} + \frac{1}{2^{k_1}} + \frac{1}{2^{k_2}} + \cdots + \frac{1}{2^{k_p}}$$

for some integers $k < k_1 < k_2 < \cdots < k_p$. So, for certain constants C_1 and C_2,

$$\left|\mathbf{X}(t) - \mathbf{X}\left(\frac{j}{2^k}\right)\right| \leq \sum_{q=1}^{p}\left(\frac{1}{2^{k_q}}\right)^{\gamma} \leq \sum_{q=k+1}^{\infty}\left(\frac{1}{2^q}\right)^{\gamma} \leq C_1 2^{-\gamma(k+1)}$$

$$\leq C_1 2^{-\gamma(k+1)(1-\delta)} = C_2(t-s)^{\gamma}.$$

In the same way, for some integers $k < k_1 < k_2 < \cdots < k_p$,

$$s = \frac{i}{2^k} - \frac{1}{2^{k_1}} - \frac{1}{2^{k_2}} - \cdots - \frac{1}{2^{k_p}},$$

and we get

$$\left| \mathbf{X}\left(\frac{i}{2^k}\right) - \mathbf{X}(s) \right| \leq C_3 (t - s)^\gamma.$$

So, combining these inequalities, we obtain $|\mathbf{X}(t) - \mathbf{X}(s)| \leq C(t - s)^\gamma$, for a certain constant C.   ☺

To summarize, we have a random function $\mathbf{X} \in \mathfrak{C}([0, 1])$ with a statistically self-affine graph. See Figure 5.5.14—three instances have been generated. To illustrate the self-affinity, imagine dividing one of these into two parts. The first part where the x-coordinate belongs to $[0, 1/2]$ and the second part where the x-coordinate belongs to $[1/2, 1]$. If we take one of these parts, magnify it horizontally by q factor of 2 and vertically by a factor of $\sqrt{2}$, and translate it so that the left-hand endpoint is at $(0, 0)$, then we will have another instance of the original random graph. In addition, these two magnified and translated graphs are independent of each other.

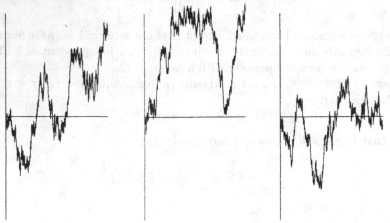

Figure 5.5.14. One-dimensional Brownian motion on $[0, 1]$.

(5.5.15) Corollary.   *There is a random function* $\mathbf{X} \colon [0, 1] \to \mathbb{R}$ *such that:*

(a) $\mathbf{X}(0) = 0$.

(b) *If* $0 \leq t_0 < t_1 < \cdots < t_n \leq 1$, *then the random variables* $\mathbf{X}(t_1) - \mathbf{X}(t_0)$, $\mathbf{X}(t_2) - \mathbf{X}(t_1)$, $\cdots$, $\mathbf{X}(t_n) - \mathbf{X}(t_{n-1})$ *are independent.*

(c) *If* $0 \leq s < t \leq 1$, *then the random variable* $\mathbf{X}(t) - \mathbf{X}(s)$ *is normally distributed with mean* 0 *and variance* $t - s$.

(d) *The function* $\mathbf{X}$ *is continuous.*

Proof. The function $\mathbf{X}$ we want is the one constructed above, extended to all $t \in [0, 1]$ as in Theorem 5.5.13. We have already proved (a) and (d).

(c) Suppose $0 \leq s < t \leq 1$ are not dyadic rationals. Choose dyadic rationals s_n, t_n with $s_n \to s$, $t_n \to t$. Then (with probability one) $\mathbf{X}(s_n) \to \mathbf{X}(s)$ and $\mathbf{X}(t_n) \to \mathbf{X}(t)$. For any $\lambda \in \mathbb{R}$,

$$\mathbb{P}\{\mathbf{X}(t) - \mathbf{X}(s) < \lambda\} \leq \lim_{n \to \infty} \mathbb{P}\{\mathbf{X}(t_n) - \mathbf{X}(s_n) < \lambda\}$$

$$= \lim_{n \to \infty} \int_{-\infty}^{\lambda} \frac{1}{\sqrt{2\pi(t_n - s_n)}} \exp\left(-\frac{x^2}{2(t_n - s_n)}\right) dx$$

$$= \int_{-\infty}^{\lambda} \frac{1}{\sqrt{2\pi(t - s)}} \exp\left(-\frac{x^2}{2(t - s)}\right) dx.$$

Similarly,

$$\mathbb{P}\{\mathbf{X}(t) - \mathbf{X}(s) \leq \lambda\} \geq \int_{-\infty}^{\lambda} \frac{1}{\sqrt{2\pi(t - s)}} \exp\left(-\frac{x^2}{2(t - s)}\right) dx.$$

These inequalities are true for all λ, so in fact we have equality. Thus $\mathbf{X}(t) - \mathbf{X}(s)$ has normal distribution with mean 0 and variance $t - s$.

(b) Approximate by dyadic rationals again. Independence in this case follows from the case of dyadic rationals using the Dominated Convergence Theorem. ☺

Brownian Bridge. Let us consider briefly a slight variant of the construction just treated. The "Brownian bridge" is a random function, constructed in almost the same way as before. We start with $\mathbf{Y}(0) = 0$ and $\mathbf{Y}(1) = 0$. Then we proceed in exactly the same way as before.

When we compare this to the random function $\mathbf{X}$ constructed before, we see that the random function $\mathbf{Y}$ may be obtained in two different ways from the random function $\mathbf{X}$. One way to think of it is this: $\mathbf{Y}(t) = \mathbf{X}(t) - t\mathbf{X}(1)$. This is because the term $t\mathbf{X}(1)$ is the first approximation to $\mathbf{X}$, and subtracting that part but leaving all the rest (due to the midpoint displacements) will match the description of $\mathbf{Y}$. The second way to think of this is as a conditional distribution:

$$\mathcal{D}_{\mathbf{Y}} = \mathcal{D}_{\mathbf{X}|\mathbf{X}(1)=0}.$$

Once we know how the first step (choice of $\mathbf{X}(1)$) comes out, the rest is done as before, so if we postulate that $\mathbf{X}(1) = 0$, the result is the Brownian bridge.

So we conclude that the two descriptions are the same. If $\mathbf{X}$ is the standard Brownian motion constructed here, then the random function $\mathbf{X}(t) - t\mathbf{X}(1)$ has the distribution $\mathcal{D}_{\mathbf{X}|\mathbf{X}(1)=0}$.

What is the statistical self-similarity for the Brownian bridge? The first half (or the second half) is to be magnified horizontally by a factor of 2, vertically by a factor of $\sqrt{2}$, translated so that the left-hand endpoint is at $(0,0)$, then vertically skewed so that the right-hand endpoint is at $(1,0)$. In Figure 5.5.16 the Brownian bridge is shown, together with the rectangle $[0,1] \times [-1,1]$; then the two parts are shown with corresponding affine images of that rectangle.

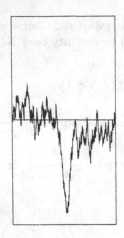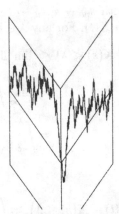

Figure 5.5.16. A Brownian bridge and its statistical division in two.

(5.5.17) Exercise. As we can see in one of the pictures, the Brownian bridge does not necessarily stay within the rectangle $[0, 1] \times [-1, 1]$. Estimate the probability that it stays within the rectangle.

Lévy Flights.* Next is another example that may be thought of as a statistically self-affine graph. We want to imitate the ideas for Brownian motion, but replace the normal distribution with a Lévy stable distribution. Mathematics texts call this sort of random function a **(Lévy) stable process**. Mandelbrot's picturesque term "Lévy flights" has become common in the physics literature.[4]

Here, we will discuss only one particular example, where the Lévy stable distribution used is the Cauchy distribution. (So it is a "symmetric stable process," not an "increasing stable process.") But the drawback (and the main difference from the Brownian motion case) is that a random variable with Cauchy distribution is not integrable.

(5.5.18) Definition. The **Cauchy flights** function should be a random function $\mathbf{Y} \colon [0, 1] \to \mathbb{R}$ with independent stationary Cauchy-distributed increments. We want (1) $\mathbf{Y}(0) = 0$; (2) for $0 \leq t_0 \leq t_1 \leq \cdots \leq t_n \leq 1$, the increments $\mathbf{Y}(t_1) - \mathbf{Y}(t_0)$, $\mathbf{Y}(t_2) - \mathbf{Y}(t_1)$, $\cdots$, $\mathbf{Y}(t_n) - \mathbf{Y}(t_{n-1})$, are independent; (3) for $0 \leq s \leq t \leq 1$, the increment $\mathbf{Y}(t) - \mathbf{Y}(s)$ has Cauchy distribution with parameters $0, t - s$.

This function may be constructed using midpoint displacements. But note (4.3.22) that the distribution to be used at the midpoints is not itself a Cauchy

---

* The rest of §5.5 is optional material.
[4] But perhaps it is not always used consistently. Sometimes the name "Lévy flights" means the increasing stable process; sometimes it means the symmetric stable process; and other times it may mean any stable process at all.

distribution. To prepare for the construction, let us recall a few details and fix some notation. The standard Cauchy distribution has density

$$\gamma_1(z) = \frac{1}{\pi(1 + z^2)},$$

and the Cauchy distribution with parameters $0, 1/2$ has density

$$\gamma_{1/2}(x) = \frac{1/2}{\pi(1/4 + x^2)} = \frac{2}{\pi(1 + 4x^2)}.$$

If X and Y are independent random variables with density $\gamma_{1/2}$, then $Z = X + Y$ has density γ_1; the conditional distribution $\mathcal{D}_{X|Z=z}$ has density

$$\theta_z(x) = \frac{z + 2x}{\pi z(1 + 4x^2)} + \frac{3z - 2x}{\pi z(1 + 4(z - x)^2)}.$$

Also note the symmetry, $\theta_z(x) = \theta_z(z - x)$, so that $\mathcal{D}_{Y|Z=z}$ has density $\theta_z(y)$. The joint distribution of (X, Z) has density

$$\theta_z(x)\gamma_1(z).$$

(5.5.19) **Exercise.** Verify the following:

(a) $\displaystyle\int_{-\infty}^{\infty} \theta_z(x)\gamma_1(z)\, dz = \gamma_{1/2}(x).$

(b) $\displaystyle\int_{-\infty}^{\infty} \theta_z(z - y)\gamma_1(z)\, dz = \gamma_{1/2}(y).$

(c) $\theta_{x+y}(x)\gamma_1(x + y) = \gamma_{1/2}(x)\gamma_{1/2}(y).$

Here is our construction of the Cauchy flights random function $\mathbf{Y}\colon [0, 1] \to \mathbb{R}$. We begin with a recursive construction on the dyadic rationals. Begin with $\mathbf{Y}(0) = 0$. Generate a standard Cauchy random variable and call it $\mathbf{Y}(1)$. Recursively, suppose that $\mathbf{Y}((i - 1)/2^k)$ and $\mathbf{Y}(i/2^k)$ have been defined. Write $V = \mathbf{Y}(i/2^k) - \mathbf{Y}((i - 1)/2^k)$ for the increment, and then generate a random variable T with density $\theta_{2^k V}$, independent of all random choices other than V, and define the midpoint value $\mathbf{Y}((2i - 1)/2^{k+1}) = \mathbf{Y}((i - 1)/2^k) + 2^{-k}T$. This completes the recursive definition of $\mathbf{Y}(t)$ for dyadic rationals t. For other values $x \in [0, 1]$, let

$$\mathbf{Y}(x) = \limsup_{\substack{t \searrow x \\ t \text{ dyadic rational}}} \mathbf{Y}(t).$$

(5.5.20) **Proposition.** *The random function* $\mathbf{Y}$ *defined here satisfies:* (a) *increments on nonoverlapping dyadic intervals are independent;* (b) *increments* $\mathbf{Y}(i/2^k) - \mathbf{Y}((i - 1)/2^k)$ *have Cauchy distribution with parameters* $0, 1/2^k$.

Proof. (a) Independence of an infinite set of random variables is by definition independence of all finite subsets. So we may assume that we have only finitely many intervals. Let $I_1, I_2, \cdots, I_n$ be nonoverlapping dyadic intervals. Let 2^k be the largest denominator among the endpoints of these intervals. Each of the intervals is a nonoverlapping union of dyadic intervals of length $1/2^k$, and increments on the longer intervals are sums of increments on these short intervals. So it is enough to prove the independence under the assumption that all intervals are of the form $I_i = \left[(i-1)/2^k, i/2^k\right]$ for a fixed value of k.

This independence is proved by induction on k. For $k = 0$, independence of a set with one element is automatic. Suppose, for a certain value of k, that the increments on the intervals $\left[(i-1)/2^k, i/2^k\right]$ are independent. Write $V_i = \mathbf{Y}(i/2^k) - \mathbf{Y}((i-1)/2^k)$, and T_i for the new random variable added in the definition of $\mathbf{Y}((2i-1)/2^{k+1})$, so that T_i has conditional density $\theta_{2^k V_i}$ and the new increments are $X_i = \mathbf{Y}((2i-1)/2^{k+1}) - \mathbf{Y}((i-1)/2^k) = 2^{-k} T_i$ and $Y_i = \mathbf{Y}(i/2^k) - \mathbf{Y}((2i-1)/2^{k+1}) = V_i - 2^{-k} T_i$. By the induction hypothesis the V_i are independent. The joint density of

$$\left(X_1, X_2, \cdots, X_{2^k}, V_1, V_2, \cdots, V_{2^k}\right)$$

at the point $(x_1, x_2, \cdots, x_{2^k}, v_1, v_2, \cdots, v_{2^k}) \in \mathbb{R}^{2 \cdot 2^k}$ is

$$\prod_{i=1}^{2^k} 2^k \theta_{2^k v_i}(2^k x_i)\, 2^k \gamma_1(2^k v_i).$$

So (substitute $v_i = x_i + y_i$) the joint density of

$$\left(X_1, X_2, \cdots, X_{2^k}, Y_1, Y_2, \cdots, Y_{2^k}\right)$$

at $(x_1, x_2, \cdots, x_{2^k}, y_1, y_2, \cdots, y_{2^k})$ is

$$\prod_{i=1}^{2^k} 2^k \theta_{2^k(x_i+y_i)}(2^k x_i)\, 2^k \gamma_1(2^k(x_i+y_i)) = \prod_{i=1}^{2^k} 2^{k+1} \theta_1(2^{k+1} x_i)\, 2^{k+1} \theta_1(2^{k+1} y_i).$$

Thus, these $2 \cdot 2^k$ random variables are independent.

Assertion (b) is also proved by induction on k. For $k = 0$, we require that $\mathbf{Y}(1) - \mathbf{Y}(0)$ has Cauchy distribution with parameters $0, 1$; this is the definition of $\mathbf{Y}(1)$. Suppose the result is true for a certain value of k. Consider the two sub-intervals of a dyadic interval $\left[(i-1)/2^k, i/2^k\right]$. In the notation used in the definition, the two increments are $\mathbf{Y}((2i-1)/2^{k+1}) - \mathbf{Y}((i-1)/2^k) = 2^{-k} T$ and $\mathbf{Y}(i/2^k) - \mathbf{Y}((2i-1)/2^{k+1}) = V - 2^{-k} T$. Now, T was chosen with conditional density $\theta_{2^k V}$, where by induction $2^k V$ has standard Cauchy distribution, so T itself (not conditioned) has density

$$\int_{-\infty}^{\infty} \theta_z(t)\gamma_1(z)\, dz = \gamma_{1/2}(t).$$

Thus T has Cauchy distribution with parameters $0, 1/2$, and the increment $2^{-k} T$ has Cauchy distribution with parameters $0, 1/2^{k+1}$. In the same way, the other half $2^k V - T$ has density

$$\int_{-\infty}^{\infty} \theta_z(z-y)\gamma_1(z)\, dz = \gamma_{1/2}(y).$$

Therefore, $V - 2^{-k}T$ also has Cauchy distribution with parameters $0, 1/2^{k+1}$. This completes the proof by induction.   ☺

(5.5.21) Exercise. Complete the proof that the random function $\mathbf{Y}$ constructed above satisfies the conditions in Definition 5.5.18.

The continuity argument does not work this time, however. In fact, with probability one, the function $\mathbf{Y}$ has jumps. In fact, the jumps occur at a dense set of values in $[0, 1]$. (That set of values is random: each given number in $[0, 1]$ occurs as a jump only with probability 0.) We can think of the graph as being nothing but jumps (and their limit points). These jumps are the "flights" in the picturesque name for this random function. We think of a series of flights from one locality to another; the period of time the function stays within a local area is called a "sojourn." But within each sojourn (if we magnify it) we will see that it contains smaller flights, interspersed with smaller sojourns. And so on.

Figure 5.5.22 is meant to illustrate Cauchy flights. If we plot only the function values, we get a totally disconnected set in the plane. We can think of the entire picture as consisting of two parts: the part with $0 \le x \le 1/2$ and the part with $1/2 \le x \le 1$. If each part is enlarged (vertically and horizontally) by a factor of 2 and translated so that the left-hand endpoint is at $(0,0)$, then each part will be another instance of the original Cauchy flight. (This graph is statistically self-similar, not merely statistically self-affine.)

The instance shown in the figure was chosen from several that were generated by my computer: many of the instances generated by the computer

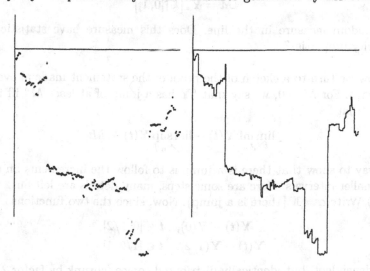

Figure 5.5.22. Cauchy flights.

contained sojourns so far above or below 0 that the pictures were mostly blank. I suppose this is because the flights making up this graph have *infinite* expected size. In the figure I have shown first (an approximation to) the graph of the actual function. Then I have plotted only the flights (the jumps in the function). Each jump is represented as a vertical line segment. The statistically self-similar set in the plane consists of the endpoints of this countable set of line segments, together with their accumulation points.

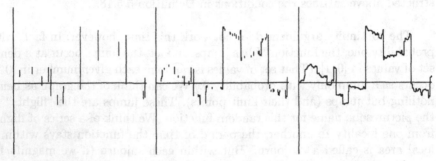

Figure 5.5.23. Flights of various sizes.

In Figure 5.5.23 I have taken (another instance of) the Cauchy flights function and graphed in turn (a) all the flights of size at least 0.04; (b) all the flights of size at least 0.01; (c) all the flights.

(5.5.24) Exercise. Cauchy flights (and Lévy flights) yield random measures: If $\mathbf{Y}$ is the random function defined here and $\mathcal{L}{\upharpoonright}[0,1]$ is the restriction of Lebesgue measure to $[0,1]$, then the image measure

$$\mathcal{M} = \mathbf{Y}_{*}[\mathcal{L}{\upharpoonright}[0,1]]$$

is a random measure in the line. Does this measure have statistical self-similarity properties?

Now we turn to a sketch of the proof of the statement made above about the jumps. For $M > 0$, we say that "$\mathbf{Y}$ has a jump of at least M" iff there is t_0 such that

$$\liminf_{t \searrow t_0} \mathbf{Y}(t) - \limsup_{t \nearrow t_0} \mathbf{Y}(t) \geq M.$$

One way to show that there is a jump is to follow the increments on smaller and smaller intervals. Here are some steps; many details are left out.

(a) Write $\alpha = \mathbb{P}\{\text{there is a jump}\}$. Now, since the two functions

$$\mathbf{Y}(t) - \mathbf{Y}(0), \quad t \in [0, 1/2],$$
$$\mathbf{Y}(t) - \mathbf{Y}(1/2), \quad t \in [1/2, 1]$$

are independent, but identically distributed copies (shrunk by factor 2) of the whole thing, we have $\mathbb{P}\{\text{there is a jump in } [0, 1/2]\} = \alpha$ and $\mathbb{P}\{\text{there is a jump}$

in $[1/2, 1]\} = \alpha$. By the independence, $\mathbb{P}\{$there is a jump$\} \geq 1 - (1 - \alpha)^2$. So either $\alpha = 0$ or $\alpha = 1$. Once we show that $\alpha = 1$, the self-similarity will show that (with probability one) there is a jump in every interval.

(b) We will need an estimate like this:

$$\int_\lambda^{z-\lambda} \theta_z(x)\,dx < \frac{1}{2\pi\lambda}$$

for z large enough and $0 < \lambda < z/2$. See Exercise 4.3.24.

(c) Given $z \geq 20$, then

$$\mathbb{P}\{1 \leq \mathbf{Y}(1/2) \leq 19 | \mathbf{Y}(1) = z\} = \int_1^{19} \theta_z(x)\,dx < \frac{1}{2\pi},$$

and if $1 \leq \mathbf{Y}(1/2) \leq 19$ fails, then one of the two increments $\mathbf{Y}(1/2) - \mathbf{Y}(0)$, $\mathbf{Y}(1) - \mathbf{Y}(1/2)$ is > 19.

(d) For intervals of half the length, the same method may be used, but to convert it so that estimate (b) may be used, we must multiply or divide by 2 in certain places. Given that an increment on an interval of length $1/2$ is at least 19 (say $\mathbf{Y}(1/2) - \mathbf{Y}(0) > 19$),

$$\mathbb{P}\{1 + 2^2/2 \leq \mathbf{Y}(1/4) - \mathbf{Y}(0) \leq 20 - 1 - 2^2/2 | \mathbf{Y}(1/2) - \mathbf{Y}(0) > 19\} < \frac{1}{2\pi 2^2},$$

and if $1 + 2^2/2 \leq \mathbf{Y}(1/4) - \mathbf{Y}(0) \leq 20 - 1 - 2^2/2$ fails, then one of the two increments $\mathbf{Y}(1/4) - \mathbf{Y}(0)$, $\mathbf{Y}(1/2) - \mathbf{Y}(1/4)$ is $> 20 - 1 - 2^2/2$.

(e) Let $A_k = \{$some increment on an interval of length $1/2^k$ is at least $20 - 1 - 2^2/2 - 3^2/2^2 - \cdots - k^2/2^{k-1}\}$; then as before,

$$\mathbb{P}[A_{k+1} | A_k] \geq 1 - \frac{1}{2\pi k^2}.$$

(f) The values of the series used here are

$$\sum_{k=1}^\infty \frac{1}{2\pi k^2} = \frac{\pi}{12}, \qquad \sum_{k=1}^\infty k^2/2^{k-1} = 12.$$

(g) We have

$$\mathbb{P}[\text{there is a jump of at least } 8 | \mathbf{Y}(1) > 20] > 1 - \frac{\pi}{12},$$

a positive number. And $\mathbb{P}\{\mathbf{Y}(1) > 20\} = \int_{20}^\infty \gamma_1(z)\,dz$ is also a positive number. So $\mathbb{P}\{$there is a jump$\} > (1 - \pi/12)\mathbb{P}\{\mathbf{Y}(1) > 20\} > 0$.

Multidimensional Lévy Flights. There is no reason to limit consideration to random variables with real values. We may use random elements of $\mathbb{R}^d$. The increments of a random function $Y: [0, 1] \to \mathbb{R}^d$ are to be independent and have Lévy stable distribution.

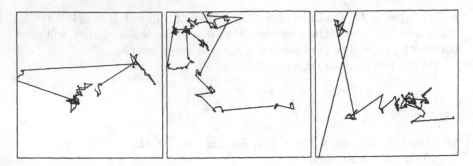

Figure 5.5.25. Trajectories of Lévy flights.

The case of stable distribution with exponent 2 (the normal distribution) is Brownian motion. Multidimensional Brownian motion is considered in §5.6.

Suppose the increments are constructed by choosing a direction at random (uniformly) and then using a Cauchy distribution to determine the size of the vector. A few pictures of this process have been included here. In Figure 5.5.25, the jumps in the process are shown as line segments. These represent the "flights," and the clusters are the "sojourns." But (as you may expect) when the sojourns are magnified, they consist of smaller flights and smaller sojourns. In Figure 5.5.26 I have not drawn the flights. Instead, a dot has been placed in any pixel where the process remains for a positive amount of time. Mandelbrot calls a random set K obtained in this manner a "Lévy dust." (In order for K to be a compact set, we need to include both the starting point and the ending point of each jump.) Because of the way we constructed this, the resulting set is statistically self-similar. If we remember the random function $\mathbf{Y} \colon [0,1] \to \mathbb{R}^d$ involved, we may obtain a random measure $\mathcal{M} = \mathbf{Y}_*(\mathcal{L} \upharpoonright [0,1])$. It is a statistically self-similar measure. Another way to think of this measure (when we think of the parameter set $[0,1]$ as time) is that $\mathcal{M}(U)$ is the total amount of time the particle spends in the set U. The jumps are thought of as happening instantaneously. The support of $\mathcal{M}$ is the statistically self similar set K.

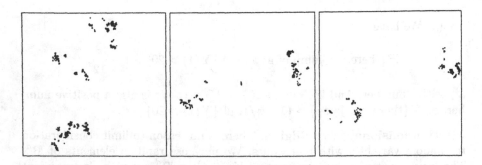

Figure 5.5.26. Lévy dusts.

5.6 Brownian Motion

Brownian motion refers to a natural phenomenon observed by a botanist (Robert Brown, 1773-1858)—fluctuations of small particles suspended in a fluid. And it refers to a stochastic process constructed by Norbert Wiener to model those fluctuations, which is also called the **Wiener process**.

One-Dimensional Brownian Motion. The stochastic process to be considered has both continuous time and continuous state space. Let us consider the version that uses the semi-infinite interval $[0, \infty)$ as "time" and the line $\mathbb{R}$ as "state space." The sample paths are continuous, so it may be considered a random element of the space $\mathfrak{C}([0, \infty), \mathbb{R})$ of continuous real-valued functions defined on $[0, \infty)$. Recall (4.2.9) that the Borel sets in $\mathfrak{C}([0, \infty), \mathbb{R})$ are generated by the cylinder sets depending on finitely many values.

(5.6.1) Definition. A (one-dimensional) **Brownian motion process** is a random element $\mathbf{X}$ of $\mathfrak{C}([0, \infty), \mathbb{R})$ satisfying:

(a) $\mathbf{X}(0) = 0$.
(b) If $0 \leq t_0 < t_1 < \cdots < t_n$, then the real random variables $\mathbf{X}(t_1) - \mathbf{X}(t_0)$, $\mathbf{X}(t_2) - \mathbf{X}(t_1), \cdots, \mathbf{X}(t_n) - \mathbf{X}(t_{n-1})$ are independent.
(c) If $0 \leq s < t$, then the real random variable $\mathbf{X}(t) - \mathbf{X}(s)$ is normally distributed with mean 0 and variance $t - s$.

(5.6.2) Proposition. *There is a random function $\mathbf{X}$ with the properties listed in Definition 5.6.1.*

Proof. The hard part of the construction was already done in Section 5.5. Let $\mathbf{X}_1, \mathbf{X}_2, \cdots$ be independent, identically distributed copies of the random function in Corollary 5.5.15. Let $\mathbf{X}$ be obtained by placing the graphs of $\mathbf{X}_k$ end to end, translating the graph of each $\mathbf{X}_k$ so that its left endpoint matches the right endpoint of $\mathbf{X}_{k-1}$.

Here is a more precise formulation. Define a random function $\mathbf{X} \colon [0, \infty) \to \mathbb{R}$ as follows. Given $t \in [0, \infty)$, let $n = \lfloor t \rfloor$ be the integer part and $s = t - n$ the fractional part. Let

$$\mathbf{X}(t) = \mathbf{X}_1(1) + \mathbf{X}_2(1) + \cdots + \mathbf{X}_n(1) + \mathbf{X}_{n+1}(s).$$

The continuity follows from the continuity of each of the parts $\mathbf{X}_n$, together with the matching endpoints: when n is an integer,

$$\lim_{t \nearrow n} \mathbf{X}(t) = \lim_{s \nearrow 1}\big(\mathbf{X}(n-1) + \mathbf{X}_n(s)\big) = \mathbf{X}(n-1) + \mathbf{X}_n(1) = \mathbf{X}(n);$$

$$\lim_{t \searrow n} \mathbf{X}(t) = \lim_{s \searrow 0}\big(\mathbf{X}(n) + \mathbf{X}_{n+1}(s)\big) = \mathbf{X}(n) + \mathbf{X}_{n+1}(0) = \mathbf{X}(n).$$

(a) If $t = 0$, then $n = 0, s = 0$, so $\mathbf{X}(0) = \mathbf{X}_1(0) = 0$.

(b) Increments on intervals that are contained in different integer blocks $[n-1, n]$ are independent, since the $\mathbf{X}_n$ are independent of each other. Increments on disjoint intervals in the same integer block $[n-1, n]$ are independent by the corresponding property of $\mathbf{X}_n$. Increments on intervals that are made up of parts of two or more different integer blocks $[n-1, n]$ are sums of increments on shorter intervals, each contained in only one block.

(c) If $[s, t]$ is contained in one integer block $[n-1, n]$, then $\mathbf{X}(t) - \mathbf{X}(s) = \mathbf{X}_n(t-n-1) - \mathbf{X}_n(s-n-1)$, so it has the required distribution. If $[s, t]$ intersects several integer blocks, say $s < n < n+1 < \cdots < m < t$, then

$$\mathbf{X}(t) - \mathbf{X}(s)$$
$$= \big(\mathbf{X}(n) - \mathbf{X}(s)\big) + \big(\mathbf{X}(n+1) - \mathbf{X}(n)\big) + \big(\mathbf{X}(n+2) - \mathbf{X}(n+1)\big)$$
$$+ \cdots + \big(\mathbf{X}(m) - \mathbf{X}(m-1)\big) + \big(\mathbf{X}(t) - \mathbf{X}(m)\big).$$

This is a sum of independent normal summands, with variances that sum to $t - s$, so the entire increment is normal with variance $t - s$. ☺

Because of (4.2.9), the Brownian motion process is unique in the sense that any two such processes $\mathbf{X}$ have the same distribution. That distribution $\mathcal{D}_\mathbf{X}$ is a Borel probability measure on $\mathfrak{C}\big([0, \infty), \mathbb{R}\big)$. It is often called the **Wiener measure**.

Elementary Properties of Brownian Motion. Let us consider first a simple application of the maximal inequality. The time set $[0, \infty)$ is not compact, and we expect that $\sup_t |\mathbf{X}(t)| = \infty$. But an interval $[0, T]$ is compact, so $\sup_{t \in [0,T]} |\mathbf{X}(t)| < \infty$. The maximal inequality provides more information about the distribution of this maximal function.

(5.6.3) Proposition. *Let $T, \lambda > 0$ be given. Then*

$$\mathbb{P}\left\{ \sup_{t \in [0,T]} |\mathbf{X}(t)| \geq \lambda \right\} \leq \frac{T}{\lambda^2}.$$

Proof. First, consider a finite subset A of $[0, T]$. Write

$$V(A) = \left\{ \sup_{t \in A} |\mathbf{X}(t)| > \lambda \right\}.$$

If we write the elements of A in order, $0 \leq t_1 < t_2 < \cdots < t_n$, then (writing for convenience $t_0 = 0$) the random variables $\mathbf{X}(t_i) - \mathbf{X}(t_{i-1})$ are independent. In fact, $\mathbf{X}(t_i) - \mathbf{X}(t_{i-1})$ has mean 0 and variance $t_i - t_{i-1}$. And for each $j \leq n$, we have the partial sums

$$\mathbf{X}(t_j) = \sum_{i=1}^{j} \big(\mathbf{X}(t_i) - \mathbf{X}(t_{i-1})\big).$$

But

$$V(A) = \left\{ \sup_{j \leq n} |\mathbf{X}(t_j)| > \lambda \right\},$$

so by the maximal inequality (4.3.11),

$$\mathbb{P}\big(V(A)\big) \leq \frac{1}{\lambda^2} \sum_{i=1}^{n} (t_i - t_{i-1}) = \frac{t_n}{\lambda^2} \leq \frac{T}{\lambda^2}.$$

Now let A_m be an increasing sequence of finite sets with union dense in $[0, T]$. With probability one, $\mathbf{X}(t)$ is a continuous function of t, so the sequence $V(A_m)$ increases to the set

$$\bigcup_{m=1}^{\infty} V(A_m) = \left\{ \sup_{t \in [0,T]} |\mathbf{X}(t)| > \lambda \right\}.$$

Therefore,

$$\mathbb{P}\left\{ \sup_{t \in [0,T]} |\mathbf{X}(t)| > \lambda \right\} = \lim_m \mathbb{P}\big(V(A_m)\big) \leq \frac{T}{\lambda^2}.$$

For the weak inequality,

$$\mathbb{P}\left\{ \sup_{t \in [0,T]} |\mathbf{X}(t)| \geq \lambda \right\} = \mathbb{P}\left(\bigcap_{n=1}^{\infty} \left\{ \sup_{t \in [0,T]} |\mathbf{X}(t)| > \lambda - \frac{1}{n} \right\} \right)$$
$$\leq \lim_{n \to \infty} \frac{T}{(\lambda - 1/n)^2} = \frac{T}{\lambda^2}. \quad \text{☺}$$

(5.6.4) Proposition. $\mathbb{E}[\mathbf{X}(s)\mathbf{X}(t)] = s \wedge t.$

Proof. Given $s < t$, the two random variables $\mathbf{X}(s), \mathbf{X}(t)$ have mean 0, so

$$\mathbb{E}[\mathbf{X}(s)\mathbf{X}(t)] = \mathbb{E}\big[\mathbf{X}(s)\big(\mathbf{X}(s) + \mathbf{X}(t) - \mathbf{X}(s)\big)\big]$$
$$= \mathbb{E}\big[\mathbf{X}(s)^2\big] + \mathbb{E}\big[\mathbf{X}(s)\big(\mathbf{X}(t) - \mathbf{X}(s)\big)\big]$$
$$= \mathrm{Var}\,[X(s)] + 0 = s. \quad \text{☺}$$

Because of the characterizations of the Brownian motion process, it is sometimes possible to prove that some stochastic process "is" Brownian motion. In particular, some of these are processes obtained in terms of Brownian motion $\mathbf{X}(t)$ itself.

(5.6.5) Proposition. (Translation) *Let* $\mathbf{X}(t)$ *be a Brownian motion. Let* $p > 0$. *Then*

$$\mathbf{X}'(t) = X(t + p) - X(p)$$

is also a Brownian motion.

Proof. Simply verify that $\mathbf{X}'(t)$ satisfies the conditions of Definition 5.6.1.

(a) $\mathbf{X}'(0) = \mathbf{X}(p) - \mathbf{X}(p) = 0.$

(b) If $0 \leq t_0 < t_1 < \cdots < t_n$, then of course $0 \leq t_0 + p < t_1 + p < \cdots < t_n + p$, so the real random variables $X'(t_i) - X'(t_{i-1}) = X(t_i + p) - X(t_{i-1} + p)$ are independent.

(c) If $0 \leq s < t$, then $X'(t) - X'(s) = X(t + p) - X(s + p)$ is normally distributed with mean 0 and variance $(t + p) - (s + p) = t - s$.   ☺

The next result deals with scaling for Brownian motion. If we have a movie of a Brownian motion particle moving in the line and we show it at half speed and magnify the picture by the factor $\sqrt{2}$, then the result should be indistinguishable from Brownian motion. This is a reformulation of the statistical self-affine property of the graph. But in fact, the graph is statistically self-affine at scales other than $1/2$.

(5.6.6) Proposition. (Scaling) *Let $X(t)$ be a Brownian motion. Let $a > 0$. Then*

$$X'(t) = a^{-1/2}X(at)$$

is also a Brownian motion.

Proof. Again, we must verify the conditions of Definition 5.6.1. (a) and (b) are easy. For (c), consider the distribution of $X'(t) - X'(s) = a^{-1/2}(X(at) - X(as))$. Now of course, $X(at) - X(as)$ is normally distributed with mean 0 and variance $at - as$, so $a^{-1/2}(X(at) - X(as))$ is normally distributed with mean 0 and variance $(a^{-1/2})^2(at - as) = t - s$, as required.   ☺

The third example has to do with reversing the time of Brownian motion. When we replace t by $1/t$, we need to scale space at a variable rate to recover Brownian motion. We do not use the result, so we leave it to the reader.

(5.6.7) Exercise. (Time inversion) Let $X(t)$ be a Brownian motion. Then

$$X'(t) = tX(1/t)$$

is also a Brownian motion.

Multidimensional Brownian Motion. Brownian motion in the physical sense comes from a particle moving in three dimensions when suspended in a fluid. It is modeled by 3-dimensional Brownian motion. We will consider next Brownian motion in d dimensions.

(5.6.8) Definition. Let d be a positive integer. A **d-dimensional Brownian motion process** is a random continuous function $X \colon [0, \infty) \to \mathbb{R}^d$ satisfying:

(a) $X(0) = 0$.

(b) If $0 \leq t_0 < t_1 < \cdots < t_n$, then the random variables $X(t_1) - X(t_0)$, $X(t_2) - X(t_1), \cdots, X(t_n) - X(t_{n-1})$ are independent.

(c) If $0 \leq s < t$, then the random variable $\mathbf{X}(t) - \mathbf{X}(s)$ is normally distributed in $\mathbb{R}^d$ with mean 0 and variance $t - s$ in every direction.

Recall that (c) means that $\mathbf{X}(t) - \mathbf{X}(s)$ is absolutely continuous with density

$$(2\pi(t-s))^{-d/2} \exp\left(-\frac{|x|^2}{2(t-s)}\right) \qquad \text{for } x \in \mathbb{R}^d.$$

Once we have constructed 1-dimensional Brownian motion, the d-dimensional case is easy.

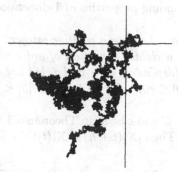

Figure 5.6.9. Range of 2-dimensional Brownian motion, $\mathbf{X}[0,1]$.

(5.6.10) Proposition. *Let* $\mathbf{X}_1, \mathbf{X}_2, \cdots, \mathbf{X}_d$ *be independent* 1-*dimensional Brownian motions. Then the random function* $\mathbf{X}: [0,\infty) \to \mathbb{R}^d$ *defined by* $\mathbf{X}(t) = (\mathbf{X}_1(t), \mathbf{X}_2(t), \cdots, \mathbf{X}_d(t))$ *is a d-dimensional Brownian motion.*

Proof. First, each component $\mathbf{X}_i(t)$ of $\mathbf{X}(t)$ is a continuous function of t, so $\mathbf{X}$ itself is continuous.

(a) $\mathbf{X}(0) = (\mathbf{X}_1(0), \mathbf{X}_2(0), \cdots, \mathbf{X}_d(0)) = (0, 0, \cdots, 0) = 0$.

(b) Let $0 \leq t_0 < t_1 < \cdots < t_n$. Then the entire family

$$\{\mathbf{X}_i(t_j - t_{j-1}) : 1 \leq i \leq d, 1 \leq j \leq n\}$$

is independent. So the vector increments $\mathbf{X}(t_j) - \mathbf{X}(t_{j-1})$ are independent.

(c) Let $0 \leq s < t$. Each component $\mathbf{X}_i(t) - \mathbf{X}_i(s)$ has normal distribution with mean 0 and variance $t - s$. Thus, $\mathbf{X}_i(t) - \mathbf{X}_i(s)$ has density

$$f_i(x_i) = (2\pi(t-s))^{-1/2} \exp\left(-\frac{x_i^2}{2(t-s)}\right).$$

These components are independent, so the density for the vector $\mathbf{X}(t) - \mathbf{X}(s)$ is their product. At the point $x = (x_1, x_2, \cdots, x_d)$ the density is

$$\prod_{i=1}^{d} f_i(x_i) = \prod_{i=1}^{d} (2\pi(t-s))^{-1/2} \exp\left(-\frac{x_i^2}{2(t-s)}\right)$$

$$= (2\pi(t-s))^{-d/2} \exp\left(\frac{1}{2(t-s)} \sum_{i=1}^{d} x_i^2\right)$$

$$= (2\pi(t-s))^{-d/2} \exp\left(-\frac{|x|^2}{2(t-s)}\right).$$

So $\mathbf{X}$ is a d-dimensional Brownian motion.   ☺

Most of the elementary properties of d-dimensional Brownian motion follow from the corresponding properties of 1-dimensional Brownian motion.

(5.6.11) Proposition. *Let d be a positive integer. The d-dimensional Brownian motion satisfies a Hölder condition of order γ for all $\gamma < 1/2$. That is, there is a constant c (depending on γ and d, but not on t or h) and $\varepsilon > 0$ such that $|\mathbf{X}(t+h) - \mathbf{X}(t)| \le c|h|^\gamma$ for all t, h with $|h| < \varepsilon$, $t \ge 0$, and $t + h \ge 0$.*

Proof. For given γ, let C and ε be as in Theorem 5.3.13. Let t, h satisfy $|h| < \varepsilon$, $t \ge 0$, and $t + h \ge 0$. Then $|\mathbf{X}_i(t+h) - \mathbf{X}_i(t)| \le C|h|^\gamma$ for all i. Then

$$|\mathbf{X}(t+h) - \mathbf{X}(t)|^2 = \sum_{i=1}^{d} |\mathbf{X}_i(t+h) - \mathbf{X}_i(t)|^2 \le \sum_{i=1}^{d} C^2 |h|^{2\gamma} = dC^2 |h|^{2\gamma}.$$

So we may use $c = C\sqrt{d}$.   ☺

(5.6.12) Exercise. The random variable $|\mathbf{X}(1)|^{-\alpha}$ is integrable if and only if $\alpha < d$.

(5.6.13) Proposition. *Let $\alpha < d$ be given. There is a constant c (which may depend on α and on d, but not on h or t) such that*

$$\mathbb{E}\left[|\mathbf{X}(t+h) - \mathbf{X}(t)|^{-\alpha}\right] = c|h|^{-\alpha/2}.$$

Proof. Let $c = \mathbb{E}\left[|\mathbf{X}(1)|^{-\alpha}\right]$. Then since $\mathbf{X}(t+h) - \mathbf{X}(t)$ has the same distribution as $h^{1/2}\mathbf{X}(1)$, we have

$$\mathbb{E}\left[|\mathbf{X}(t+h) - \mathbf{X}(t)|^{-\alpha}\right] = \mathbb{E}\left[||h|^{1/2}\mathbf{X}(1)|^{-\alpha}\right]$$

$$= |h|^{-\alpha/2}\mathbb{E}\left[|\mathbf{X}(1)|^{-\alpha}\right] = c|h|^{-\alpha/2}.$$ ☺

Dimensions Determined by Brownian Motion. Many random sets are associated with stochastic processes, such as Brownian motion. We will now consider the fractal dimensions of a few of these sets.

Statistical Self-Similarity. Consider the Brownian motion process $\mathbf{X}(t)$ in $\mathbb{R}^d$, with $d \geq 2$. It has statistical self-similarity properties, such as:

(a) For any $t_0 \geq 0$, the process $\left(\mathbf{X}(t_0 + t) - \mathbf{X}(t_0)\right)_{t \geq 0}$ has the same (joint) distribution as $\left(\mathbf{X}(t)\right)_{t \geq 0}$.

(b) For any $a > 0$, the process $\left(a^{-1/2}\mathbf{X}(at)\right)_{t \geq 0}$ has the same distribution as $\left(\mathbf{X}(t)\right)_{t \geq 0}$.

Consider the (random) set K consisting of the points that Brownian motion process visits during the first unit interval of time $[0, 1]$. We will write $K = \mathbf{X}[0, 1] = \{\, \mathbf{X}(t) : 0 \leq t \leq 1 \,\}$. As we will see, K has fractal dimension 2.

Now, K is the union (possibly not disjoint) of the points visited during the first half-unit of time and the points visited during the second half-unit of time: $K = \mathbf{X}[0, 1/2] \cup \mathbf{X}[1/2, 1]$. But by the above self-similarity properties, $\mathbf{X}[0, 1/2]$ has the same distribution as the range of $2^{-1/2}\mathbf{X}$ on the entire interval $[0, 1]$. So $\mathbf{X}[0, 1/2]$ has the same distribution as $f_0[K]$, where f_0 is the similarity defined by $f_0(x) = 2^{-1/2}x$. Also, $\mathbf{X}[1/2, 1]$ has the same distribution as $\mathbf{X}[0, 1/2]$ translated by the vector $\mathbf{X}(1/2)$. So $\mathbf{X}[1/2, 1]$ has the same distribution as $f_1[K]$, where f_1 is the (random) similarity defined by $f_1(x) = 2^{-1/2}x + \mathbf{X}(1/2)$. So $K = f_0[K] \cup f_1[K]$. Thus K is statistically self-similar; the similarity dimension is the solution s of

$$2 \cdot \left(2^{-1/2}\right)^s = 1,$$

so $s = 2$. Then by Theorem 5.4.11, $\dim K \leq 2$ (and by (5.4.14) $\mathrm{Dim}\, K \leq 2$).

Proof of the lower bound $\dim K \geq 2$ would require an open set condition (which we have not discussed). But we will follow another route to the lower bound. First, we note another proof for the upper bound.

Recall the Hölder condition for Brownian motion (5.6.11): If $0 < \alpha < 1/2$ is given, then there is a constant b such that

$$\left|\mathbf{X}(t + h) - \mathbf{X}(t)\right| \leq b|h|^\alpha \qquad \text{for } |h| < \varepsilon$$

for some $\varepsilon > 0$ and all $t > 0$.

Recall (Exercise 1.2.4 and 1.2.14) the estimates for the dimension of an image under a map with Hölder condition: Let S_1 and S_2 be metric spaces. Suppose the function $f \colon S_1 \to S_2$ maps S_1 onto S_2 and there exist positive numbers b, α, ε such that

$$\rho\big(f(x), f(y)\big) \leq b\rho(x, y)^\alpha$$

for all $x, y \in S_1$ with $\rho(x, y) \leq \varepsilon$. Then $\dim S_1 \geq \alpha \dim S_2$ and $\mathrm{Dim}\, S_1 \geq \alpha \mathrm{Dim}\, S_2$.

This may be used to prove the upper bound on the dimension of the Brownian path $K = \mathbf{X}[0, 1]$, namely, $\dim K \leq (1/\alpha)\dim[0, 1] = 1/\alpha$ for all $\alpha < 1/2$, so $\dim K \leq 2$.

(5.6.14) Exercise. Let $M \subseteq [0,1]$ be a Borel set, and let $\mathbf{X}[M] = \{\mathbf{X}(t) : t \in M\}$ be its image under Brownian motion. Then $\dim \mathbf{X}[M] \leq 2 \dim M$ and $\operatorname{Dim} \mathbf{X}[M] \leq 2 \operatorname{Dim} M$.

Lower bound. For the lower bound estimate $\dim K \geq 2$ we will use the formula (5.6.13) for the distribution of the length of an increment of Brownian motion.

(5.6.15) Theorem. *Let $\mathbf{X}$ be d-dimensional Brownian motion ($d \geq 2$), and let $K = \mathbf{X}[0,1]$ be its path. Then $\dim K = \operatorname{Dim} K = 2$ with probability one.*

Proof. Let $\mathcal{M}$ be the image under $\mathbf{X}$ of Lebesgue measure on $[0,1]$, that is,

$$\mathcal{M}(E) = \mathcal{L}\big(\mathbf{X}^{-1}[E] \cap [0,1]\big)$$

for Borel sets $E \subseteq \mathbb{R}^d$. Thus $\mathcal{M}$ is a (random) probability measure concentrated on the compact set $K = \mathbf{X}[0,1]$. Let $s < 1$, so that $2s < 2 \leq d$. Then by (5.6.13) there is a $c > 0$ (which depends on d and s, but not on t or h) such that

$$\mathbb{E}\left[|\mathbf{X}(t+h) - \mathbf{X}(t)|^{-2s}\right] = ch^{-s}.$$

Now compute with Fubini's Theorem

$$\mathbb{E}\left[\iint \frac{\mathcal{M}(dx)\,\mathcal{M}(dy)}{|x-y|^{2s}}\right] = \mathbb{E}\left[\int_0^1 \int_0^1 |\mathbf{X}(t) - \mathbf{X}(u)|^{-2s}\,dt\,du\right]$$

$$= \int_0^1 \int_0^1 \mathbb{E}\left[|\mathbf{X}(t) - \mathbf{X}(u)|^{-2s}\right]\,dt\,du$$

$$= c\int_0^1 \int_0^1 |t-u|^{-s}\,dt\,du < \infty.$$

Since the expectation is finite, we have with probability one,

$$\iint \frac{\mathcal{M}(dx)\,\mathcal{M}(dy)}{|x-y|^{2s}} < \infty.$$

Therefore, by (3.2.5), $\dim K \geq 2s$. This is true for all $s < 1$, so $\dim K \geq 2$.

So we conclude that $2 \leq \dim K \leq \operatorname{Dim} K \leq 2$, and thus equality holds everywhere.   ☺

(5.6.16) Exercise. Let $M \subseteq [0,1]$ be a compact set. Then

(a) $\dim \mathbf{X}[M] = 2 \dim M$.
(b) if $\dim M = \operatorname{Dim} M$, then $\dim \mathbf{X}[M] = \operatorname{Dim} \mathbf{X}[M]$.
(c) $\operatorname{Dim} \mathbf{X}[M] = 2 \operatorname{Dim} M$.

*5.7 A Multifractal Decomposition

It has been argued in the physics literature (for example Halsey et al. [125]) that certain fractals carrying a natural measure may be analyzed in terms of the scaling properties of the measure. The fractal K should contain "singularities of strength α" for certain values of a parameter α; a fractal dimension $f(\alpha)$ describes how densely those singularities are distributed. Computations show a typical concave shape for this function $f(\alpha)$, sometimes known as a "dimension spectrum."[5]

The mathematical theory for this study is known as a "multifractal formalism." Such formalisms have been described from the mathematical point of view in recent years (for example, [203]).

Begin with a (possibly fractal) nonempty compact set K and a measure $\mathcal{M}$ on K. For example, we might have a dynamical system where in the limit, the trajectories approach a (strange) attractor K, and the ergodic time-averages along the process approach a corresponding measure $\mathcal{M}$. Or we might imagine an iterated function system, which approaches its attractor K as points are chosen according to the chaos game; the time-averages again converge to a natural measure on the attractor (as we saw in Theorem 5.1.5).

When a set K has fractal dimension s and supports a "natural" finite measure $\mathcal{M}$, we may expect "typically," for $x \in K$ and $\varepsilon > 0$, that the measure $\mathcal{M}(\overline{B}_\varepsilon(x))$ of the ball of radius ε centered at x is roughly equal to $(2\varepsilon)^s$, the sth power of the diameter of the ball. This might mean that

$$0 < \lim_{\varepsilon \searrow 0} \frac{\mathcal{M}(\overline{B}_\varepsilon(x))}{(2\varepsilon)^s} < \infty$$

or, more generally, that

$$\lim_{\varepsilon \searrow 0} \frac{\log \mathcal{M}(\overline{B}_\varepsilon(x))}{\log \varepsilon} = s$$

for all $x \in K$. (Recall that this limit is called the pointwise dimension of $\mathcal{M}$ at x.) Multifractal decomposition will be interesting exactly when this does not happen—the pointwise dimension takes different values at different points; for many different values of the parameter α, the set

$$K_{(\alpha)} = \left\{ x \in K : \lim_{\varepsilon \searrow 0} \frac{\log \mathcal{M}(\overline{B}_\varepsilon(x))}{\log \varepsilon} = \alpha \right\}$$

is nontrivial.[6]

---

* An optional section.
[5] The letters f and α are used in the physics literature, so I will follow that custom here. So in this section α will not denote a string.
[6] But it is sometimes argued that the pointwise dimension does not exist at all in typical cases.

The sets $K_{(\alpha)}$ may be thought of as the multifractal components of K. The Hausdorff dimension of $K_{(\alpha)}$ is called $f(\alpha)$; this function describes the dimension spectrum.

It is important to note that we are using the *Hausdorff dimension* of $K_{(\alpha)}$ and not the box dimension. Typically, all the sets $K_{(\alpha)}$ are dense in K, so they all have box dimension equal to the box dimension of K itself. But the Hausdorff dimension $f(\alpha)$ varies with α.

There are some special cases when all of the Hausdorff dimension computations can be carried out explicitly. We will do it in the case of a string model for a ratio list with weights. They do, indeed, show the characteristic concave multifractal dimension spectrum curve $f(\alpha)$.

String Space. Let E be a finite alphabet (with at least two letters); let ratios with $0 < r_e < 1$ be given; and let weights be given with $0 < p_e$, $\sum_e p_e = 1$. As usual, we will use the space $E^{(\omega)}$ of infinite strings from the alphabet E, together with the metric ρ such that $\operatorname{diam}[\gamma] = r(\gamma)$ for all finite strings γ.

Now, $E^{(\omega)}$ itself is the attractor of our iterated function system consisting of right shifts. So I will write $K = E^{(\omega)}$. Starting with the weights (p_e), a measure $\mathcal{M}$ is defined on this string space such that $\mathcal{M}([\gamma]) = p(\gamma)$ for all finite strings γ. Now (as argued in the proof of Theorem 5.2.1),

$$K_{(\alpha)} = \left\{ \sigma \in K : \lim_{\varepsilon \searrow 0} \frac{\log \mathcal{M}(\overline{B}_\varepsilon(\sigma))}{\log \varepsilon} = \alpha \right\}$$

$$= \left\{ \sigma \in K : \lim_{k \to \infty} \frac{\log p(\sigma \restriction k)}{\log r(\sigma \restriction k)} = \alpha \right\}.$$

The problem is the computation of

$$f(\alpha) = \dim K_{(\alpha)}$$

for all real numbers $\alpha > 0$. When $K_{(\alpha)} = \varnothing$, we write by convention $f(\alpha) = -\infty$.[7] We know that $\log p(\sigma \restriction k)/\log r(\sigma \restriction k)$ converges to the similarity dimension of the measure $\mathcal{M}$ for $\mathcal{M}$-almost all $\sigma \in K$; so here we are interested in the Hausdorff dimension of certain sets of $\mathcal{M}$-measure zero.

Auxiliary Functions. The investigation of the dimensions will involve a few auxiliary functions. This discussion is taken from [43].

Recall that the similarity dimension of the ratio list (r_e) is the solution s of $\sum_e r_e^s = 1$; the weights r_e^s are called the "uniform weights." The similarity dimension for the ratio list with (possibly nonuniform) weights (p_e) is

$$s_0 = \frac{\sum p_e \log p_e}{\sum p_e \log r_e}.$$

We will define a function $\beta \colon \mathbb{R} \to \mathbb{R}$ as follows. For a fixed $q \in \mathbb{R}$, the function

---

[7] Mandelbrot has proposed a method of assigning negative dimensions $f(\alpha)$ in these cases [179, 180].

$$\sum_{e \in E} p_e^q r_e^z$$

of z is continuous, strictly decreasing, goes to ∞ as $z \to -\infty$, and goes to 0 as $z \to \infty$. So there is a unique z for which it is 1. We accordingly may define $\beta(q)$ by the equation

$$\sum_{e \in E} p_e^q r_e^{\beta(q)} = 1.$$

Note that $\beta(1) \doteq 0$ and $\beta(0) = s$.

By implicit differentiation,

$$\beta'(q) = -\frac{\sum (\log p_e) p_e^q r_e^{\beta(q)}}{\sum (\log r_e) p_e^q r_e^{\beta(q)}}$$

and

$$\beta''(q) = -\frac{\sum (\log p_e + \beta'(q) \log r_e)^2 p_e^q r_e^{\beta(q)}}{\sum (\log r_e) p_e^q r_e^{\beta(q)}}.$$

So we may conclude that $\beta'(q) < 0$ for all q, so that β is strictly decreasing, and that $\beta''(q) \geq 0$, so that β is a convex function. Now in fact, the only way to have $\beta''(q) = 0$ at some point q is $\beta'(q) = -\log p_e / \log r_e$ for all e; therefore, $p_e = r_e^{\beta'(q)}$ and $p_e = r_e^s$. That is, the weights are the uniform weights.

(5.7.1) Proposition. *Define $\beta(q)$ by $\sum_{e \in E} p_e^q r_e^{\beta(q)} = 1$. If the weights are the uniform weights, then $\beta(q) = -sq + s$ for all q. Otherwise, β is strictly decreasing, and $\beta''(q) > 0$ for all q.*

Next, define

$$\alpha(q) = -\beta'(q) = \frac{\sum (\log p_e) p_e^q r_e^{\beta(q)}}{\sum (\log r_e) p_e^q r_e^{\beta(q)}}.$$

If the weights are the uniform weights, $\alpha(q)$ is the constant s. Otherwise, $\alpha(q)$ is strictly decreasing. Write $\alpha(-\infty)$ and $\alpha(\infty)$ for the limits as q goes to $-\infty$ or ∞.

Now define $f(q) = q\alpha(q) + \beta(q)$. In particular, $f(0) = s$ and $f'(q) = -q\beta''(q)$. So if the weights are the uniform weights, then f is the constant s, and otherwise f is strictly increasing from $-\infty$ to 0 and strictly decreasing from 0 to ∞.

Unless the weights are the uniform weights, the function $\alpha(q)$ is one-to-one, and we can express f as a function of α: for each real number α between $\alpha(\infty)$ and $\alpha(-\infty)$ set $f(\alpha) = f(q)$, where $\alpha(q) = \alpha$. The graph of $f(\alpha)$ is smooth and concave down.

We can describe the asymptotic behavior of these auxiliary functions. From the definition

$$1 = \sum p_e^q r_e^{\beta(q)}$$

we have $\lim_{q \to \infty} \beta(q) = -\infty$ and $\lim_{q \to -\infty} \beta(q) = \infty$. Set $\lambda_e = \log p_e / \log r_e$, for $e \in E$ and $\lambda_{\min} = \min \lambda_e$, $\lambda_{\max} = \max \lambda_e$.

(5.7.2) Exercise. $\alpha(\infty) = \lambda_{\min}$, $\alpha(-\infty) = \lambda_{\max}$.

In Figure 5.7.3 we show the graphs of these functions in a typical case. The ratio list used was $(1/2, 1/3, 1/3, 1/4, 1/6)$ and the corresponding weights $(1/4, 1/9, 2/9, 1/4, 1/6)$. The similarity dimension for the ratio list is $s = 1.4316$, and the similarity dimension for the ratio list with weights is $s_0 = 1.3254$. We may compute $\lambda_{\min} = 1$ and $\lambda_{\max} = 2$. Note certain special values:

$$
\begin{aligned}
&q \to -\infty, &&\beta \to \infty, &&\alpha \to \lambda_{\max}, &&f \to 0.78788; \\
&q = 0, &&\beta = s, &&\alpha = 1.5465, &&f = s; \\
&q = 1, &&\beta = 0, &&\alpha = s_0, &&f = s_0; \\
&q \to \infty, &&\beta \to -\infty, &&\alpha \to \lambda_{\min}, &&f \to 0.438694.
\end{aligned}
$$

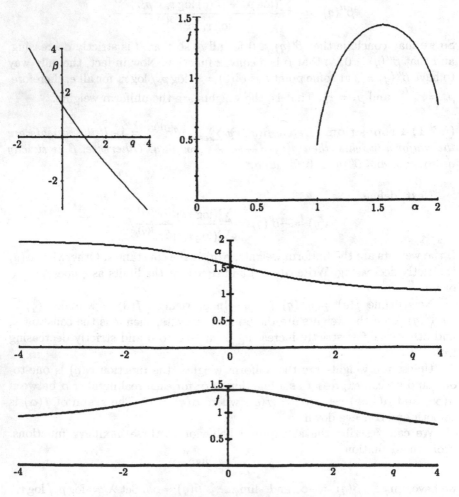

Figure 5.7.3. Graphs of the auxiliary functions.

(5.7.4) Exercise. In this example, the number $f(\alpha(-\infty)) = 0.78788$ is the solution of $(1/2)^z + (1/3)^z = 1$, and the number $f(\alpha(\infty)) = 0.438694$ is the solution of $(1/4)^z + (1/6)^z = 1$. What is 1.5465?

(5.7.5) Theorem. *Let (r_e) be a ratio list and (p_e) weights. Let K be the string space defined above, and let $\mathcal{M}$ be the self-similar measure on K. Let s be the similarity dimension for the ratio list (r_e).*

(a) *If $p_e = r_e^s$ for all $e \in E$, then $\dim K_{(s)} = s$ and $K_{(\alpha)} = \varnothing$ for all $\alpha \neq s$.*
(b) *If $p_e \neq r_e^s$ for at least one e, then for any α between $\alpha(\infty)$ and $\alpha(-\infty)$, we have $\dim K_{(\alpha)} = \operatorname{Dim} K_{(\alpha)} = f(\alpha)$.*

Proof. (a) We have $\log p(\sigma{\restriction}k)/\log r(\sigma{\restriction}k) = s$ for every k and σ, so $K_{(s)} = K$, and $K_{(\alpha)} = \varnothing$ for all other α. The dimension of K was computed to be s in [MTFG, Theorem 6.3.7].

So consider part (b). Let α be in the specified range. There is $q \in \mathbb{R}$ with $\alpha = \alpha(q)$. We also have corresponding numbers $\beta = \beta(q)$ and $f = q\alpha + \beta$. We will define a new measure $\mathcal{M}_q$ on $K = E^{(\omega)}$.

We know (by the definition of β) that

$$\sum_{e \in E} p_e^q r_e^\beta = 1.$$

Let $p_e' = p_e^q r_e^\beta$, and let $\mathcal{M}_q$ be the self-similar measure with weights (p_e'). I claim that $\mathcal{M}_q(K_{(\alpha)}) = 1$. The similarity dimension associated with these weights is

$$\frac{\sum p_e' \log p_e'}{\sum p_e' \log r_e} = \frac{\sum p_e^q r_e^\beta \log(p_e^q r_e^\beta)}{\sum p_e^q r_e^\beta \log r_e}$$

$$= \frac{\sum p_e^q r_e^\beta q \log p_e}{\sum p_e^q r_e^\beta \log r_e} + \frac{\sum p_e^q r_e^\beta \beta \log r_e}{\sum p_e^q r_e^\beta \log r_e}$$

$$= q\alpha + \beta = f.$$

Therefore (as in Theorem 5.2.1), for $\mathcal{M}_q$-almost all σ, we have

$$q\alpha + \beta = f = \lim_k \frac{\log p'(\sigma{\restriction}k)}{\log r(\sigma{\restriction}k)}$$

$$= \lim_k \frac{q \log p(\sigma{\restriction}k) + \beta \log r(\sigma{\restriction}k)}{\log r(\sigma{\restriction}k)}$$

$$= q \lim_k \frac{\log p(\sigma{\restriction}k)}{\log r(\sigma{\restriction}k)} + \beta.$$

As long as $q \neq 0$, we conclude that $\log p(\sigma{\restriction}k)/\log r(\sigma{\restriction}k) \to \alpha$ for $\mathcal{M}_q$-almost all σ. That is, $\mathcal{M}_q(K_{(\alpha)}) = 1$.

For $q = 0$ we have $\beta = f = s$, so that $p_e' = r_e^s$ are the uniform weights and $\mathcal{M}_0$ is the uniform measure. Also,

$$\alpha = \alpha(0) = \frac{(\log p_e)r_e^s}{(\log r_e)r_e^s}.$$

But then by the Strong Law of Large Numbers, $\mathcal{M}_0$-almost every σ satisfies

$$\lim_k \frac{\log p(\sigma\restriction k)}{k} = \sum_e p_e' \log p_e,$$

$$\lim_k \frac{\log r(\sigma\restriction k)}{k} = \sum_e p_e' \log r_e,$$

so that

$$\lim_k \frac{\log p(\sigma\restriction k)}{\log r(\sigma\restriction k)} = \frac{\sum_e p_e' \log p_e}{\sum_e p_e' \log r_e} = \frac{\sum_e (\log p_e)r_e^s}{\sum_e (\log r_e)r_e^s} = \alpha.$$

So in all cases, $\mathcal{M}_q(K_{(\alpha)}) = 1$.

Now, for any $\delta > 0$, the upper pointwise dimension $\overline{L}_{\mathcal{M}_q}(\sigma) < f + \delta$ for every $\sigma \in K_{(\alpha)}$. By Proposition 3.3.13 the lower density $\underline{D}_{\mathcal{M}_q}^{f+\delta}(\sigma)$ is bounded below on $K_{(\alpha)}$. Also, $\mathcal{M}_q(K_{(\alpha)}) < \infty$, so[8] by Theorem 1.5.11, $\mathcal{P}^{f+\delta}(K_{(\alpha)}) < \infty$, so that $\operatorname{Dim} K_{(\alpha)} \leq f + \delta$. This is true for all $\delta > 0$, so $\operatorname{Dim} K_{(\alpha)} \leq f$.

On the other hand, the lower pointwise dimension $\underline{L}_{\mathcal{M}}(\sigma) > f - \delta$ for every $\sigma \in K_{(\alpha)}$. By Proposition 3.3.13 the upper density $\overline{D}_{\mathcal{M}_q}^{f-\delta}(\sigma)$ is bounded above on $K_{(\alpha)}$. Also, $\mathcal{M}_q(K_{(\alpha)}) > 0$, so by Theorem 1.5.13, $\mathcal{C}^{f-\delta}(K_{(\alpha)}) > 0$, so that $\dim K_{(\alpha)} \geq f - \delta$. This is true for all $\delta > 0$, so $\dim K_{(\alpha)} \geq f$.

Combining these results, we get $f \leq \dim K_{(\alpha)} \leq \operatorname{Dim} K_{(\alpha)} \leq f$, and the claimed result follows.   ☺

(5.7.6) Exercise. If $\alpha > \lambda_{\max}$ or $\alpha < \lambda_{\min}$, then $K_{(\alpha)} = \varnothing$.

*5.8 Remarks

The "chaos game" is the colorful name proposed by M. Barnsley [13. p. 2] for this random method of generating the attractor of an iterated function system. He also discusses [13, p. 385] a version of the chaos game that applies to graph-directed iterated function systems (known there as recurrent iterated function systems).

An interesting article dealing with the Sierpiński gasket is by I. Stewart [245].

The use of the Law of Large Numbers to compute the dimension of a self-similar measure goes back at least to Billingsley [30].

---

[8] Again, since this space is ultrametric, the strong Vitali condition is automatically satisfied.
* An optional section.

The random Cantor set is a particular example of a statistically self-similar set. A variant of this example is also found in [87, §15.1]. Statistical self-similarity in general is treated in [118, 143, 188, 175, 176].

Note that when $\Phi(s_0) = 1$ for some s_0, we may use a martingale argument like the one outlined in §5.3 (Exercise 5.4.13). But if Φ jumps from ∞ to a value below 1 (as in the example following Definition 5.4.9), the corresponding random sequence is not a martingale, but a supermartingale. Although there is a supermartingale convergence theorem, in this case the limit is 0, so it is not useful for computation of the lower bound for the dimension. Other methods are required in such cases [118].

See [262][9] for an interesting discussion of the Poisson process, used, for example, as a model for the random clicking of a Geiger counter.

Statistically self-similar measures are found in [4, 89, 209].

Random midpoint displacement (especially in two dimensions) has been used to model rugged terrain; many realistic-seeming pictures are shown in [176].

The random homeomorphism comes from [117, 63] with reference to [64].

The rigorous mathematical definition of Brownian motion is attributed to N. Wiener. The midpoint construction given here is due to P. Lévy. For more on Brownian motion, in particular its applications, see [68].

Lévy stable processes are due to P. Lévy; see [147]. Historical remarks may be found in [176], applications to physics in [214, 237, 238].

Historical information on the Hausdorff dimension and packing dimension for the range of Brownian processes (and other related computations) may be found in [255]. More recent developments may be found in [186, 271].

Multifractal decomposition (expecially in the physics literature) often begins with the assertion (or assumption) that the pointwise dimension exists:

$$\lim_{\varepsilon \searrow 0} \frac{\log \mathcal{M}\left(\overline{B}_\varepsilon(x)\right)}{\log \varepsilon}$$

at least for typical points x. This is used to define a dimension spectrum $f(\alpha)$, the dimension of the set where the pointwise dimension is α. On the other hand, mathematicians have shown that there are measures where this sort of thing does not work. One side of the dispute asserts that these counterexamples may be dismissed as "unphysical" and that measures that arise in practice exhibit a dimension spectrum. The other side asserts that the dimension spectrum exists only in rare artificial cases (like self-similar measures). This argument remains to be played out.

The computation of a multifractal decomposition in §5.7 is from [43]. See also [4, 78, 81, 89, 125, 208, 209, 220, 226, 246] and especially [203].

An interesting discussion of fractal sets (and measures) that arise in connection with "balanced random sampling" is in [106].

Brownian Zero Set. Let $\mathbf{X} \colon [0, \infty) \to \mathbb{R}$ be one-dimensional Brownian motion. The **Brownian zero set** is the random set

---

[9] A book written for high-school students.

$$Z = \{t \geq 0 : \mathbf{X}(t) = 0\}.$$

With probability one, Z is a closed set, Z has no isolated points, and Z has Hausdorff (and packing) dimension $1/2$. (The value $1/2$ is due to Taylor [254].)

The main feature of interest here is the statistical self-similarity. Consider the set $Z_1 = Z \cap [0,1]$. It turns out that Z_1 is a random Cantor set.

Certainly, $0 \in Z$, and as we have seen above, there are other elements of Z greater than 0 but as close as we like to 0. Now with probability one, $\mathbf{X}(1) \neq 0$. Let $\beta_\Lambda = \sup(Z \cap [0,1])$. Because Z is closed, $\beta_\Lambda \in Z$. Our random Cantor set begins with the random interval $C_0 = [a_\Lambda, b_\Lambda]$, where $a_\Lambda = 0$.

Next, with probability one, the value $\mathbf{X}((b_\Lambda + a_\Lambda)/2) \neq 0$ at the midpoint. So we let

$$
\begin{aligned}
a_{\mathsf{L}} &= a_\Lambda, \\
b_{\mathsf{L}} &= \sup(Z \cap [a_\Lambda, (b_\Lambda + a_\Lambda)/2]), \\
a_{\mathsf{R}} &= \inf(Z \cap [(b_\Lambda + a_\Lambda)/2, b_\Lambda]), \\
b_{\mathsf{R}} &= b_\Lambda.
\end{aligned}
$$

The next approximation is $C_1 = [a_{\mathsf{L}}, b_{\mathsf{L}}] \cup [a_{\mathsf{R}}, b_{\mathsf{R}}]$. We continue in this way. If a_α, b_α have been defined for all strings α of length k, then for the successors $\alpha\mathsf{L}$ and $\alpha\mathsf{R}$ use the smallest intervals in the left half or right half of $[a_\alpha, b_\alpha]$ that contain all points of Z in that half. Because of the self-similarity properties (at all scales, not just $1/2$), these subsequent steps are statistically similar to the original step. (Strictly speaking, we need a more technical argument here, since the endpoints are random: the "strong Markov property.")

In order to compute the exact dimension of Z_1, we would need to determine the distribution of the (i.i.d.) random variables

$$(x_\alpha, y_\alpha) = \left(\frac{b_{\alpha\mathsf{L}} - a_{\alpha\mathsf{L}}}{b_\alpha - a_\alpha}, \frac{b_{\alpha\mathsf{R}} - a_{\alpha\mathsf{R}}}{b_\alpha - a_\alpha} \right).$$

That requires a more careful study of Brownian motion than we have done in this book. See, for example, [68, Chapter 1].

On the Exercises. *(5.3.3):* (a) $(3/8)^k$. Note that (x_α, y_α) is independent of u_α. (b) $\left(2\left((1/4)^s + (1/8)^s\right)\right)^k$. (c) There is a critical value s_0 such that the expectation goes to infinity when $s < s_0$, goes to zero when $s > s_0$, and is equal to 1 when $s = s_0$. This critical value is $s_0 = -\log \lambda / \log 2 \approx 0.8232$, where $\lambda \approx 0.5652$ is the solution of $\lambda^3 + \lambda^2 = 1/2$:

$$\lambda = \frac{\sqrt[3]{46 + 6\sqrt{57}}}{6} + \frac{\sqrt[3]{46 - 6\sqrt{57}}}{6} - \frac{1}{3}.$$

(5.3.4): $2 \int_T (x^s + y^s)\, dx\, dy = 4/(s+1)(s+2)$, so this is 1 when $s = (-3 + \sqrt{17})/2 \approx 0.5616$.

(5.4.1): Use the fact that each node in the tree has only finitely many children.

(5.4.10): (1) "Keep squares at random (3 of 4)": Similarity dimension is $\log 3/\log 2$, the same as keeping 3 of 4 squares using a nonrandom choice.

(3) "Uniform distribution on an interval": 0.

(2) "Keep squares at random (independent)": $\Phi(s) = (1/2)^s(4p)$. If $p \leq 1/4$, then $\Phi(0) \leq 1$, so the similarity dimension is 0. If $p \geq 1/4$, then the similarity dimension is the solution of $(1/2)^s(4p) = 1$, or $s = 2 + \log p/\log 2$.

(5.4.12): Here we may rely on a theory of so-called "birth and death processes" [97, p. 479]. The process eventually dies out (so that $K = \varnothing$) even when $\Phi(0) = 1$.

(5.5.12): Choose a real number $u \in (0,1)$ at random with uniform distribution. Define

$$h(t) = \begin{cases} 0, & \text{if } t < u, \\ 1, & \text{if } t \geq u. \end{cases}$$

(5.6.7): Time Inversion. For (a), since $\mathbf{X}'(0)$ is not defined by the formula, we may simply define $\mathbf{X}'(0) = 0$. But this does not complete the requirement, since Brownian motion is supposed to be continuous. So in fact, we must show that $\lim_{t \to 0} t\mathbf{X}(1/t) = 0$ with probability one. For this, we will use the maximal inequality 4.3.11. For a positive integer n, we have

$$\mathbb{P}\left\{ \max_{t \in [1/(n+1)^4, 1/n^4]} |t\mathbf{X}(1/t)| \geq \frac{1}{n} \right\}$$

$$\leq \mathbb{P}\left\{ \max_{t \in [1/(n+1)^4, 1/n^4]} |\mathbf{X}(1/t)| \geq n^4 \frac{1}{n} \right\}$$

$$\leq \mathbb{P}\left\{ \max_{s \in [0, (n+1)^4]} |\mathbf{X}(s)| \geq n^3 \right\}$$

$$\leq \frac{(n+1)^4}{n^6}.$$

Now, the series $\sum (n+1)^4/n^6$ converges, so by the easy direction of the Borel–Cantelli Lemma 4.1.2, with probability one, only finitely many of the events occur. So for all n beyond some point,

$$\max_{t \in [1/(n+1)^4, 1/n^4]} \left| t\mathbf{X}\left(\frac{1}{t}\right) \right| < \frac{1}{n},$$

and thus $\lim_{t \to 0} |t\mathbf{X}(1/t)| = 0$.

For (c), let $0 < s < t$, so that $0 < 1/t < 1/s$, and consider the distribution of $\mathbf{X}'(t) - \mathbf{X}'(s) = t\mathbf{X}(1/t) - s\mathbf{X}(1/s) = (t-s)\mathbf{X}(1/t) - s(\mathbf{X}(1/s) - \mathbf{X}(1/t))$. Now, $(t-s)\mathbf{X}(1/t)$ and $-s(\mathbf{X}(1/s) - \mathbf{X}(1/t))$ are independent normal random variables, so their sum is also normal. The mean of the sum is the sum of the means, 0. The variance of the sum is the sum of the variances, $(t-s)^2(1/t) + (-s)^2(1/s - 1/t) = t - s$.

For (b), write down a joint density for the increments required. This will need some careful algebra to get it right.

(5.6.16): For (a), imitate the proof in 5.6.15: Let $s < \dim M$, so there is a probability measure $\mathcal{N}$ on M with $\int \int |t - u|^{-s} \mathcal{N}(dt)\,\mathcal{N}(du) < \infty$. Let $\mathcal{M}$ be the image of $\mathcal{N}$ under $\mathbf{X}$. For (b), combine (a) with (5.6.14). The published proof of (c) is [212]. If any student manages to prove it, I would be interested in hearing about it. A student proof is likely to be simpler than the published proof.

(5.7.2): [43].

References

1. Althoen, S. C., Schilling, K. E., Wyneken, M. F.: Cutting corners: A four-gon conclusion. *College Mathematics Journal* **25** (1994) 266–279.
2. Anchorage Math Solutions Group: A minimal angle of a constrained triangle (Solution to Problem 1472). *Mathematics Magazine* **69** (1996) 225–226.
3. Arbeiter, M.: Random recursive constructions of self-similar fractal measures: The noncompact case. *Probability Theory and Related Fields* **88** (1991) 497–520.
4. Arbeiter, M., Patzschke, N.: Random self-similar multifractals. *Mathematische Nachrichten* **181** (1996) 5–42.
5. Austin, D. G., Edgar, G. A., Ionescu-Tulcea, A.: Pointwise convergence in terms of expectations. *Zeitschrift für Wahrscheinlichkeitstheorie und verwandte Gebiete* **30** (1974) 17–26.
6. Banach, S.: *Théorie des Opérations Linéaires*. Warsaw, 1932 [English translation: North-Holland, Amsterdam, 1987].
7. Bandt, C.: Self-similar sets 1: Topological Markov chains and mixed self-similar sets. *Mathematische Nachrichten* **142** (1989) 107–123.
8. Bandt, C.: Self-similar sets 3: Constructions with sofic systems. *Monatshefte für Mathematik* **108** (1989) 89–102.
9. Bandt, C.: Deterministic fractals and fractal measures. *Rendiconti dell'Istituto di Matematica dell'Università di Trieste* **23** (1991) 1–40 (1993).
10. Bandt, C.: Self-similar sets 5: Integer matrices and fractal tilings of $\mathbb{R}^n$. *Transactions of the American Mathematical Society* **112** (1991) 549–562.
11. Bandt, C., Graf, S.: Self-similar sets 7: A characterization of self-similar fractals with positive Hausdorff measure. *Proceedings of the American Mathematical Society* **114** (1992) 995–1001.
12. Barnsley, M. F.: Fractal functions and interpolation. *Constructive Approximation* **2** (1986) 303–329.
13. Barnsley, M. F.: *Fractals Everywhere*. Academic Press, Boston, 1988.
14. Barnsley, M. F., Demko, S. G., Elton, J. H., Geronimo, J. S.: Invariant measures for Markov processes arising from iterated function systems with place-dependent probabilities. *Annales de l'Institut Henri Poincaré* **24** (1988) 367–394.
15. Barnsley, M. F., Elton, J., Hardin, D., Massopust, P.: Hidden variable fractal interpolation functions. *S.I.A.M. Journal of Mathematical Analysis* **20** (1989) 1218–1242.
16. Barnsley, M. F., Hardin, D. P.: A Mandelbrot set whose boundary is piecewise smooth. *Transactions of the American Mathematical Society* **315** (1989) 641–657.
17. Barnsley, M. F., Hurd, L. P.: *Fractal Image Compression*. A K Peters, Wellesley, Massachussetts, 1993.
18. Bartle, R. G.: Return to the Riemann integral. *American Mathematical Monthly* **103** (1996) 625–632.

19. Bedford, T.: Dimension and dynamics for fractal recurrent sets. *Journal of the London Mathematical Society* **33** (1986) 89–100.

20. Bedford, T.: On Weierstrass-like functions and random recurrent sets. *Mathematical Proceedings of the Cambridge Philosophical Society* **106** (1989) 325–342.

21. Bedford, T.: The box dimension of self-affine graphs and repellers. *Nonlinearity* **2** (1989) 53–71.

22. Bedford, T, Urbański, M.: The box and Hausdorff dimension of self-affine sets. *Ergodic Theory and Dynamical Systems* **10** (1990) 627–644.

23. Besicovitch, A. S.: On the sum of digits of real numbers represented in the dyadic system. *Mathematische Annalen* **110** (1934) 321–330 {Selection 9 in [74]}.

24. Besicovitch, A. S.: On rational approximation to real numbers. *Journal of the London Mathematical Society* **9** (1934) 126–131 {Selection 10 in [74]}.

25. Besicovitch, A. S.: A general form of the covering principle and relative differentiation of additive functions. *Proceedings of the Cambridge Philosophical Society* **41** (1945) 103–110.

26. Besicovitch, A. S.: On existence of subsets of finite measure of sets of infinite measure. *Indagationes Mathematicae* **14** (1952) 339–344.

27. Besicovitch, A. S., Moran, P. A. P.: The measure of product and cylinder sets. *Journal of the London Mathematical Society* **20** (1945) 110–120.

28. Besicovitch, A. S., Taylor, S. J.: On the complementary intervals of a linear closed set of zero Lebesgue measure. *Journal of the London Mathematical Society* **29** (1954) 449–459 {Selection 15 in [74]}.

29. Besicovitch, A. S., Ursell, H. D.: On dimensional numbers of some continuous curves. *Journal of the London Mathematical Society* **12** (1937) 18–25 {Selection 11 in [74]}.

30. Billingsley, P.: *Ergodic Theory and Information*. Wiley, New York, 1965.

31. Billingsley, P.: *Convergence of Probability Measures*. Wiley, New York, 1968.

32. Billingsley, P.: *Probability and Measure*. Wiley, New York, 1979.

33. Blumenthal, L. M., Menger, K.: *Studies in Geometry*. Freeman, New York, 1970.

34. Boor, C. de: Cutting corners always works. *Computer Aided Geometric Design* **4** (1987) 125–131.

35. Bossomaier, T. R. J., Green, D. G. (eds.): *Complex Systems*. Cambridge University Press, Cambridge, to appear.

36. Bouligand, G.: Ensembles impropres et nombre dimensionnel. *Bulletin des Sciences Mathématiques* **52** (1928) 320–344, 361–376 {Selection 7 in [74]}.

37. Bourbaki, N.: *Topologie Générale*. Chap. 9 (Actualités Scientifiques et Industrielles 1045). Hermann, Paris, 1958.

38. Bourbaki, N.: *Intégration*. Chap. 9 (Actualités Scientifiques et Industrielles 1343). Hermann, Paris, 1969.

39. Briggs, J.: *Fractals: The Patterns of Chaos*. Simon & Schuster, New York, 1992.

40. Brown, G., Moran, W., Pearce, C. E. M.: Riesz products, Hausdorff dimension and normal numbers. *Mathematical Proceedings of the Cambridge Philosophical Society* **101** (1987) 529–540.

41. Cantor, G.: De la puissance des ensembles parfait de points. *Acta Mathematica* **4** (1884) 381–392 {Selection 2 in [74]}.

42. Carathéodory, C.: Über das lineare Maß von Punktmengen—eine Verallgemeinerung des Längenbegriffs. *Nachrichten der K. Gesellschaft der Wissenschaften zu Göttingen, Mathematisch-physikalische Klasse* (1914) 404–426 {Selection 4 in [74]}.

43. Cawley, R., Mauldin, R. D.: Multifractal decompositions of Moran fractals. *Advances in Mathematics* **92** (1992) 196–236.

44. Cherbit, G. (ed.): *Fractals: Non-Integral Dimensions and Applications*. Wiley, Chichester, 1991.
45. Choquet, G.: Ensembles singulières et structure des ensembles mesurables pour les mesures de Hausdorff. *Bulletin de la Société Mathématique de France* **74** (1946) 1–14.
46. Chung, K. L.: *A Course in Probability*. Academic Press, Boston, 1974.
47. Cinlar, E., Chung, K. L., Getoor, R. K. (eds.): *Seminar on Stochastic Processes* (Progress in Probability and Statistics 12). Birkhäuser, Boston, 1987.
48. Cohn, D. L.: *Measure Theory*. Birkhäuser, Boston, 1980.
49. Crilly, A. J., Earnshaw, R. A., Jones, H. (eds.): *Applications of Fractals and Chaos*. Springer-Verlag, Berlin, 1993.
50. Cutler, C. D.: The Hausdorff dimension distribution of finite measures in Euclidean space. *Canadian Journal of Mathematics* **38** (1986) 1459–1484.
51. Cutler, C. D.: A review of the theory and estimation of fractal dimension. {pp. 1–107 in [261]}.
52. Cutler, C. D.: The density theorem and Hausdorff inequality for packing measure in general metric spaces. *Illinois Journal of Mathematics* **39** (1995) 676–694.
53. Darst, R.: Hausdorff dimension of sets of non-differentiability points of Cantor functions. *Mathematical Proceedings of the Cambridge Philosophical Society* **117** (1995) 185–191.
54. Davies, R. O.: Measures not approximable or not specifiable by means of balls. *Mathematika* **18** (1971) 157–160.
55. Davies, R. O., Rogers, C. A.: The problem of subsets of finite positive measure. *Bulletin of the London Mathematical Society* **1** (1969) 47–54.
56. Dekking, F. M.: Recurrent sets. *Advances in Mathematics* **44** (1982) 78–104.
57. Dekking, F. M.: Replicating superfigures and endomorphisms of free groups. *Journal of Combinatorial Theory*, Series A **32** (1982) 315–320.
58. Devaney, R. L.: *An Introduction to Chaotic Dynamical Systems*. Benjamin/Cummings, Menlo Park, California, 1986.
59. Devaney, R. L.: *Chaos, Fractals, and Dynamics: Computer Experiments in Mathematics*. Addison-Wesley, Menlo Park, California, 1990.
60. Devaney, R. L.: *A First Course in Chaotic Dynamical Systems*. Addison-Wesley, Reading, Massachussetts, 1992.
61. Devaney, R. L., Keen, L.: *Chaos and Fractals: The Mathematics Behind the Computer Graphics*. American Mathematical Society, Providence, 1980.
62. Doob, J. L.: The development of rigor in mathematical probability (1900–1950). *American Mathematical Monthly* **103** (1996) 586–595.
63. Downarowicz, T., Mauldin, R. D., Warnock, T. T.: Random circle homeomorphisms. *Ergodic Theory and Dynamical Systems* **12** (1992) 441–458.
64. Dubins, L. E., Freedman, D. A.: Random distribution functions. {pp. 183–214 in [164]}.
65. Dubuc, S.: Une équation fonctionnelle pour diverses constructions géométriques. *Annales des Sciences Mathématiques du Québec* **9** (1985) 151–174.
66. Dudley, R. M.: Convergence of Baire measures. *Studia Mathematica* **27** (1966) 251–268.
67. Dunford, N., Schwartz, J. T.: *Linear Operators, Part I*. Wiley, New York, 1958.
68. Durrett, R.: *Brownian Motion and Martingales in Analysis*. Wadsworth, Belmont, California, 1984.
69. Durrett, R.: *Probability: Theory and Examples*. Wadsworth & Brooks/Cole, Pacific Grove, California, 1991.
70. Duvall, P., Keesling, J.: The Hausdorff dimension of the bounary of the Lévy Dragon. (preprint).
71. Edgar, G. A.: Kiesswetter's fractal has Hausdorff dimension 3/2. *Real Analysis Exchange* **41** (1989) 215–223.

72. Edgar, G. A.: *Measure, Topology, and Fractal Geometry*. Springer-Verlag, New York, 1990.
73. Edgar, G. A.: A fractal puzzle. *Mathematical Intelligencer* **13** (1991) 44–50.
74. Edgar, G. A.: *Classics on Fractals*. Addison-Wesley, Menlo Park, California, 1992
75. Edgar, G. A.: Fractal dimension of self-affine sets: some examples. {pp. 341–358 in [156]}.
76. Edgar, G. A.: Packing measure as a gauge variation. *Proceedings of the American Mathematical Society* **122** (1994) 167–174.
77. Edgar, G. A.: Fine variation and fractal measures. *Real Analysis Exchange* **20** (1995) 256–280.
78. Edgar, G. A., Mauldin, R. D.: Multifractal decompositions of digraph recursive fractals. *Proceedings of the London Mathematical Society* **65** (1992) 604–628.
79. Edgar, G. A., Sucheston, L.: *Stopping Times and Directed Processes*. Cambridge University Press, Cambridge, 1992.
80. Elton, J. H.: An ergodic theorem for iterated maps. *Ergodic Theory and Dynamical Systems* **7** (1987) 481–488.
81. Evertsz, C. J. G., Mandelbrot, B. B.: Multifractal measures. {Appendix B in [210]}.
82. Ewing, J.: Can we see the Mandelbrot set? *College Mathematics Journal* **26** (1995) 90–99.
83. Falconer, K. J.: *The Geometry of Fractal Sets*. Cambridge University Press, Cambridge, 1985.
84. Falconer, K. J.: Random fractals. *Mathematical Proceedings of the Cambridge Philosophical Society* **100** (1986) 559–582.
85. Falconer, K. J.: The Hausdorff dimension of some fractals and attractors of overlapping construction. *Journal of Statistical Physics* **47** (1987) 123–132.
86. Falconer, K. J.: The Hausdorff dimension of self-affine fractals. *Mathematical Proceedings of the Cambridge Philosophical Society* **103** (1988) 339–350.
87. Falconer, K. J.: *Fractal Geometry: Mathematical Foundations and Applications*. Wiley, Chichester, 1990.
88. Falconer, K. J.: The dimension of self-affine fractals II. *Mathematical Proceedings of the Cambridge Philosophical Society* **111** (1992) 169–179.
89. Falconer, K. J.: The multifractal spectrum of statistically self-similar measures. *Journal of Theoretical Probability* **7** (1994) 681–702.
90. Falconer, K. J.: On the Minkowski measurability of fractals. *Proceedings of the American Mathematical Society* **123** (1995) 1115–1124.
91. Falconer, K. J.: Sub-self-similar sets. *Transactions of the American Mathematical Society* **347** (1995) 3121–3129.
92. Falconer, K. J., Howroyd, J. D.: Projection theorems for box and packing dimensions. *Mathematical Proceedings of the Cambridge Philosophical Society* **119** (1996) 287–295.
93. Falconer, K. J., March, D. T.: The dimension of affine-invariant fractals. *Journal of Physics* A **21** (1988) L121–125.
94. Falconer, K. J., O'Neil, T. C.: Vector-valued multifractal measures. *Proceedings of the Royal Society of London* Series A **452** (1996) 1433–1457.
95. Farmer, J. D., Ott, E., Yorke, J. A.: The dimension of chaotic attractors. *Physica* D **7** (1983) 153–180.
96. Federer, H.: *Geometric Measure Theory*. Springer-Verlag, Berlin, 1969.
97. Feller, W.: *An Introduction to Probability Theory and Its Applications* (2nd ed., vol. II). Wiley, New York, 1971.
98. Ferhau, H.: *Iterierte Funktionen, Sprachen und Fraktale*. BI-Wissenschaftsverlag, Mannheim, 1994.
99. Fickel, N.: Selbstaffine Interpolation. Diplomarbeit, Universität Erlangen, 1988.
100. Fisher, Y.: A discussion of fractal image compression. {Appendix A in [210]}.

101. Fisher, Y. (ed.): *Fractal image compression: Theory and application.* Springer-Verlag, New York, 1995.
102. Folland, G. B.: *Real Analysis: Modern Techniques and Their Applications.* Wiley-Interscience, New York, 1984.
103. Frostman, O.: Potential d'équilibre et capacité des ensembles avec quelques applications à la théorie des fonctions. *Meddelanden fran Lunds Universitest Matimatiska Seminarum* **3** (1935) 1–118.
104. Garcia, L.: *The Fractal Explorer.* Dynamic Press, Santa Cruz, California, 1991.
105. Gatzouras, D., Lalley, S. P.: Statistically self-affine sets: Hausdorff and box dimensions. *Journal of Theoretical Probability* **7** (1994) 437–468.
106. Gerow, K., Holbrook, J.: Statistical sampling and fractal distributions. *Mathematical Intelligencer* **18** (1996) no. 2, 12–22.
107. Gilbert, W. J.: Fractal geometry derived from complex bases. *Mathematical Intelligencer* **4** (1982) 78–86.
108. Gilbert, W. J.: The fractal dimension of sets derived from complex bases. *Canadian Mathematical Bulletin* **29** (1986) 495–500.
109. Gilbert, W. J.: Complex bases and fractal similarity. *Annales des Sciences Mathématiques du Québec* **11** (1987) 65–77.
110. Gillman, L., Jerison, M.: *Rings of Continuous Functions.* Van Nostrand, New York, 1960.
111. Gleason, A. M.: *Fundamentals of Abstract Analysis.* Addison-Wesley, Reading, MA, 1966
112. Gleick, J.: *Chaos: Making a New Science.* Viking, New York, 1987.
113. Gordon, R. A.: *The Integrals of Lebesgue, Denjoy, Perron, and Henstock.* American Mathematical Society, 1994.
114. Graf, S.: Random fractals. *Rendiconti dell'Istituto di Matematica dell'Università di Trieste* **23** (1991) 81–144 (1993).
115. Graf, S.: On the characterization of the Hausdorff measure on a self-similar set. *Atti del Seminario Matematico e Fisico dell'Università di Modena* **42** (1994) 189–197.
116. Graf, S.: On Bandt's tangential distribution for self-similar measures. *Monatshefte für Mathematik* **120** (1995) 223–246.
117. Graf, S., Mauldin, R. D., Williams, S. C.: Random homeomorphisms. *Advances in Mathematics* **60** (1986) 239–359.
118. Graf, S., Mauldin, R. D., Williams, S. C.: *The exact Hausdorff dimension in random recursive constructions* (Memoirs of the American Mathematical Society 381). American Mathematical Society, Providence, 1988.
119. Guzmán, M. de: *Real Variable Methods in Fourier Analysis* (North-Holland Mathematics Studies 46). North-Holland, Amsterdam, 1981.
120. Haase, H.: Non-σ-finite sets for packing measure. *Mathematika* **33** (1986) 129–136.
121. Haase, H.: Packing measures on ultrametric spaces. *Studia Mathematica* **91** (1988) 189–203.
122. Haase, H.: The packing theorem and packing measure. *Mathematische Nachrichten* **146** (1990) 77–84.
123. Haeseler, F. v., Peitgen, H.-O., Skordev, G.: Multifractal decompositions of rescaled evolution sets of equivariant cellular automata. *Random and Computational Dynamics* **3** (1995) 93–119.
124. Halmos, P. R.: *Measure Theory.* Van Nostrand, New York, 1950.
125. Halsey, T., Jensen, M., Kadanoff, L., Procaccia, I., Shraiman, B.: Fractal measures and their singularities: the characterization of strange sets. *Physical Review* A **33** (1986) 1141–1151.
126. Hardy, G. H.: Weierstrass's non-differentiable function. *Transactions of the American Mathematical Society* **17** (1916) 301–325.

127. Harrison, J.: Continued fractals and the Seifert conjecture. *Bulletin of the American Mathematical Society* **13** (1985) 147–153.
128. Hastings, H. M., Sugihara, G.: *Fractals, A User's Guide for the Natural Sciences.* Oxford University Press, Oxford, 1993.
129. Hata, M.: Fractals in mathematics. {pp. 259–278 in [202]}.
130. Hausdorff, F.: Dimension und äußeres Maß. *Mathematische Annalen* **79** (1918) 157–179 {Selection 5 in [74]}.
131. Hausdorff, F.: *Mengenlehre* (3 Aufl.). Dover, New York, 1944.
132. Hayes, C. A., Pauc, C. Y.: *Derivation and Martingales* (Ergebnisse der Mathematik und Ihrer Grenzgebiete 49). Springer-Varlag, New York, 1970.
133. Henstock, R.: A Riemann-type integral of Lebesgue power. *Canadian Journal of Mathematics* **20** (1968) 79–87.
134. Henstock, R.: *The General Theory of Integration.* Oxford University Press, Oxford, 1991.
135. Hewitt, E., Stromberg, K.: *Real and Abstract Analysis.* Springer-Verlag, New York, 1965.
136. Hofbauer, F., Urbański, M.: Fractal properties of invariant subsets for piecewise monotonic maps on the interval. *Transactions of the American Mathematical Society* **343** (1994) 659–673.
137. Hofstadter, D. R.: Energy levels and wave functions of Bloch electrons in rational and irrational magnetic fields. *Physical Review* B **14** (1976) 2239–2249.
138. Howroyd, J.: On dimension and on the existence of sets of finite positive Hausdorff measure. *Proceedings of the London Mathematical Society* **70** (1995) 581–604.
139. Howroyd, J.: On Hausdorff and packing dimension of product spaces. *Mathematical Proceedings of the Cambridge Philosophical Society* **119** (1996) 715–727.
140. Hu, X., Taylor, S. J.: Fractal properties of products and projections of measures in $\mathbb{R}^d$. *Mathematical Proceedings of the Cambridge Philosophical Society* **115** (1994) 527–544.
141. Hurewicz, W., Wallman, H.: *Dimension Theory.* Princeton University Press, Princeton, 1941.
142. Hutchinson, J. E.: Fractals and self similarity. *Indiana University Mathematics Journal* **30** (1981) 713–747.
143. Hutchinson, J. E.: Deterministic and Random Fractals. to appear in [35].
144. Ikeda, S., Tamashiro, M.: Frostman theorem for the packing measure. *Comptes Rendus, Academie des Sciences, Paris,* Série I **320** (1995) 1445–1448.
145. Jarník, V.: Über die simultanen diophantischen Approximationen. *Mathematische Zeitschrift* **33** (1931) 505–543.
146. Joyce, H., Preiss, D.: On the existence of subsets of finite positive packing measure. *Mathematika* **42** (1995) 14–24.
147. Kahane, J.-P.: Definition of stable laws, infinitely divisible laws, and Lévy processes. {pp. 99–109 in [238]}.
148. Kakutani, S.: Concrete representation of abstract (M)-spaces. *Annals of Mathematics* **42** (1941) 994–1024.
149. Keesling, J.: The boundary of a self-similar tile in $\mathbb{R}^n$. (preprint).
150. Kelly, J. D.: A method for constructing measures apporpriate for the study of Cartesian products. *Proceedings of the London Mathematical Society* **26** (1973) 521–546.
151. Kießwetter, K.: Ein einfaches Beispiel für eine Funktion, welche überall stetig und nicht differenzierbar ist. *Mathematisch-Physikalische Semesterberichte zur Pflege des Zusammenhangs von Schule und Universität* **13** (1966) 216–221 {Selection 18 in [74]}.

152. Klafter, J., Shlesinger, M. F., Zumofen, G.: Beyond Brownian motion. *Physics Today* **49** (Feb. 1996) 33–39.
153. Knopp, K.: Ein einfaches Verfahren zur Bildung stetiger nirgends differenzierbar Funktionen. *Mathematische Zeitschrift* **2** (1918) 1–26.
154. Koch, H. v.: Sur une courbe continue sans tangente obtenue par une construction géométrique élémentaire. *Arkiv för Matematik, Astronomi och Fysik* **1** (1904) 681–702 {Selection 3 in [74]}
155. Kolmogorov, A. N., Tihomirov, V. M.: ε-энтропия и ε-емкость множеств в функциональных пространствах. *Успехи Математических Наук* **14** (1959) № 2, 3–86 [translation published in *American Mathematical Society Translations*, Series 2, Volume 17, pp. 277–364] {Selection 17 in [74]}.
156. Kölzow, D. (ed.): Measure Theory Oberwolfach 1990. *Supplemento ai Rendiconti del Circolo Matematico di Palermo* **28** (1992).
157. Kuratowski, K.: *Topology* (New edition, vol. I). Academic Press, New York, 1966.
158. Kurzweil, J.: Generalized ordinary differential equations and continuous dependence on a parameter. *Czechoslovak Mathematics Journal* **7** (1957) 418–446.
159. Lalley, S. P.: The packing and covering functions of some self-similar fractals. *Indiana University Mathematics Journal* **37** (1988) 699–709.
160. Lalley, S. P., Gatzouras, D.: Hausdorff and box dimensions of certain self-affine fractals. *Indiana University Mathematics Journal* **41** (1992) 533–568.
161. Lau, K.: Fractal measures and mean p-variations. *Journal of Functional Analysis* **108** (1992) 427–457.
162. Lau, K.: Dimension of a family of singular Bernoulli convolutions. *Journal of Functional Analysis* **116** (1993) 335–358.
163. Lau, K.: Self-similarity, L^p-spectrum and multifractal formalism. *Progress in Probability* **37** (1995) 55–90.
164. LeCam, L. M., Neyman, J. (eds.): *Proceedings of the Fifth Berkeley Symposium on Mathematical Statistics and Probability* (Volume II, Part 1). University of California Press, Berkeley, 1967.
165. Le Gall, J.-F., Taylor, S. J.: The packing measure of planar Brownian motion. {pp. 139–147 in [47]}.
166. Le Méhauté, A.: *Fractal Geometrics: Theory and Applications*. Hermes, Paris, 1990.
167. Lefschetz, S.: On compact spaces. *Mathematische Annalen* **32** (1931) 521–538.
168. Lévy, P.: *Calcul des probabilités*. Gauthier-Villars, Paris, 1925.
169. Lévy, P.: *Théorie de l'addition des variables aléatoirs*. Gauthier-Villars, Paris, 1937.
170. Lévy, P.: Les courbes planes ou gauches et les surfaces composée de parties semblables au tout. *Journal de l'École Polytechnique* **81** (1938) 227–247, 249–291 {Selection 12 in [74]}.
171. Lindstrøm, T.: *Brownian Motion on Nested Fractals* (Memoirs of the American Mathematical Society 420). American Mathematical Society, Providence, 1990.
172. Littlewood, J. E.: *Lectures on the Theory of Functions*. Oxford University Press, Oxford, 1944.
173. Littlewood, J. E.: The Dilemma of Probability Theory. In *A Mathematician's Miscellany*. Methuen & Co., London, 1953.
174. Lubin, A.: Extensions of measures and the von Neumann selection theorem. *Proceedings of the American Mathematical Society* **43** (1974) 118–122.
175. Mandelbrot, B. B.: How long is the coast of Britain: Statistical self-similarity and fractional dimension. *Science* **156** (1967) 636–639 {Selection 19 in [74]}.
176. Mandelbrot, B. B.: *The Fractal Geometry of Nature*. Freeman, New York, 1982.
177. Mandelbrot, B. B.: Self-affine fractals and fractal dimension. *Physica Scripta* **32** (1985) 257–260.

178. Mandelbrot, B. B.: Self-affine sets I: The basic fractal dimensions. {pp. 3–28 in [216]}.

179. Mandelbrot, B. B.: Negative fractal dimensions and multifractals. *Physica* A **163** (1990) 306–315.

180. Mandelbrot, B. B.: Random multifractals: Negative dimensions and the resulting limitations of the thermodynamic formalism. *Proceedings of the Royal Society of London* Series A **434** (1991) 79–88.

181. Mandelbrot, B. B., Gefen, Y., Aharony, A., Peyrière, J.: Fractals, their transfer matrices and their eigen-dimensional sequences. *Journal of Physics* A **18** (1985) 335–354.

182. Mandelbrot, B. B., Ness, J. W. van: Fractional Brownian motion, fractional noises and applications. *SIAM Review* **10** (1968) 422–437.

183. Marstrand, J. M.: The dimension of Cartesian product sets. *Proceedings of the Cambridge Philosophical Society* **50** (1954) 198–202 {Selection 14 in [74]}.

184. Marstrand, J. M.: The (φ, s)-regular subsets of n-space. *Transactions of the American Mathematical Society* **113** (1964) 369–392.

185. Massopust, P.: *Fractal Functions, Fractal Surfaces, and Wavelets.* Academic Press, San Diego, 1994.

186. Mattila, P.: *Geometry of Sets and Measures in Euclidean Spaces: Fractals and Rectifiability.* Cambridge Univerisity Press, Cambridge, 1995.

187. Mauldin, R. D.: Infinite iterated function systems: Theory and applications. *Progress in Probability* **37** (1995) 91–110.

188. Mauldin, R. D., Williams, S. C.: Random recursive constructions: Asymptotic geometric and topological properties. *Transactions of the American Mathematical Society* **295** (1986) 325–346.

189. Mauldin, R. D., Williams, S. C.: Hausdorff dimension in graph directed constructions. *Transactions of the American Mathematical Society* **309** (1988) 811–829.

190. McGuire, M.: *An Eye for Fractals: A Photographic Essay.* Addison-Wesley, Redwood City, California, 1991.

191. McMullen, C.: The Hausdorff dimension of general Sierpiński carpets. *Nagoya Mathematics Journal* **96** (1984) 1–9.

192. Meinershagen, S.: The symmetric derivation basis measure and the packing measure. *Proceedings of the American Mathematical Society* **103** (1988) 813–814.

193. Mendès France, M., Shallit, J. O.: Wire bending. *Journal of Combinatorial Theory,* Series A **50** (1989) 1–23.

194. Menger, K.: Allgemeine Räume und Cartesische Räume. *Proceedings of the Section of Sciences, Koniklijke Akademie van Wetenschappen te Amsterdam* **29** (1926) 476–482, 1125–1128 {Selection 6 in [74]}.

195. Menger, K.: Zur allgemeinen Kurventheorie. *Fundamenta Mathematicae* **10** (1927) 96–115.

196. Menger, K.: *Dimensionstheorie.* B. G. Teubner, Leipzig, 1928.

197. Moore, E. H.: On certain crinkly curves. *Transactions of the American Mathematical Society* **1** (1900) 72–90.

198. Moran, M.: Dimension functions for fractal sets associated to series. *Proceedings of the American Mathematical Society* **120** (1994) 749–754.

199. Moran, P. A. P.: Additive functions of intervals and Hausdorff measure. *Proceedings of the Cambridge Philosophical Society* **42** (1946) 15–23 {Selection 13 in [74]}.

200. Munroe, M. E.: *Introduction to measure and integration.* Addison-Wesley, Cambridge, Massachusetts, 1953.

201. Nikodým, O.: Sur une généralisation des intégrales de M. J. Radon. *Fundamenta Mathematicae* **15** (1930) 131–179.

202. Nishida, T., Mimura, M., Fuji, H. (eds.): *Patterns and Waves—Qualitative Analysis of Differential Equations.* North-Holland, Amsterdam, 1986.
203. Olsen, L.: A multifractal formalism. *Advances in Mathematics* **116** (1995) 82–196.
204. Olsen, L.: *Random geometrically graph directed self-similar multifractals.* Wiley, New York, 1994.
205. Olsen, L.: Multifractal dimensions of product measures. *Mathematical Proceedings of the Cambridge Philosophical Society* **120** (1996) 709–734.
206. Ott, E., Sauer, T., Yorke, J. A. (eds.): *Coping with Chaos.* Wiley-Interscience, New York, 1994.
207. Oxtoby, J. C., Ulam, S. M.: On the existence of a measure invariant under a transformation. *Annals of Mathematics* **40** (1939) 560–566.
208. Patzschke, N.: Self-conformal multifractal measures. *Advances in Mathematics* (to appear).
209. Patzschke, N.: The strong open set condition in the random case. *Proceedings of the American Mathematical Society* **125** (1997) 2119–2125.
210. Peitgen, H.-O., Jürgens, H., Saupe, D.: *Chaos and Fractals.* Springer-Verlag, New York, 1992.
211. Peres, Y., Solomyak, B.: Absolute continuity of Bernoulli convolutions, a simple proof. *Mathematical Research Letters* **3** (1996) 231–239.
212. Perkins, E. A., Taylor, S. J.: Uniform measure results for the image of subsets under Brownian motion. *Probability Theory and Related Fields* **76** (1987) 257–289.
213. Peruggia, M.: *Discrete Iterated Function Systems.* A K Peters, Wellesley, Massachussetts, 1993.
214. Peterson, I.: Trails of the wandering albatross: Applying the mathematics of haphazard motion. *Science News* **150** (1996) 104–105.
215. Pfeffer, W. F.: *The Riemann Approach to Integration.* Cambridge University Press, Cambridge, 1993.
216. Pietronero, L., Tosatti, E. (eds.): *Fractals in Physics* (Proceedings of the sixth Trieste international symposium held in Trieste, July 9–12, 1985). North-Holland, Amsterdam, 1986.
217. Pontrjagin, L., Schnirelmann, L.: Sur une propriété métrique de la dimension. *Annals of Mathematics* **33** (1932) 156–162 {Selection 8 in [74]}.
218. Prusinkiewicz, P., Lindenmayer, A.: *The Algorithmic Beauty of Plants.* Springer-Verlag, New York, 1990.
219. Przytycki, F., Urbański, M.: On the Hausdorff dimension of some fractal sets. *Studia Mathematica* **93** (1989) 155–186.
220. Rand, D.: The singularity spectrum $f(\alpha)$ for cookie-cutters. *Ergodic Theory and Dynamical Systems* **9** (1989) 527–541.
221. Read, A. H.: The solution of a functional equation. *Proceedings of the Royal Society of Edinburgh,* Section A **63** (1952) 336–345.
222. Rham, G. de: Sur une courbe plane. *Journal de Mathématiques Pures et Appliquées* **35** (1956) 25–42.
223. Rham, G. de: Sur quelques courbes definies par des equations fonctionnelles. *Rendiconti del Seminario Matematico dell'Università e del Politecnico di Torino* **16** (1957) 101–113 {Selection 16 in [74]}.
224. Rham, G. de: Sur les courbes limites de polygones obtenus par trisection. *L'Enseignement Mathématique,* Ser. II **5** (1959) 29–43.
225. Riedi, R. H.: An improved multifractal formalism and self-similar measures. *Journal of Mathematical Analysis and Applications* **189** (1995) 462–490.
226. Riedi, R. H., Mandelbrot, B. B.: Multifractal formalism for infinite multinomial measures. *Advances in Applied Mathematics* **16** (1995) 132–150.
227. Rogers, C. A.: *Hausdorff Measures.* Cambridge University Press, Cambridge, 1970.

228. Rogers, C. A.: Dimension prints. *Mathematika* **35** (1988) 1–27.
229. Royden, R. L.: *Real Analysis*. Macmillan, New York, 1968.
230. Rudin, W.: *Real and Complex Analysis*. Third Edition, McGraw-Hill, New York, 1966.
231. Ruelle, D.: Repellers for real analytic maps. *Ergodic Theory and Dynamical Systems* **2** (1982) 99–107.
232. Saint Raymond, X., Tricot, C.: Packing regularity of sets in n-space. *Mathematical Proceedings of the Cambridge Philosophical Society* **103** (1988) 133–145.
233. Saks, S.: *Theory of the Integral* (2nd rev. ed.). Stechert, Warsaw, 1937.
234. Schief, A.: Separation properties for self-similar sets. *Proceedings of the American Mathematical Society* **122** (1994) 111-115.
235. Schief, A.: Self-similar sets in complete metric spaces. *Proceedings of the American Mathematical Society* **124** (1996) 481–490.
236. Shishikura, M.: The boundary of the Mandelbrot set has Hausdorff dimension two. Complex analytic methods in dynamical systems (Rio de Janeiro, 1992). *Astérisque* **222** (1994) 389–405.
237. Shlesinger, M. F.: Lévy flights: Variations on a theme. *Physica D* **38** (1989) 304–309.
238. Shlesinger, M. F., Zaslavsky, G. M., Frisch, U. (eds.): *Lévy Flights and Related Topics in Physics* (Proceedings of the International Workshop Held at Nice, France, 27–30 June 1994). Springer-Verlag, Berlin, 1995.
239. Sierpiński, W.: Sur une courbe cantorienne qui contient une image biunivoque et continue de toute courbe donnée. *Comptes Rendus, Academie des Sciences, Paris* **162** (1916) 629–632.
240. Snell, J. L.: Gambling, probability and martingales. *Mathematical Intelligencer* **4** (1982) 118–124.
241. Solomyak, B.: On the random series $\sum \pm \lambda^n$ (an Erdös problem). *Annals of Mathematics* **142** (1995) 611–625.
242. Spear, D. W.: Measures and self similarity. *Advances in Mathematics* **91** (1992) 143–157.
243. Sprott, J. C.: Automatic generation of iterated function systems. *Computers & Graphics* **18** (1994) 417–425.
244. Stewart, I.: *Does God Play Dice? The Mathematics of Chaos*. Blackwell, Cambridge, Massachussetts, 1989.
245. Stewart, I.: Four encounters with Sierpiński's gasket. *Mathematical Intelligencer* **17** (1995) 52–64.
246. Strichartz, R. S.: Self-similar measures and their Fourier transforms, I. *Indiana University Mathematics Journal* **39** (1990) 798–817.
247. Strichartz, R. S.: Self-similar measures and their Fourier transforms, II. *Transactions of the American Mathematical Society* **336** (1993) 335–361.
248. Strichartz, R. S.: Self-similar measures and their Fourier transforms, III. *Indiana University Mathematics Journal* **42** (1993) 367–411.
249. Strichartz, R. S.; Taylor, A.; Zhang, T.: Densities of self-similar measures on the line. *Experimental Mathematics* **4** (1995) 101–128.
250. Stromberg, K. R.: *An Introduction to Classical Real Analysis*. Wadsworth, Belmont, California, 1981.
251. Sullivan, D.: Entropy, Hausdorff measures old and new, and limit sets of finite Kleinian groups. *Acta Mathematica* **153** (1984) 259–277.
252. Szpilrajn, E.: La dimension et la mesure. *Fundamenta Mathematicae* **28** (1937) 81–89.
253. Talagrand, M., Xiao, Y.: Fractional Brownian motion and packing dimension. *Journal of Theoretical Probability* **9** (1996) 579–593.
254. Taylor, S. J.: The α-dimensional measure of the graph and set of zeros of a Brownian path. *Proceedings of the Cambridge Philosophical Society* **51** (1955) 265–274.

255. Taylor, S. J.: The measure theory of random fractals. *Mathematical Proceedings of the Cambridge Philosophical Society* **100** (1986) 383–406.
256. Taylor, S. J., Tricot, C.: Packing measure and its evaluation for a Brownian path. *Transactions of the American Mathematical Society* **288** (1985) 679–699.
257. Taylor, S. J., Tricot, C.: The packing measure of rectifiable subsets of the plane. *Mathematical Proceedings of the Cambridge Philosophical Society* **99** (1986) 285–296.
258. Thomson, B. S.: Construction of measures in metric spaces. *Journal of the London Mathematical Society* (2) **14** (1976) 21–24.
259. Thomson, B. S.: Measures generated by a differentiation basis. *Bulletin of the London Mathematical Society* **9** (1977) 279–282.
260. Thomson, B. S.: Derivation bases on the real line (I). *Real Analysis Exchange* **8** (1982) 67–207.
261. Tong, H. (ed.): *Nonlinear Time Series and Chaos* (vol. 1). World Scientific, Singapore, 1993.
262. Topsøe, F.: *Spontaneous Phenomena*. Academic Press, New York, 1990.
263. Tricot, C.: Douze définitions de la densité logarithmique. *Comptes Rendus, Academie des Sciences, Paris*, Série I **293** (1981) 549–552.
264. Tricot, C.: Two definitions of fractional dimension. *Mathematical Proceedings of the Cambridge Philosophical Society* **91** (1982) 57–74.
265. Urbański, M.: The probability distribution and Hausdorff dimension of self-affine functions. *Probability Theory and Related Fields* **84** (1990) 377–391.
266. Varadarajan, V.: Groups of automorphisms of Borel spaces. *Transactions of the American Mathematical Society* **109** (1963) 191–220.
267. Wallin, H.: Hausdorff measures and generalized differentiation. *Mathematische Annalen* **183** (1969) 275–286.
268. Wegmann, H.: Die Hausdorff-Dimension von kartesischen Produkten metrischer Räume. *Journal für die reine und angewandte Mathematik* **246** (1971) 46–75.
269. Weierstrass, K.: Über continuirliche Functionen eines reellen Arguments, die für keinen Werth des Letzteren einen bestimmten Differentialquotienten besitzen. In *Mathematische Werke*, Band II. Mayer & Müller, Berlin, 1895, pp. 71–74 {Selection 1 in [74]}.
270. Whitney, H.: A function not constant on a connected set of critical points. *Duke Mathematical Journal* **1** (1935) 514–517.
271. Xiao, Y.: Fractal Measures and Related Properties of Gaussian Random Fields. Ph.D. thesis, The Ohio State University, 1996.
272. Xiao, Y.: Packing dimension, Hausdorff dimension and Cartesian product sets. *Mathematical Proceedings of the Cambridge Philosophical Society* **120** (1996) 535–546.
273. Xiao, Y.: Packing measure of the sample paths of fractional Brownian motion. *Transactions of the American Mathematical Society* **348** (1996) 3193–3213.

Notation

An index of notations used in the book.

dim A—Hausdorff dimension of set A, 16.

$\mathcal{P}^s$—packing measure, 19.

$\widetilde{\mathcal{P}}^s_\varepsilon$, $\widetilde{\mathcal{P}}^s$, $\widetilde{\mathcal{P}}^s$, $\overline{\mathcal{P}}^s$—construction of $\mathcal{P}^s$, 19.

Dim A—packing dimension of set A, 20.

$\mathcal{C}^s$—covering measure, 29.

$\widetilde{\mathcal{C}}^s_\varepsilon$, $\widetilde{\mathcal{C}}^s$, $\overline{\mathcal{C}}^s$—construction of $\mathcal{C}^s$, 29.

$\mathcal{J}^s$—lower entropy measure, 33.

$\widetilde{\mathcal{J}}^s_\varepsilon$, $\widetilde{\mathcal{J}}^s$, $\overline{\mathcal{J}}^s$—construction of $\mathcal{J}^s$, 33.

led E—lower entropy dimension, 34.

lei E—lower entropy index, 34.

$\mathcal{K}^s$—upper entropy measure, 34.

$\widetilde{\mathcal{K}}^s_\varepsilon$, $\widetilde{\mathcal{K}}^s$, $\overline{\mathcal{K}}^s$—construction of $\mathcal{K}^s$, 34.

$S_\varepsilon(A)$—neighborhood, Minkowski sausage, parallel body, 36.

$\overline{D}^s_{\mathcal{M}}$—upper density, 46.

$\underline{D}^s_{\mathcal{M}}$—lower density, 46.

$D^s_{\mathcal{M}}$—density, 46.

$A_n \searrow A$ strongly—convergence of sets, 61.

$\mathbb{N}^{\mathbb{N}}$—Baire's space of sequences, 62.

$\mathcal{F} \otimes \mathcal{G}$—product σ-algebra, 69.

$h_*(\mathcal{M})$—image measure, 73.

$\overline{\mathcal{M}_1 \otimes \mathcal{M}_2}$—product outer measure, 73.

$\mathcal{M}_1 \otimes \mathcal{M}_2$—product measure, 74.

$\int f(x)\, \mathcal{M}(dx)$—integral, 77.

Lip_b—space of bounded Lipschitz functions, 98.

ρ_γ—metric on measures, 105.

$\dim^* \mathcal{M}$—upper Hausdorff dimension of a measure, 125.

$\dim_* \mathcal{M}$—lower Hausdorff dimension of a measure, 125.

$\mathrm{Dim}^* \mathcal{M}$—upper packing dimension of a measure, 126.

$\mathrm{Dim}_* \mathcal{M}$—lower packing dimension of a measure, 126.

$\overline{L}_{\mathcal{M}}(x)$—upper pointwise dimension, 126.

$\underline{L}_{\mathcal{M}}(x)$—lower pointwise dimension, 126.

$\mathrm{Var}\,[X]$—variance of a random variable, 164.

$\mathbb{E}\,[X]$—expectation of a random variable, 164.

$\mathcal{D}_X$—distribution (law) of a random variable, 166.

$\mathbb{1}_E$—indicator function, characteristic function, 168.

$\mathbb{P}\,[A\,|\,K]$—conditional probability, 175.

$\mathcal{D}_{Y|E}$—conditional distribution, 183.

$\mathbb{E}\,[Y\,|\,K]$—conditional expectation, 184.

Index